Advances in
PARASITOLOGY

Global Mapping of Infectious Diseases
Methods, Examples and Emerging Applications

VOLUME 62

Advances in
PARASITOLOGY

Global Mapping of Infectious Diseases

Methods, Examples and Emerging Applications

Guest Edited by

SIMON I. HAY, ALASTAIR GRAHAM
and DAVID J. ROGERS

TALA Research Group
Department of Zoology
University of Oxford
Oxford
UK

VOLUME 62

ELSEVIER

Amsterdam Boston Heidelberg London New York Oxford
Paris San Diego San Francisco Singapore Sydney Tokyo

ACADEMIC PRESS

ELSEVIER B.V.
Radarweg 29
P.O. Box 211
1000 AE Amsterdam
The Netherlands

ELSEVIER Inc.
525 B Street, Suite 1900
San Diego
CA 92101-4495
USA

ELSEVIER Ltd
The Boulevard, Langford
Lane, Kidlington
Oxford OX5 1GB
UK

ELSEVIER Ltd
84 Theobalds Road
London
WC1X 8RR
UK

First edition 2006

British Library Cataloguing in Publication Data
A catalogue record is available from the British Library.

ISBN-10: 0-12-031762-1
ISBN-13: 978-0-12-031762-2
ISBN-10: 0-12-031766-4 (DVD)
ISBN-13: 978-0-12-031766-0 (DVD)
ISSN: 0065-308X (series)

∞ The paper used in this publication meets the requirements of ANSI/NISO Z39.48-1992 (Permanence of Paper). Printed in Great Britain.

CONTRIBUTORS TO VOLUME 62

P. M. Atkinson, *School of Geography, University of Southampton, Highfield, Southampton SO17 1BJ, UK*

D. L. Balk, *CIESIN, Columbia University, Palisades, NY 10964, USA*

S. Brooker, *Department of Infectious and Tropical Diseases, London School of Hygiene & Tropical Medicine, Keppel Street, London WC1E 7HT, UK*

D. A. P. Bundy, *Human Development Division, The World Bank, 1818 H Street NW, Washington DC 20433, USA*

A. C. A. Clements, *Department of Infectious and Tropical Diseases, London School of Hygiene & Tropical Medicine, Keppel Street, London WC1E 7HT, UK*

U. Deichmann, *Development Research Group, The World Bank, 1818 H Street NW, Washington, DC 20433, USA*

S. J. Goetz, *The Woods Hole Research Center, Woods Hole, MA 02543-0296, USA*

A. J. Graham, *TALA Research Group, Tinbergen Building, Department of Zoology, University of Oxford, South Parks Road, Oxford OX1 3PS, UK*

C. A. Guerra, *TALA Research Group, Tinbergen Building, Department of Zoology, University of Oxford, South Parks Road, Oxford OX1 3PS, UK*

ISSN: 0065-308X $35.00
DOI: 10.1016/S0065-308X(05)62011-8

S. I. Hay, *TALA Research Group, Tinbergen Building, Department of Zoology, University of Oxford, South Parks Road, Oxford OX1 3PS, UK*

A. Nelson, *JRC (Joint Research Centre of the European Commission), Global Environment Monitoring Unit, TP 440, Via Enrico Fermi 1, I-21020 Ispra (VA), Italy*

F. Pozzi, *Università Cattolica del Sacro Cuore, sede di Brescia, via dei Musei 41, 25121, Brescia, Italy*

S. E. Randolph, *Oxford Tick Research Group, Tinbergen Building, Department of Zoology, University of Oxford, Oxford OX1 3PS, UK*

D. J. Rogers, *TALA Research Group, Tinbergen Building, Department of Zoology, University of Oxford, South Parks Road, Oxford OX1 3PS, UK*

R. W. Snow, *Malaria Public Health & Epidemiology Group, Centre for Geographic Medicine, KEMRI/Wellcome Trust Research Laboratories, PO Box 43640, 00100 Nairobi, Kenya*

A. J. Tatem, *TALA Research Group, Tinbergen Building, Department of Zoology, University of Oxford, South Parks Road, Oxford OX1 3PS, UK*

A. J. Wilson, *TALA Research Group, Tinbergen Building, Department of Zoology, University of Oxford, South Parks Road, Oxford OX1 3PS, UK*

G. Yetman, *CIESIN, Columbia University, Palisades, NY 10964, USA*

Series Editors' Preface

Nearly six years ago, a special volume of *Advances in Parasitology* (vol. 47) dealt with the uses of remote sensing and geographical information systems in the study of disease epidemiology. In a sense, this volume is a follow-on to that publication, dealing as it does with some practical applications of those techniques to the study of parasitic and infectious diseases.

We are once again fortunate in having Simon Hay, David Rogers and—a newcomer this time—Alastair Graham, of the University of Oxford in the United Kingdom, as guest editors. They have assembled a formidable array of talented research workers from the UK and the USA as contributors to what we are sure will be a valuable source of both technical and epidemiological data in this rapidly growing field.

We are sincerely grateful to the guest editors, authors and all those who have contributed to the production of this volume.

John Baker
Ralph Muller
David Rollinson

ADVANCES IN PARASITOLOGY VOL 62
ISSN: 0065-308X $35.00
DOI: 10.1016/S0065-308X(05)62017-9

Guest Editors' Preface

It has been five years since an earlier special issue of *Advances in Parasitology*, Volume 47, outlined the advances that remote sensing (RS) and geographical information systems (GIS) could bring to epidemiology. During this interval a vast amount of work has been undertaken in this area and these RS data and GIS tools have moved from the novel to part of the mainstream of spatial epidemiology. Data availability has continued to limit the engagement of many potential users, however. This has been most obvious in continental and global scale public-health applications, and predictably these limitations have been particularly acute in regions with low bandwidth internet connections, often where the public health need is greatest. The primary reason for compiling this new volume was to enable a wider range of epidemiologists to have access to the global environmental data (satellite and demographic), which we have been collectively working with for over a decade. The second reason for devising this special issue was to demonstrate that RS and GIS do not simply create pretty maps, but biologically informative information and ultimately pragmatic control tools. That being said, we also hope you like the front cover!

This special issue of *Advances in Parasitology*, Volume 62, "*Global mapping of infectious diseases: methods, examples and emerging applications*" comprises 10 reviews and a DVD of global environmental and population data. There are four introductory reviews: one on the various methods used to predict disease distributions (Rogers, this volume, pp. 1–35); another on the global environmental datasets that can be used for disease mapping (Hey *et al.*, this volume, pp. 37–77); a further one exploring the concepts of spatial resolution, accuracy and uncertainty measures in disease mapping based on remote sens-

ADVANCES IN PARASITOLOGY VOL 62
ISSN: 0065-308X $35.00
DOI: 10.1016/S0065-308X(05)62012-X

ing (Atkinson and Graham, this volume, pp. 79–118) and a final one on predicting the global distribution of human population (Balk *et al.*, this volume, pp. 119–156). Three reviews follow dealing with defining the global distribution limits of *Plasmodium falciparum* and *P. vivax* malaria (Guerra *et al.*, this volume, pp. 157–179), first attempts to map the environmental limits of dengue and yellow fever at the global scale (Rogers *et al.*, this volume, pp. 181–220) and continued efforts to map the geo-helminths for the targeting of control activities (Brooker *et al.*, this volume, pp. 221–262). Finally, three application reviews discuss current research topics that have emerged from our ability to consider epidemiological phenomena at the global scale. Tick-borne diseases are used as an example of how to coincidentally map geographic and phylogenetic space (Randolph and Rogers, this volume, pp. 263–291). The penultimate review investigates the spread of disease vectors and the pathogens they transmit via global transport networks (Tatem *et al.*, this volume, pp. 293–343). The final review discusses methods and evidence required to evaluate the impacts of climate change on vector-borne disease (Rogers and Randolph, this volume, pp. 345–381).

The public health environment, as well as our research, has evolved significantly in the last five years. The millennium development goals (http://www.un.org/millenniumgoals/) have been signed and largely drive the global development agenda. More recently these have been emphasized and augmented by the commission for Africa (http://www.commissionforafrica.org/), which helped to highlight the increasing importance of public health interventions in achieving international development goals. Our ability to document the "epidemiological state of the Earth" is therefore a crucial underpinning to measuring, planning, costing and ultimately delivering on these promises. We hope that the information and methods outlined in these reviews will in some small way contribute.

Furthermore, global environmental change has continued apace throughout the past five years. These changes for example to climate, transport networks, disease pathogens and their vectors do not respect administrative boundaries and their influences and impacts are best addressed at the global scale. With increasing computing power and ever cheaper data storage capacity, a wider variety of researchers can start to evaluate changes using the baseline data provided here. Our final reviews aim to convey some of the research opportunity that, we believe, these data facilitate.

Finally, we would like to note the passing of Byron Woods and Louisa Beck. Both participated in the previous special issue and were valued colleagues, whose contributions will be missed. We would also like to thank the series editors and development editors at Academic Press for their continued support and their help in making this collection of reviews, a book.

S.I. Hay
A.J. Graham
D.J. Rogers

Contents

Contributors to Volume 62 v

Series Editors' Preface vii
Guest editiors' Preface ix

Models for Vectors and Vector-Borne Diseases

D.J. Rogers

Abstract 1
1. A Brief History of Distribution Modelling 2
2. Families of Distribution Models 4
3. Predictor Variable Selection in Distribution Models 12
4. What to Do With Sparse Datasets? 14
5. Incorporating Spatial Information Into Models 19
6. Model Selection and Multi-Model Inference 21
7. Conclusion 30
Acknowledgements 33
References 33

Global Environmental Data for Mapping Infectious Disease Distribution

S.I. Hay, A.J. Tatem, A.J. Graham, S.J. Goetz and D.J. Rogers

Abstract 38
1. Introduction 38
2. The AVHRR Sensor 39
3. Temporal Fourier Analysis (TFA) 49
4. Future Global Environmental Data 57
5. Conclusions 70
Acknowledgements 70
References 71

ISSN: 0065-308X $35.00
DOI: 10.1016/S0065-308X(05)62015-5

Issues of Scale and Uncertainty in the Global Remote Sensing of Disease

P.M. Atkinson and A.J. Graham

Abstract 80
1. Introduction 80
2. Issues of Scale and Spatial Resolution 86
3. Issues of Uncertainty 96
4. Summary 107
Acknowledgements 108
References 108

Determining Global Population Distribution: Methods, Applications and Data

D.L. Balk, U. Deichmann, G. Yetman, F. Pozzi, S.I. Hay and A. Nelson

Abstract 120
1. Introduction 120
2. Data 124
3. Methodology 133
4. Health Applications 138
5. Discussion 144
6. Data Dissemination 148
Acknowledgements 151
References 151

Defining the Global Spatial Limits of Malaria Transmission in 2005

C.A. Guerra, R.W. Snow and S.I. Hay

Abstract 157
1. Introduction 158
2. The Distribution of Malaria Risk from Travel Guidelines 159
3. The Biological Limits of Transmission 161
4. Distinguishing *P. falciparum* and *P. vivax* risk 169
5. Regional Analysis 169
6. Discussion 173

7. Conclusions 174
Acknowledgements 175
References 175

The Global Distribution of Yellow Fever and Dengue

D.J. Rogers, A.J. Wilson, S.I. Hay and A.J. Graham

Abstract 182
1. Introduction 182
2. The Pathogens 183
3. Materials and Methods 193
4. Results 200
5. Discussion 208
6. Conclusion 209
Acknowledgements 210
References 211

Global Epidemiology, Ecology and Control of Soil-Transmitted Helminth Infections

S. Brooker, A.C.A. Clements and D.A.P. Bundy

Abstract 221
1. Introduction 222
2. Transmission Dynamics and the Environment 224
3. Ecological Correlates 227
4. Predicting Distributions 231
5. Urbanization. 232
6. Global Control Strategies. 234
7. Control Applications of GIS/RS. 237
8. Global Distributions 243
9. Predicted Numbers of Infections. 246
10. The Future 250
Acknowledgements 252
References 252

Tick-borne Disease Systems: Mapping Geographic and Phylogenetic Space

S.E. Randolph and D.J. Rogers

Abstract 263
1. Predicting Changing Risk of Infection on Evolutionary Time Scales 264
2. The Evolutionary Time Scale for Vector-Borne Flaviviruses . . 267
3. Correlates of Phylogenetic Patterns 269
4. Testing the Role of Climate in the Evolution of Tick-borne Flaviviruses 276
Acknowledgements 285
References 285

Global Transport Networks and Infectious Disease Spread

A.J. Tatem, D.J. Rogers and S.I. Hay

Abstract 294
1. Introduction 294
2. Global Transport Networks and Pandemics 295
3. Global Transport Networks and Disease Vector Invasions . . . 306
4. Global Transport Networks and Vector-borne Diseases 319
5. Conclusions 332
Acknowledgements 332
References 333

Climate Change and Vector-Borne Diseases

D.J. Rogers and S.E. Randolph

Abstract 346
1. The Mathematics and Biology of Changes in Vector-Borne Diseases 346
2. Defining the Criteria for Claiming Climate Impacts on Vector-Borne Diseases 351
3. Models for Climate Change Impacts on Vector-Borne Diseases 353

4. Biological and Statistical Approaches to Vector-Borne Disease Futures 355
5. Recent Changes in Vector-Borne Diseases: Has Climate Change Already had an Impact? 366
6. Conclusions 376
Acknowledgements 377
References 377

Index 383
Contents of Volumes in This Series 391

Colour Plate Section can be found at the back of this book

Models for Vectors and Vector-Borne Diseases

D.J. Rogers

TALA Research Group, Tinbergen Building, Department of Zoology, University of Oxford, South Parks Road, Oxford 0X1 3PS, UK

Abstract 1
1. A Brief History of Distribution Modelling 2
2. Families of Distribution Models 4
2.1. Logistic Models: The Theory 4
2.2. Discriminant Analysis Models: The Theory 10
3. Predictor Variable Selection in Distribution Models 12
4. What to Do With Sparse Datasets? 14
4.1. Bootstrap Sampling 15
4.2. Environmental Envelope Expansion 17
5. Incorporating Spatial Information Into Models 19
6. Model Selection and Multi-Model Inference 21
6.1. Application to Vector and Disease Mapping 26
7. Conclusion 30
Acknowledgements 33
References 33

ABSTRACT

The development of models for species' distributions is briefly reviewed, concentrating on logistic regression and discriminant analytical methods. Improvements in each type of modelling approach have led to increasingly accurate model predictions. This review addresses several key issues that now confront those wishing to choose the "right" sort of model for their own application. One major issue is the number of predictor variables to retain in the final model.

ADVANCES IN PARASITOLOGY VOL 62
ISSN: 0065-308X $35.00
DOI: 10.1016/S0065-308X(05)62001-5

Another is the problem of sparse datasets, or of data reported to administrative levels only, not to points. A third is the incorporation of spatial co-variance and auto-covariance in the modelling process. It is suggested that many of these problems can be resolved by adopting an information-theoretic approach whereby a group of reasonable potential models is specified in advance, and the "best" candidate model is selected among them. This approach of model selection and multi-model inference, using various derivatives of the Kullback–Leibler information or distance statistic, puts the biologist, with her or his insight, back in charge of the modelling process that is usually the domain of statisticians. Models are penalized when they contain too many variables; careful specification of the right set of candidate models may also be used to identify the importance of each predictor variable individually; and finally the degree to which the current "best" model improves on all the other models in the candidate set may be quantified. The ability definitely to exclude some models from the realm of all possible models appropriate for any particular distribution problem may be as important as the ability to identify the best current model.

1. A BRIEF HISTORY OF DISTRIBUTION MODELLING

Human beings, first as hunter-gatherers and later as scientists, have always appreciated the value of spatial information. Simple observations and collections of museum specimens reinforce the notion that no single species occurs everywhere. Maps of plant and animal distributions that were made for a variety of purposes record the known distribution of these living organisms as points or within polygons specifying some local region, either ecological (e.g. within a forest) or political (e.g. within a country or region). It is clear that the development of some of these polygon boundaries also benefited from the increasing knowledge of climatic constraints to species' distributions. For example, the historical map of the distribution of tsetse flies (*Glossina* spp.) in Zimbabwe is based on very few point records (tsetse were collected at known points by farmers, missionaries, District Officers and others) supplemented with the entomologists' increasing understanding that to complete development successfully tsetse require certain minimum temperatures that are found only at lower altitudes. In many places, therefore, the historical tsetse map follows elevation contours.

This example suggests that, as was said of models by Box, "All maps are wrong, but some are useful" (Box, 1979). The real question is, how useful are such maps, and can we improve upon them with current technology? This technology comes in four guises. The first is simply the increased information that we now have for many species that is available both digitally and online; the second is the increasingly powerful computers that we have, not just to hold but also to process the data; the third is the new generations of satellites that are providing us with information about our natural environments at unprecedented spectral, spatial and temporal resolutions; and finally the fourth is the increasingly sophisticated mathematical models that can be applied to distribution data. This combination of information, knowledge and understanding is available at a critical moment in human history, when the distributions of many other animals, plants and diseases are being affected, directly and indirectly, by the most ubiquitous of all species—our own. It seems we are beginning to understand the distributions of plants and animals only at the point when we are about to change them dramatically.

Biologists require maps for many different purposes. Here we concentrate on the use of maps to increase our understanding of the biological and other processes that determine the distribution and abundance of species in space and time. Which are the important variables; how do they act; and how do they differ in places of disease presence and absence, or in places of different disease abundance? We therefore eschew mapping methods that simply reproduce the distribution map, or points drawn from the distribution map (the "training set" data), through one or other pure "pattern-matching" approach. Thus both neural network models and *k*-nearest neighbour techniques will be ignored (Williams *et al.*, 1992), as will be tree-based classification methods (Green, 1978) and those hybrids, such as GARP, that use genetic algorithms to improve on initial approaches using more traditional methods (Stockwell and Peters, 1999). Each of these appears to make the assumption that what is required is a description, as accurate as possible, of the training set data in terms of the available suite of predictor variables, from which a distribution prediction is made. While the end results of these alternative approaches can be maps with high levels of statistical accuracy, the biological insight they provide is minimal.

2. FAMILIES OF DISTRIBUTION MODELS

In a recent review of predictive mapping of species' distribution, Rushton *et al.* (2004) show that logistic regression methods are by far the most commonly used approach. In fact, none of the examples reviewed used discriminant analysis, which is curious given the pervasive assumption of some or other form of normality (the key assumption of discriminant analysis) throughout the ecological literature (see below). This section shows that, despite the great accuracy of many logistic regression models, biologically their assumptions are essentially unrealistic. If we are to use statistics to increase our biological understanding we should start with a model type that potentially allows us to do so.

2.1. Logistic Models: The Theory

Logistic regression models are often applied to the relatively simple problem of the distribution of an organism or disease and whether or not it is absent or present in any area. The heart of logistic regression is a relationship between a response variable y and sets of predictor variables x_1, x_2 etc. as follows:

$$y = \beta_0 + \beta_1 x_1 + \beta_2 x_2 + \cdots + \varepsilon \quad (1)$$

where β_0 is the intercept, β_1, β_2 etc. are partial coefficients and ε is a residual error term (Dobson, 1990; Cramer, 2003).

Equation (1) is not used directly for the simple reason that the response variable y may take any value, and could certainly be well outside the theoretical limits (for binary presence/absence modelling) of 0 to 1.0. Even if the range of observed values of x on which the equation is based limits y to this range, there is no *a priori* reason why the model should not be extrapolated using a wider range of values of x, giving "impossible" values of y. Instead, the variable y from Eq. (1) is used in the following logistic equation that predicts the probability of presence or absence of a species:

$$\mathrm{Pr}_{\mathrm{presence}} = \frac{\exp(y)}{(1 + \exp(y))}$$

and therefore

$$\mathrm{Pr}_{\mathrm{absence}} = 1 - \frac{\exp(n)}{(1 + \exp(n))} \tag{2}$$

Equation (2) has the convenient property that $0 \leqslant \mathrm{Pr}_{\mathrm{presence}} \leqslant 1$, i.e. the response variable $\mathrm{Pr}_{\mathrm{presence}}$ is bounded within the only logical possible range of 0 to 1. Equation (2) is simply one of a number of alternatives that has this desirable property of confining an unbounded prediction (Eq. (1)) to a fixed range, but there seems to have been little investigation in the literature of these alternatives to date.

One of the "problems" with Eqs (1) and (2) is illustrated in Figure 1, where the simplest form of Eq. (1) has been used:

$$y = \beta_0 + \beta_1 x_1 \tag{3}$$

(i.e. also dropping the error term for illustration purposes). Figure 1 shows that logistic regression allows only a single transition from 0 (absence, or "off") to 1.0 (presence, or "on") across the entire range of any single predictor variable x_1, in the situation where y is an increasing function of x_1 (i.e. positive β_1, Figure 1a), or a single transition from 1.0 (presence) to 0 (absence) when y is a decreasing function of x_1 (i.e. negative β_1, Figure 1b). Many biological responses are non-linear with respect to increases in critical driving variables; for example, as temperature increases from very low levels, where species are absent, to intermediate levels (species present) to very high levels (species absent again) we clearly have two transitions (effectively "off" to "on", and then back to "off" again) on a single environmental variable. Without specifying the form of this biological relationship it is clear that a simple logistic model using this single environmental variable could not capture the entire species'

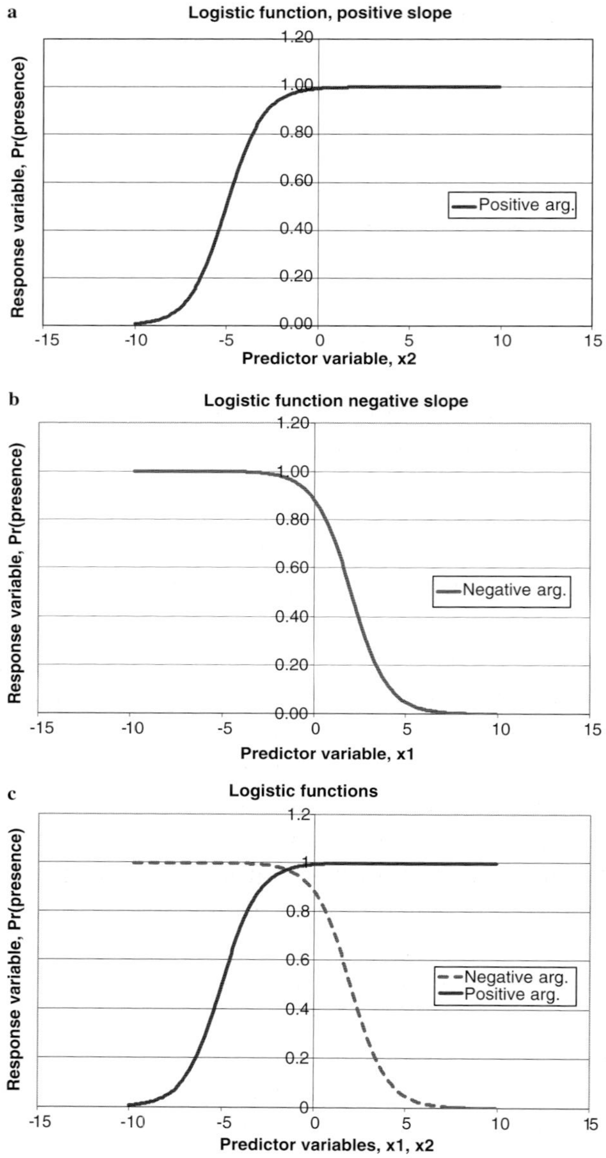

Figure 1 Examples of logistic curves where the dependent variable (y in text Eq. (3)) (a) increases and (b) decreases with the independent variable, x. There can be only one "on"–"off" or "off"–"on" transition with any single variable when y is linearly related to x, so that two independent variables are required to define an "off"–"on"–"off" response (e.g. c).

distribution, although it may be satisfactory for some parts of this distribution.

To capture the "off"–"on"–"off" nature of species' distributions, simple logistic regression requires two variables, one capturing the "off" to "on" transition and the other the "on" to "off" transition. Put together, these can give a reasonable semblance of what is happening on the ground (Figure 1c), but biologically, in a rather unrealistic way.

It may be deduced from the above that a non-linear form of Eq. (3) may in fact be the solution to the "off"–"on"–"off" problem if it can be arranged that y is an increasing function of x_1 over part of the range of x_1 and a decreasing function over other parts of the range of x_1. This may be achieved by the following quadratic expression for y in Eq. (1) (again omitting the error term):

$$y = \beta_0 + \beta_1 x_1 + \beta_2 x_1^2 \tag{4}$$

Incorporating Eq. (4) into Eq. (2) gives the so-called Gaussian logistic equation, which has rarely been used in distribution studies, despite its apparent applicability.

While the various forms of Eq. (2) will produce numbers that range between zero and one the distribution of these outcomes is usually non-normal. This helps to explain why the optimal cut-off threshold in logistic models (i.e. the level of the output variable that is taken to separate "presence" from "absence" in any predictive distribution map) is rarely exactly 0.5. Sometimes authors take an arbitrary cut-off value that gives the best-looking output map. At other times more objective criteria are used (such as ROC or area-under-the-curve methods—see Table 1 and later) that essentially maximize the discriminatory ability of the logistic model (i.e. as judged by its ability correctly to describe the input data).

While logistic regressions are usually applied to simple binary situations they can be extended to situations with more than two categories or outcomes, when they are known as multinomial logistic regressions (Cramer, 2003). This extension of logistic regression models is rarely used in distribution studies, even when obvious categorical variables, or continuous variables made into categories (such as various levels of the abundance of vectors or the prevalence of a disease), are to be modelled.

Table 1 Various accuracy metrics applicable to distribution modelling

Accuracy metric	Range of values	Description	Advantages	Disadvantages
Correct (%)	0–100%	Overall percentage accuracy, all categories combined	Simple and easy to calculate	Presence and absence sites given equal weight. Metric usually affected by prevalence
False positives (%)	0–100%	% of total training set sample wrongly predicted as 'presence'	Simple and easy to calculate	Should be considered with its complement—false negatives
False negatives (%)	0–100%	% of total training set sample wrongly predicted as 'absence'	Simple and easy to calculate	Should be considered with its complement—false positives
Sensitivity	0–1	Ability to identify positives correctly	Derived from diagnostics. Useful measure of positive test accuracy	Concentrates on positives only. Should be considered with its complement—specificity
Specificity	0–1	Ability to identify negatives correctly	Derived from diagnostics. Useful measure of negative test accuracy	Concentrates on negatives only. Should be considered with its complement—sensitivity
Producer's accuracy	0–100%	Ability to predict the training set data correctly	A guide to the modeller to identify where current models are wrong	Not particularly useful to users
Consumer's accuracy	0–100%	Accuracy of model predictions	A guide to the user to indicate the probability with which each model prediction is correct	An important metric for operational use, but not particularly useful to the modeller in identifying model errors
κ (kappa)	−1 to 1	Index of agreement for positive and negative samples combined	Adjusts for chance model agreement with training set data (for which $\kappa = 0$). Applicable to multiple categories of presence/absence or abundance	Sensitive to overall prevalence at high and low prevalence levels
AUC	0–1	AUC is the area under the curve of	Effectively combines sensitivity and	Rather more time consuming to

Table 1 (*continued*)

Accuracy metric	Range of values	Description	Advantages	Disadvantages
		a plot of sensitivity (y-axis) against (1-specificity) (x-axis), sometimes called the receiver operating characteristics (ROC) plot	specificity to assess model accuracy. Commonly used in logistic regression analyses where probability thresholds to achieve best fit (for presence/absence) are often NOT 0.5. Less affected than κ by high/low overall prevalence	calculate than other methods, and more difficult to interpret. Works only for binary (presence/absence) situations
AIC	0–∞	AIC is Akaike's Information Criterion used in information-theoretic models	Estimates the difference between a model's performance and some unknown, ultimate truth. Models with lower AICs are better than those with higher AICs	AIC is used to compare models on an arbitrary scale. Absolute and relative differences between models are more informative, and can determine which models to drop from a candidate set of 'possible' models

A little thought will show that the intercept term in the basic logistic Eq. (1) determines the position of the logistic curve on the predictor-variable axis, while the coefficients determine how quickly the curve rises or falls as the predictor variables change. In other words they reflect the sensitivity of the organism's distribution to changes in the associated environmental variables. Small values of the coefficients result in a slow transition from "off" to "on" or "on" to "off"; large values produce a rapid transition. In practice, of course, the coefficients are estimated by least squares and other methods applied to real datasets, and their inspection can reveal the likely sensitivity of the species to variations of the predictor variables. Needless to say, the greater the number of predictor variables, the less easy it is to draw conclusions about either the contribution of any single one of them to describing distributions, or the sensitivity of the species concerned to small variations in their values.

2.2. Discriminant Analysis Models: The Theory

Details of the discriminant analytical approach to distribution models have been given elsewhere (Rogers, 2000). Initially these models used linear discriminant functions with equal co-variance matrices (Rogers, 1993; Rogers and Randolph, 1993); the use of non-linear discriminant functions and unequal covariance matrices later allowed the description of more complex distributions over larger areas (Robinson *et al.*, 1997; Rogers, 1998). Here we highlight the important differences between discriminant analytical and logistic regression model concepts and conclusions.

Discriminant analysis is one of the techniques that assume a multivariate normal distribution both of predictor datasets and also of the response variable, which is an estimate of the probability of species presence or absence. Briefly, the areas of presence of a species are assumed to experience a range of conditions described by a multivariate normal distribution, and the areas of absence are assumed to be described also by a multivariate normal distribution with a different multivariate mean or "centroid" and (usually) different co-variances of the same set of predictor variables. These two distributions therefore exist in multivariate environmental space and together define a multivariate surface on which it is possible to locate the environmental conditions of any point on a map, and to calculate the probability with which this point "belongs" to the cluster defining presence or, alternatively, to the cluster defining absence. These probabilities are more correctly described as such than is the case for logistic regression probabilities, since they assume some underlying normal frequency distribution, although a normalization step is usually required to calculate them (thus they are linear functions of the exact probabilities rather than the exact probabilities themselves). The use of observed or other more appropriate prior probabilities produces (Bayesian) maximum likelihood output predictions in such discriminant analysis models (Swain, 1978).

This brief description highlights the difference between logistic and discriminant analytical methods and identifies the critical assumption of an underlying uni- or multivariate normal frequency distribution of the predictor variables (and response variable) in the latter, but not the former method. Univariate and multivariate assumptions pervade many branches of ecology in different guises. For example, much competition theory assumes a univariate response of competing species on a continuous environmental variable (Macarthur, 1972); ideas of species packing (hence competitive exclusion etc.) along the same environmental variables were first derived assuming identical and essentially univariate normal responses of species to variation in the resource—the species' exploitation curves (Macarthur, 1972). Canonical and other correspondence analyses, widely used in community ecology, also assume underlying univariate normal responses of individual species within the community to whole sets of environmental variables that are explicitly specified in canonical analyses, but implied in the other sorts of correspondence analyses (Jongman *et al.*, 1995). And so on. While one might legitimately question whether or not the actual responses are precisely uni- or multivariate normal, most uni- and multivariate analytical methods are fairly robust to small departures from normality, and these distribution-based approaches appear to have worked well for these ecological applications. One might even argue that an ecological "model" that does not make some underlying assumption of multivariate normality is the exception rather than the rule. If this is the case why, then, do most biologists use logistic regression to model species distributions? The answer appears to be a mistaken assumption that a frequency distribution of environmental conditions (of either presence or absence points) that is clearly non-normal renders any multivariate method inapplicable. However, rather than change the model to one that does not assume any underlying distribution at all (e.g. the logistic regression model), an alternative option is first to pre-process the environmental data in some way or other to make them more nearly normal, and thus suitable for discriminant analysis or other methods. This can usually be achieved by clustering the environmental data (Rogers, 2000). When this is done, a single distribution (e.g. of environmental conditions in areas of a species' presence) is broken into a series of distributions, each of which is much more likely to be multivariate normal than is the original distribution (in fact, clustering algorithms tend to ensure this outcome).

While the early forms of discriminant analysis, which assumed common and equal multivariate normal co-variances of sets of data points around their respective centroids, performed well for certain datasets (generally involving classifications of flowers or skulls), other methods, such as logistic regression, performed better on species' distribution data. Discriminant analytical methods were, in general, ignored for such applications. This situation was not improved by the relative paucity of software for maximum likelihood discriminant analysis that newcomers to the field could use. Nevertheless, the biological insight provided by discriminant analysis admits this approach to the existing family of informative ecological models that make analogous assumptions about species' responses to environmental conditions. Discriminant analysis easily handles multiple categories of the dependent variables (whether of absence and presence clusters, or multiple categories of continuous variables such as vector abundance or disease prevalence) and thus also copes well with a variety of non-linear biological responses (when the categories or clusters are not collectively linearly dependent on any combination of environmental variables). In addition, the maximum likelihood outputs of discriminant analysis lend themselves to the information-theoretic and multi-model selection methods outlined below.

3. PREDICTOR VARIABLE SELECTION IN DISTRIBUTION MODELS

For both logistic and discriminant analysis approaches to species' distribution modelling, most workers select variables from a suite of candidate descriptors in step-wise inclusion or step-wise exclusion fashion (Rushton *et al.*, 2004). Two decisions have to be taken here. The first is to select a criterion on which to base variable selection; the second is to decide when enough variables have been selected because additional variables no longer improve the model in some way or other.

Criteria for variable selection are based either upon some strictly statistical measure such as the change in Mahalanobis Distance, residual sum of squares, odds- or variance-ratio, or else upon the improved fit of the model to the data with and without the variable in question, in comparison with all other candidate variables in the dataset. These methods will not necessarily, and usually do not, select the same variables. Step-wise inclusion methods will tend to select variables that are un-correlated with each other (because variables that are correlated with those already in the model do not improve model fit), but it follows from this that the first selected variable strongly influences the selection of all later variables. A different first variable very often results in a completely different set of selected variables in alternative models of the same distribution. These different sets of predictor variables often confuse the novice modeller, but they arise ultimately from the correlation structure of the entire predictor dataset, embedded in which are various "subsets" of predictor variables sharing the common characteristic of being least correlated with other variables in the same subset (and therefore by definition more correlated with other variables in other subsets).

Knowing when to stop adding (or subtracting) additional variables in step-wise inclusion (or exclusion) methods for distribution modelling is somewhat problematic. Strict statistical criteria are difficult to define (e.g. what absolute or relative value should be chosen for reduction in the sums of squares?), but easy to apply, once selected. Criteria based on model fits to the data are easier to define, but more difficult to apply in practice (e.g. a 5% improvement in overall model fit is easily obtained with a small change to a large patch of the species' distribution, but a larger change to a smaller patch—giving the same overall 5% improvement—might result in a better-looking or more useful predictive map).

Table 1 lists a variety of accuracy metrics as used in the remote-sensing literature, and many of these can be used for variable selection at model-building time. That is to say, whatever metric is preferred by the user is calculated for each candidate variable's inclusion into the existing model, and the variable that maximizes the metric is the one selected for inclusion next. Clearly those metrics that need to be paired to achieve the best overall accuracy (such as sensitivity and specificity) may be weighted equally when used together, or differentially if the "cost" of incorrect predictions is greater for a false absence than for a false presence prediction (Congalton, 1991; Fielding and Bell, 1997). Kappa varies from -1 (model entirely opposite to observations) through 0 (model fit no better than random) to 1 (perfect fit). Landis and Koch (1977) suggest the following ranges of agreement for the κ statistic: poor, $\kappa < 0.4$; good, $0.4 < \kappa < 0.75$ and excellent, $\kappa > 0.75$.

With a burgeoning number of potential predictor variables, derived from a variety of ground-based or satellite systems, the danger is that some of these layers, by chance alone, will have some skill in describing the observed distribution, and will therefore be selected in the final model. The greater the number of predictor variables available for modelling, the higher is the probability that this will happen. Stepwise inclusion methods will continue to select variables as long as they meet the threshold criterion for inclusion. There is no particular penalty for selecting a model with many variables over one with fewer variables other than that imposed by the residual degrees of freedom, which are usually large enough not to affect the level of significance of a final model with many or with few variables.

The practice of using as many descriptor layers as possible and selecting them in some fashion or other that improves the "fit" of the model is called "data-mining" by its proponents, and "data-dredging" by its detractors. The latter suggest that this practice leads to final models that are neither parsimonious nor informative biologically and suggest instead an information-theoretic approach that overcomes many of these problems. These methods are described later.

4. WHAT TO DO WITH SPARSE DATASETS?

Distribution models are designed to draw inferences from sparse datasets to make predictions about distributions over much larger areas. In this section, we discuss the nature of sparse datasets, and what may be done to ensure that the resulting maps use as much of the information they contain as is possible.

The training set data are a subset of the total distribution of any species and thus contain partial but not full information about the species' distribution, hereafter referred to as "reality" or "truth". A different training set, of the same sample size, would contain different points and therefore would present different partial information about the same reality. A modelling approach applied rigorously to each training set would produce two different predictions of reality. Which is correct? In fact it is very likely that neither model is correct, although one is going to be more correct than the other. The problem for the modeller is that she or he does not know which is the more correct, because she or he does not know what reality—the true distribution—actually is. All that we can know is that modelling using different training sets will give different predictions of reality, among which it is apparently impossible to choose the "best". Nevertheless, we should try to estimate from the training set how variable are our modelling estimates of reality, because users of predictive maps can rightly expect of the modeller some statement about the uncertainty of any model's predictions.

A second problem arises when we know a sparse dataset does not sample the entire geographic range of a species' distribution. When observations come from only part of a species' range, is it possible to infer from them the wider geographical distribution of which they are only a part? Thus the first problem above is one of sparse but geographically unbiased training set data while the second is one of sparse and geographically biased training sets. Two methods help us to draw as much information as possible from sparse training sets of different degrees of geographical bias: bootstrap sampling and environmental envelope expansion.

4.1. Bootstrap Sampling

Bootstrap sampling is one of the methods normally used for testing the accuracy of predictive models (Davison and Hinkley, 2003). A bootstrap sample is simply a sub-sample of a set of training data that is used to make one prediction of a species' distribution. Multiple bootstrap samples are taken, a prediction is made for each and the entire set of predictions is then combined to produce a single, average prediction. This prediction reflects the uncertainty we should have of a species' real distribution "on the ground", on the basis of any particular set of training data.

Bootstrap samples should be taken from the training set with replacement because we assume that the training set itself is a sample of reality, and the occurrence of any one observation within it is essentially random. A different training set could contain that observation more than once, or not at all. Only bootstrap sampling with replacement reflects this.

One advantage of bootstrap sampling is that within any one model the samples can be arranged to have equal numbers of presence and absence observations. Recent work suggests that this situation produces model outputs with the greatest accuracy (McPherson *et al.*, 2004).

Figure 2 (Figure 2 is Plate 1.2 in the Separate Color Plate Section) shows an example of the bootstrap approach applied to a very sparse dataset for Rift Valley Fever (RVF) in Africa and the near Middle East. Figure 2a shows the fit of a discriminant analytical model applied to the entire training set of presence ($n = 62$) and absence ($n = 2000$) points (the latter selected to be within 0.5 and 10.00 degrees of any presence point). The accuracy metrics for this model are high (overall % correct = 97.0; false positives = 2.3%, false negatives = 0.7%; $\kappa = 0.813$), although this is more due to an accurate description of absence than of presence points (sensitivity = 0.774, specificity = 0.976). Overall, the map underestimates the distribution of this disease within Africa. Figure 2b shows the mean predictions of 100 bootstrap models using the same dataset (each model with 200 presence and 200 absence points, randomly selected with replacement from the training set). The bootstrap approach extends the areas predicted to be suitable for this disease, but there is considerable uncertainty about the status of many of the additional high-risk areas of Figure 2b compared with Figure 2a. Nevertheless, the average accuracy statistics for the 100 models that contributed to Figure 2b are impressive. The mean κ value for the top 10 models was 0.895 (S.D. = 0.017) and for the bottom 10 models was 0.851 (S.D. = 0.020).

4.2. Environmental Envelope Expansion

When it is suspected that the training set is geographically biased, a different approach may be adopted. In this approach, only the presence data points are used and a selection of absence points is generated within the GIS (as in the bootstrap example above) that are within a specified geographical distance of the presence points. There should be many more absence than presence points. The approach assumes that some of these absence points are in reality presence points that have simply not yet been identified as such, and the challenge is to establish which these are. Again multiple models are run. In each model, all of the observed presence points (or bootstrap samples of them) are taken, together with an equal number (i.e. a subset) of the absence points. In the various runs of the model, any particular absence point appears with a different sample of other absence points and they collectively define the environmental envelope of absence; the presence points obviously define the environmental envelope of presence. Clearly in any one run of the model, a particular point of absence may be assigned to the category of presence with a certain probability. If this same point is always assigned a high probability of presence in all the models in which it is included, and regardless of the other absence points with which it is associated, then it seems reasonable to re-assign it to the category of presence points. The output of each model run is therefore stored in terms of the predicted probability of membership of the presence and absence categories of each point in the dataset used. When all the model runs are completed the average probabilities are calculated from the stored data. Data points for presence that are consistently predicted to belong to the presence category with very low probability are re-assigned to the absence category (there are usually very few of these) and data points for absence that are consistently predicted to belong to the presence category are re-assigned to this category. Thus the environmental envelope for presence is expanded along the environmental dimensions defined by the sparse training set. The new training set with the re-assigned data points is then used to make map predictions in the usual way (or by bootstrap sampling, previous section). Further rounds of envelope expansion can be carried out, each one starting with the output of the previous round.

This approach has been very successful in expanding predictions of the geographical distributions of diseases within the same broad climatological/vegetation type as the training set data, but it obviously cannot identify other areas of vector or disease presence that are quite different from the training set regions. An example using the Rift Valley dataset (previous section) is shown in Figure 2c. The figure shows the results after the first round of environmental envelope expansion applied to the training set data, and using exactly the same bootstrap samples as were used to produce Figure 2b. In this case, the presence/absence status of the training set data was adjusted on the basis of the mean probabilities of presence/absence determined for them in the 100 bootstrap samples; training set "positives" that were predicted to belong to the presence category with a mean probability of <0.25 were re-assigned to the absence class, while training set "negatives" with a mean probability of presence of 0.50 or greater were re-assigned to the presence class. A single model was then produced, in the same way as Figure 2a, but using this corrected training set (overall % correct = 94.4, false positives = 4.3%, false negatives = 1.4%, $\kappa = 0.757$, sensitivity = 0.767, specificity = 0. 955). In comparison with the predictions in Figure 2b, Figure 2c indicates a greater degree of model certainty of disease presence throughout the savannah zones of West Africa but is more cautious in its predictions for the Arabian Peninsula and in East Africa (Ethiopia, Somalia, Kenya, Tanzania, Zambia and Zimbabwe). RVF certainly occurs in the latter five countries, but has not yet been reported from Ethiopia (Gideon, 2005). Figure 2 shows that the two modelling approaches are making rather different predictions using the same sparse dataset. The bootstrap approach seeks to extract the maximum information from the training set as given, and without necessarily questioning the status of each data point. Environmental envelope expansion begins with the assumption that some of the training set data are wrongly assigned to the absence category, and seeks to identify these, to re-assign them to the presence category. More experience with both methods is required before we can decide which is better for handling sparse datasets.

5. INCORPORATING SPATIAL INFORMATION INTO MODELS

So far the models discussed do not explicitly take into account the spatial arrangement of the training set observations, nor are any spatial variables explicitly used in model predictions. These predictions are generally spatially coherent for the simple reason that the environmental data on which they are based exhibit spatial co-variation, as discussed by Atkinson and Graham (this volume, pp. 79–118). Many sets of training data consist simply of point records from which it is difficult to infer the spatial relationships of the underlying distribution (e.g. the RVF dataset of the previous section). In other cases, however, and certainly when the training data are in polygon form (where the polygons represent the natural boundaries of the species' distribution, not political or administrative level units), these data do contain some information on the spatial pattern of the species' distribution, and this information can be, and probably should be, included in the model. Individuals of a species occur in a spatially coherent pattern for two different reasons: firstly the individuals may be independently responding to the same set of environmental conditions. Since the individuals are likely to share environmental preferences, they are likely to co-occur regardless of any biological interaction between them. Secondly animals and plants occur in patches through the processes of natural reproduction and limited dispersal away from the parental area. Whatever the underlying causes, the pattern can be included as a predictor variable in the model by defining, for each observation of species' presence, the occurrence of the species in adjacent habitat squares (or pixels). Unfortunately, of course, the number of occupied adjacent pixels is unknown for each of the training set observations and so must be estimated. The modelling therefore proceeds in stages. Initially an ordinary (i.e. non-spatial) model is fitted to a set of training data and applied to all the pixels in the region, thus generating a first "guess" of the species' distribution, from which the pattern of adjacent occupation (i.e. the auto-covariance) can be estimated for the training set data. Next a new model is constructed, this time using the auto-covariances (plus the other predictor variables) and the training set data. This model is applied to all the pixels in the image using the current model to estimate the auto-covariances. Predicted values for the unsurveyed squares are combined with observed values of the surveyed squares to generate a new distribution prediction.

This new prediction is the starting point for further rounds of prediction, constituting what is known as the Gibbs' sampler. During these rounds, a model is made of the training set data and taking account of their auto-covariances as judged by the current map. A random starting point in the map is selected and predictions are then made only for the unsurveyed squares, again using each square's auto-covariance. This process is repeated until the predicted map does not change significantly between iterations.

In the above modelling approach, the predictions are generally probabilities but the interim and final maps record simply presence/absence, and the auto-covariances are calculated from these. Initial implementation of the Gibbs' sampler recommended probabilities be turned into presence/absence predictions stochastically (i.e. rather than by thresholding the probabilities at a certain level to ensure, for example, the same number of predicted as observed occupied squares).

When this approach was applied to describe the distribution of red deer in the Grampian Region of Scotland, Augustin *et al.* (1996) showed that this stochastic rule does not rapidly lead to convergence of parameter estimates, whereas leaving the probabilities as predicted by the models did so. In this case, the auto-covariances are calculated as a weighted function of the probabilities rather than of the 0/1 stochastic realizations of these probabilities. This alternative approach is called the "modified Gibbs' sampler" and was found by the authors to lead to rapid convergence of parameter estimates and also to give the best results of the four methods tried, which were a simple logistic model, an auto-logistic model (that stopped before the Gibbs' sampler stage), a Gibbs' sampler and a modified Gibbs' sampler model.

As pointed out at the start of this section, models which do not explicitly take account of the spatial coherence of species' occurrences, nevertheless, usually give predictions that are spatially coherent for the simple reason that the predictor environmental variables change gradually through space, not abruptly. The relative contributions of environmental variable and species' auto-covariances to species' occurrence in any area can be gauged by running models without and with the auto-logistic, Gibbs' sampler or other explicit measures of the species' distribution as predictor variables. If, for example, the Gibbs' sampler adds nothing to the descriptive power of a distribution model, it may be concluded that direct or indirect individual interactions are unimportant in determining the species' presence in any area and that any spatial coherence seen is due to environmental variables alone.

6. MODEL SELECTION AND MULTI-MODEL INFERENCE

A significant departure from the modelling approach described above is provided by the work of Burnham, Anderson and others who promote what is called an information-theoretic approach that appears to tackle a number of problems that arise with the more traditional ways of modelling species' distributions (Burnham and Anderson, 2002).

In this approach, it is assumed that there exists an *n*-dimensional and unknowable truth (the real distribution of organisms in the present case) that models can only attempt to approximate rather than describe completely. There exists, therefore, a certain distance ($I(f,g)$) between model (g) and reality (f) that is captured by the Kullback–Leibler (K–L) information or distance measure which is defined as:

$$I(f,g) = \int f(x) \log\left(\frac{f(x)}{g(x|\theta)}\right) \mathrm{d}x \tag{5}$$

for continuous functions and

$$I(f,g) = \sum_{i=1}^{k} p_i \cdot \log\left(\frac{p_i}{\pi_i}\right) \tag{6}$$

for discrete distributions such as the Poisson, binomial etc. In Eq. (5), full reality f is considered fixed while g varies over a range of models indexed by θ. In Eq. (6), there are k possible outcomes of the underlying random variable. The true probability of outcome i is p_i, while the modelled outcome is π_i, with $\sum p_i = \sum \pi_i = 1$.

These rather fearsome looking equations are really quite simple. It is obvious from both, for example, that in the unlikely event that the models perfectly describe reality, $g(x|\theta) = f(x)$ in Eq. (5) and $\pi_i = p_i$ in Eq. (6). The logarithmic terms will therefore be 0 (because $\log(1) = 0$) and the Kullback–Leibler distance, $I(f,g)$, will thus also be 0 in each case. The greater the discrepancy between model and reality, the larger will $I(f,g)$ become. Thus the K–L distance is a guide to model accuracy and may be used to select the best from a set of candidate models for any particular situation.

There is one obvious problem, however, and that is that we do not know in each case what the truth ($f(x)$ or p_i) actually is. Taking the continuous case as an example, Eq. (5) can be re-arranged as follows:

$$I(f, g) = \int f(x) \log(f(x))\, \mathrm{d}x - \int f(x) \log(g(x|\theta))\, \mathrm{d}x \qquad (7)$$

with the following statistical expectations:

$$I(f, g) = E_f[\log(f(x))] - E_f[\log(g(x|\theta))] \qquad (8)$$

each with respect to the distribution f. The first expectation on the right-hand side of Eq. (8) will be unknown (because it is the expectation of reality) but constant (reality does not change!). The second expectation on the right-hand side of Eq. (8) will vary, depending both upon the model and its current parameters. This means that although $I(f,g)$ cannot be evaluated exactly, it can be estimated up to a constant C (viz $E_f[\log(f(x))]$)

$$I(f, g) = C - E_f[\log(g(x|\theta))]$$

or

$$I(f, g) - C = -E_f[\log(g(x|\theta))] \qquad (9)$$

The left-hand side is a relative directed distance between f and g and thus the value of the right-hand side can be used to select between different candidate models. A model with a lower value of this quantity is better than one with a higher value. Because we do not know C we can never know just how good our "best" model really is, but the difference between models is a guide to how much better is our best model than any others in the candidate set.

In the discussion so far, it is assumed that the parameters of the candidate models are already known. In reality they must be estimated from a set of data. Akaike showed that in practice the K–L distance could be estimated from the empirical log-likelihood function evaluated at its maximum point (Akaike, 1973). The practical equivalent of Eq. (9) is what has since become known as the "Akaike Information Criterion" or *AIC*, defined as follows:

$$AIC = -2\log(\ell(\hat{\theta}|y)) + 2K \tag{10}$$

where $\log(\ell(\theta|y))$ is the value of the log-likelihood at its maximum point (i.e. the maximum likelihood estimate) and K is the number of estimated parameters in the model. It is clear from Eq. (10) that the first term on the right-hand side will tend to decrease as the number of parameters in the model increases (because a model with more parameters is almost bound to fit a dataset better than one with fewer parameters) while the second term ($2K$) will obviously increase. This achieves a neat balance between over-fitting a model (too many parameters, *AIC* penalized with a large value of $2K$) and under-fitting a model (too few parameters, *AIC* large because the first term is large).

A modification of the *AIC* was suggested by Hurvich and Tsai (1989) for the situation where the sample size is small in relation to the number of fitted parameters. This modification, the corrected *AIC* or AIC_c, is calculated as follows:

$$AIC_c = -2\log(\ell(\hat{\theta}|y)) + 2K\left(\frac{n}{n-K-1}\right) \tag{11}$$

where n is the sample size and all other terms are as in Eq. (10). In general, unless the sample size is large in relation to the number of estimated parameters, Eq. (11) is to be preferred over Eq. (10).

The modelling approach recommended by Burnham and Anderson (2002) involves proposing a set of candidate models for the biological situation involved, then fitting these models to the data and calculating the *AIC* or AIC_c values. As mentioned before, the absolute values of these quantities are usually of little interest, but differences between them are very informative. The *AIC* difference (Δ_i) is defined as follows:

$$\Delta_i = AIC_i - AIC_{min} \tag{12}$$

where AIC_{min} is the minimum *AIC* for any candidate model in the set of models, and the model with this minimum value is the current best one. Despite the very wide possible range of absolute values of *AIC*, *AIC* differences of approximately >10 indicate models that have very little support and therefore can be omitted from further consideration, while *AIC* differences of <2 are indicative of strong support. Given any particular set of models, the likelihood of one of the models within the set (g_i), given the data, is proportionately related to the *AIC* difference by the following:

$$\ell(g_i|x) \propto \exp\left(-\frac{1}{2}\Delta_i\right) \tag{13}$$

These likelihoods are usually normalized across the entire set, *R*, of candidate models to determine a set of Akaike weights, w_i that sum to 1.0:

$$w_i = \frac{\exp\left(-\frac{1}{2}\Delta_i\right)}{\sum_{r=1}^{R} \exp\left(-\frac{1}{2}\Delta_r\right)} \tag{14}$$

These weights are thus an effective way to scale and interpret the *AIC* difference values.

Equations (12)–(14) involve comparisons *between* models, and Eq. (14) refers to a particular set of models. Thus one can only conclude that a particular model has a Δ_i, likelihood or Akaike weight relative to some one (Δ_i) or all other models (w_i) in a particular set of models. Choice of a candidate set of models therefore becomes crucial. If a candidate model is dropped from the set, or a new model is added, the various quantities should be recalculated. However, a quantity called the evidence ratio w_i/w_j, where *i* and *j* are just two of the candidate models, is *not* affected by any other model in the candidate set, but just by the two models being compared. Evidence ratios may be used to judge how much better one model is compared with another, regardless of any other models in the candidate set.

For some biological systems where the mechanisms are fairly well understood, the set of candidate models may be easy to define. For example if we seek a model for plant growth, we might generate a series of potential models that involve the quantity of available sunlight, water or soil nutrients in various combinations. The information-theoretic approach is ideal in this situation because what we really seek is some idea of the relative importance of variables we know, or suspect, to be of importance. In the case of animal and plant distributions, however, it is much more difficult to identify in advance a set of "reasonable" environmental variables and so we tend to fall back upon the step-wise or data-mining methods described earlier in this review. Nevertheless, it seems that even here this alternative approach may be able to help. For example, we could generate a set of candidate models which described distributions using different sorts of variables (temperature, humidity, vegetation indices) and select between them. Burnham and Anderson are sympathetic to this approach, if only because it is, in their view, the lesser of two evils:

> "While we do not condone the use of information theoretic approaches to blatant data dredging, we suggest it might be a more useful tool than hypothesis testing in exploratory data analysis where little a priori knowledge is available. Data dredging has enough problems and risks without using a testing-based approach that carries its own set of substantial problems and limitations" (Burnham and Anderson, 2002).

The information-theoretic approach provides a completely different paradigm from the traditional statistical approach to model building. There are no formal levels of any test statistic that determine "significance" of one result over another, and therefore no formal hypothesis testing either. As Burnham and Anderson point out, there are many areas of life and science that involve numbers that do not readily fall within the realm of traditional statistical testing. For example, one does not ask for a formal test of significance if a soccer match is won by 3 goals to 1 or by 10 goals to 1. One infers that the winners in the second match were considerably better than their opponents, in comparison with the winners of the first match. How large should be the differences in goals scored for them to be judged "significant" is irrelevant in this case. The match results simply give us evidence for the greater superiority of the winners (compared with the losers) of the second match compared with the first, and allow us to rank the teams in a tournament situation. Model selection and multi-model inference is in many ways more like a tournament. We seek the best possible candidate from a whole suite of models to do the job we have in hand. We are able to say how much better is this model compared to all the other models we have constructed, and we are able to discard at least some models because some or other information metric (the Akaike weight, or the evidence ratio) puts them so much lower than the current best model. There are, however, no threshold values for any of these metrics, signifying "significant" in one case or "not significant" in another, because such formal statistics are inappropriate in this situation. Burnham and Anderson go so far as to say that the use of null hypothesis testing for model selection must be considered ad hoc (albeit a rather refined set of ad hoc procedures), whereas there is a sound theoretical basis to the information-theoretic approach to model selection criteria. (There remains a role for formal hypothesis testing in more experimental situations where the experimenter can define treatment and control groups that differ only in a single or limited number of variables, although even here it is not so much the significance of the effect that is of interest, but the size of the effect.)

Needless to say, the issue of the type of models we should use for distribution mapping is still debated. A recent review of distribution modelling strongly favours the information-theoretic approach (Rushton *et al.*, 2004). A later article in the same journal redresses the balance with a plea for pluralism (Stephens *et al.*, 2005).

6.1. Application to Vector and Disease Mapping

Since the output of discriminant analysis can be expressed as a probability (strictly a Bayesian posterior probability), the likelihood ℓ is simply

$$\ell = \prod_i \Pr(Y_i) \tag{15}$$

where $\Pr(Y_i)$ is the probability of the observed outcome, defined as

$$\Pr(Y_i) = P_i^{Y_i} Q_i^{1-Y_i} \tag{16}$$

where P_i is the predicted probability of presence for a presence point ($Y_i = 1$), and Q_i is the complement of P_i.

The most convenient form of the log likelihood function of Eqs. (15) and (16) is the following:

$$\log(\ell(\hat{\theta}|y)) = \sum_{i \in A_1} \log P(x_i, \theta) + \sum_{i \in A_0} \log Q(x_i, \theta) \tag{17}$$

where A_1 and A_0 denote the sets of observations with $Y = 1$ and 0, respectively (Cramer, 2003). Thus it is possible to calculate the corrected AIC_c, from Eqs. (17) and (11). One could therefore use the AIC_c to select between models and, equally importantly, to decide how much better the best model is compared with the others.

The Akaike weights are also useful in helping to determine the relative importance of the predictor variables. If the current best model contains variable x_1, say, but has only a modest Akaike weight, then it is clear that there is considerable model uncertainty and therefore only weak evidence for the importance of x_1 as a predictor variable. However, the Akaike weights can be summed for all models across the set that contains x_1, or x_2, or x_3, etc. and these summed weights reflect the relative importance of these variables across all models (Burnham and Anderson, 2002). It will generally happen that the sum of the Akaike weights for a variable will exceed the Akaike weight of the best model (in which the variable may or may not occur) and it can also happen that a variable not in the "best" model can have a summed Akaike weight that exceeds that of any other variable, even those included in the "best" model. These summed Akaike weights therefore highlight the relative importance of each variable regardless of which models the variable occurs in. This procedure can also be extended to pairs of variables, or to interaction effects between variables (if interaction terms are included in the candidate models). For the correct conclusions to be drawn about any particular variable, it is advisable to use a set of candidate models in which the variables being compared occur about the same number of times. Obviously this is more likely to be the case for important variables than for unimportant ones.

In all the models shown in Figure 2 the variables were selected for sequential step-wise inclusion using the AIC_c as the selection criterion; thus during each round the variable was selected that gave the smallest value of the AIC_c, indicating a closer approach to the "truth" or "reality" than with any other candidate variable. Virtually always the reduction in AIC_c exceeded 10 for every included variable, indicating that each additional variable made a substantial contribution to improving the fit of the model. The models always included only 10 variables, although on the criterion of AIC_c reduction it appears that many more could have been included. There were no signs of an increase in the AIC_c due to the penalty imposed by increasing numbers of predictor variables K (Eq. (11)), which is not surprising given the large total number of data points in each model (400 in the bootstrap models; >2000 in the others). When the bootstrap models were ranked from lowest AIC_c (i.e. best fit) to highest AIC_c, the mean values of the alternative kappa statistic for model accuracy correlated sensibly with them, although there was considerable scatter of individual results ($\kappa = 0.895$, S.D. $= 0.017$ for the top 10 models; $\kappa = 0.851$, S.D. $= 0.020$ for the bottom 10 models). Figure 3 (Figure 3 is Plate 1.3 in the Separate Color Plate Section) shows the sets of variables selected by each of the bootstrapped models, which are arranged in AIC_c order from lowest (at the top of the image) to highest (bottom of the image). In Figure 3 each row of the image refers to one model, the variables selected for it are shown as coloured squares, and their order of selection is indicated by the rainbow colour scale, red orange, yellow green etc. (see figure legend). The red line down the middle of the image indicates that variable 14, the annual amplitude of the Land Surface Temperature, was not only consistently selected in all of the models but was often selected as the most (or second most) important variable. No other variable had such a consistent performance or such a high rank, although the very last one, the variance of the Normalized Difference Vegetation Index, was fairly often selected second in the lower-ranking models. Within this ranked series of models the Akaike weights (Eq. (14)) indicated that the very few top models were so much better than all the others that the rest could be discarded (the weights for the top five models were 0.8074, 0.0799, 0.0710, 0.0413 and 0.0003, respectively). Table 2 shows the summed Akaike weights for each of the top 20 variables across all models in which they appeared. This allows the assessment

Table 2 Rift Valley Fever model predictor variables

Variable	Summed Akaike weight	Mean rank	N/100
wd1007a1	1.00000	1.26	100
wd1014p3	1.00000	6.28	81
wd1014a1	0.99993	7.74	69
wd1007mn	0.99951	10.33	23
DEM	0.95830	7.46	54
wd1003a1	0.92011	8.41	51
wd1007p2	0.88755	8.51	43
wd1007vr	0.84904	6.45	65
wd1003a2	0.80750	10.03	27
wd1007p3	0.80750	9.99	22
wd1007a3	0.19211	9.77	34
wd1003a0	0.15076	10.62	11
wd1014p2	0.11242	8.98	44
wd1003vr	0.07989	9.35	40
wd1014a2	0.07985	10.5	17
wd1014p1	0.07106	10.26	25
wd1003a3	0.04170	9.8	31
wd1014mn	0.04131	9.48	22
wd1014vr	0.00046	6.72	62
wd1007mx	0.00042	10.44	20

Summed Akaike weights (second column) for the top predictor variables (first column) of the 100 RVF bootstrap models used to produce Figure 2b (see text for details). The mean ranks (i.e. the order in which the variables were selected, where rank 1 = the first selected variable, rank 10 = the tenth selected variable and all non-selected variables are given a rank of 11) are given in the third column, and the number of times (out of 100 models) each variable was selected is given in the final column.

Notes: Key to variable names: *wd10* refers to NOAA-AVHRR data at 0.10 degrees resolution in the latitude/longitude format, *03* refers to the AVHRR channel 3 (MIR), *07* to Land Surface Temperature and *14* to Normalized Difference Vegetation Index (NDVI) data; *a0* is the (Fourier and arithmetic) mean, *a1*, *a2* and *a3* refer to the amplitudes of the annual, bi-annual and tri-annual cycles, respectively of temporal Fourier processed imagery and *p1*, *p2* and *p3* to their corresponding phases (timing of the first peak); mn and mx refer to the minimum and maximum and vr to the variance. DEM is a digital elevation layer (see Hay *et al.*, this volume, pp. 37–77 for more details).

of the importance of each variable independently of both the other variables, and of the particular models in which the variable appears. Given the very high weights of the very few top models in this series, these summed weights reflect whether or not the variables concerned were included in these top models. The summed Akaike weights (Table 2, second column) do not necessarily scale with the mean rank (i.e. mean order of step-wise inclusion into the models) of each predictor variable (Table 2, third column). Perhaps not surprisingly, given its occurrence in all models and often as the first variable, the annual amplitude of the Land Surface Temperature has the highest summed Akaike weight. The next variable (phase of the tri-annual NDVI cycle) had an equal summed weight, but appeared in fewer models (Table 2, last column), although in all of the top 25 models. The next three variables appeared in 20 (NDVI annual amplitude), 6 (LST minimum) and 14 (DEM) of the top 25 models, respectively. The "best" model had a combination of variables not found in any other of the 100 models. Whether or not this indicates that this unique combination of variables describes the disease better than any other, or simply the bootstrap sample that happened to have been selected during that run of the model, is a moot point. By its very nature bootstrap sampling is bound to produce, simply by chance, one set of randomly selected data that gives rise to a model that is "better" than all the rest. Only if this subset of data is an unbiased sample of the real disease situation on the ground (something that cannot be assessed; this is the unknown "truth" we are trying to model) will this best model also be the best model for the disease.

7. CONCLUSION

Despite a plethora of models available to describe vector and disease distributions, practitioners tend to favour only a few approaches. Logistic regression methods appear to have the largest number of followers (Rushton *et al.*, 2004), especially within Europe, while the GARP approach has a strong following in North America (Stockwell and Peters, 1999). Part of the reason for this dominance is the availability of logistic-regression routines within many standard statistical packages, or of the GARP program from dedicated websites. This review seeks to redress the balance, to make distribution mapping yet another branch of ecology, with underlying assumptions of normally or multivariate normally distributed predictor variables. In so doing its aim is not only to point out the numerous advantages of such assumptions (for example, their biological realism coupled with statistical tractability) but also to provide a route map towards a more unified theory combining population and community ecology. Population dynamics and the distribution of individual species are the basic ingredients of ecological communities but, to date, these subjects have been studied in isolation because, it seems, of an incompatibility between the under-pinning theoretical models. Briefly, methods that made no assumptions about underlying frequency distributions of predictor variables, or that allowed only single "off"–"on" transitions on single variables (as does simple logistic regression) were hardly compatible with the Macarthurian tradition of quantitative community ecology in which the normal distribution is an all-pervading assumption, regarded by many even as an axiom. It might indeed be a delusion that the response of any species to gradual changes in each key predictor variable is shaped like a normal, or apparently normal, curve; but it is certainly not shaped like the underlying sigmoid curve of the logistic regression model. Biological processes occur within bounds, and therefore the sum total of biological processes that ultimately determine a species' distribution is also likely to occur within bounds. The future holds many uncertainties for biological communities such as how they will respond to gradual habitat destruction, or to climate change? By unifying individual and community analysis we might be able to discover if communities are no more than the "sum of their parts", in which case we might expect communities to re-arrange themselves under the forces of global change, or if indeed they are greater that the sum of their parts, in which case environmental change may result in community disintegration, and a significant loss of ecosystem services to human-kind.

Vectors and vector-borne diseases are one component of ecosystems, and they too may show gradual or catastrophic changes as environments change. Rogers and Randolph (this volume, pp. 345–381) show that we are presently far from understanding which type of change to expect. Their great sensitivity to environmental conditions suggests that these diseases may be among the first of all diseases to show distribution and intensity changes as climates change but, as Rogers and Randolph (this volume, pp. 345–381) points out, we must not let this sensitivity lead us to conclude that *any* change in a vector-borne disease is due to climate. By modelling species' distributions in the ways outlined in the present review, it should be possible to estimate the sensitivity of any disease to climate change by examining the limits of its environmental envelope in multivariate space. When this is matched to the predicted changes in climate, also in multivariate space, it is possible to map the areas on the ground, which will fall within the environmental envelope in the future. It is these areas that are at risk of disease invasion and spread.

A common feature of many vector-borne and other diseases is the paucity of hard data we have for their precise geographic distributions. This did not particularly matter in the days of equally sparse climatic data (also from point sources) or of fairly coarse spatial "climate surfaces" that were produced from such data. One set of coarse data could be related to another set of coarse data to produce a risk map of such poor spatial resolution as to be almost useless. Today, however, environmental data from satellites are available at unprecedented spatial resolutions, and these reveal the inadequacies both of past maps and of the data on which they are based (see Hay *et al.*, this volume, pp. 37–77). This therefore presents us with a new problem of how to deal with sparse distribution data. While it is possible to produce risk maps from such data using the techniques outlined in this review, one must nevertheless ask whether the current best model is describing simply the data or, more importantly, the disease. Bootstrap sampling does not resolve this problem, since a bootstrap sample of a biased training set is itself likely to be biased. If all maps are wrong, and only some of them are useful, we must be able to distinguish the useful from the useless. It is likely that only additional, independent data collected during ground-truthing of risk map predictions will enable us to do this. It is, perhaps, time to return to the field.

ACKNOWLEDGEMENTS

I thank my fellow editors, Drs Simon Hay and Alastair Graham, for their patience during the production of this article, and for reading and commenting on the manuscript. Simon Hay kindly provided the RVF data set used in the models of Figure 2. I also thank Dr Bethan Purse, Dr Andy Tatem and Ben McCormick for their constructive comments. The research upon which this chapter is based was carried out with funding from a Wellcome Trust Showcase Award (Ref. ♯067614), from the Synergy program administered by the Raytheon Corporation, from the International Research Partnership for Infectious Diseases (INTREPID) headed by Ms Monica Myers. It was also partially funded by EU grant GOCE-2003-010284 EDEN and the paper is catalogued by the EDEN Steering Committee as EDEN0008 (http://www.eden-fp6project.net/). The contents of this publication are the sole responsibility of the authors and do not reflect the views of the European Union.

REFERENCES

Akaike, H. (1973). *Information theory as an extension of the maximum likelihood principle*. In: Second International Symposium on Information Theory (B.N. Petrov and F. Csaki, eds), pp. 267–281. Budapest: Akademiai Kiado.

Augustin, N.H., Mugglestone, M.A. and Buckland, S.T. (1996). An autologistic model for the spatial distribution of wildlife. *Journal of Applied Ecology* **33**, 339–347.

Box, G.E.P. (1979). Robustness in the strategy of scientific model building. In: *Robustness in Statistics* (R.L. Launer and G.N. Wilkinson, eds), pp. 201–236. New York: Academic Press.

Burnham, K.P. and Anderson, D.R. (2002). *Model Selection and Multimodel Inference: A Practical Information-Theoretic Approach*, 2nd edn. New York: Springer.

Congalton, R.G. (1991). A review of assessing the accuracy of classifications of remotely sensed data. *Remote Sensing of Environment* **37**, 35–46.

Cramer, J.S. (2003). *Logit Models from Economics and other Fields*. Cambridge: Cambridge University Press.

Davison, A.C. and Hinkley, D.V. (2003). *Bootstrap Methods and their Application*, 1st edn. (reprinted). Cambridge Series on Statistical and

Probabilistic Mathematics (R. Gill, B. D. Ripley, S. Ross, M. Stein and D. Williams, eds) Cambridge: Cambridge University Press.
Dobson, A.J. (1990). *An Introduction to Generalized Linear Models*. London: Chapman & Hall.
Fielding, A.H. and Bell, J.F. (1997). A review of methods for the assessment of prediction errors in conservation presence/absence models. *Environmental Conservation* **24**, 38–49.
Gideon (2005). Gideon. Gideon Informatics Inc., 6010 Wilshire Blvd, Suite 302. Los Angeles, CA 90036 USA. http://www.GIDEONonline.com.
Green, P.E. (1978). *Analyzing Multivariate Data*. Hinsdale, IL: The Dryden Press.
Hurvich, C.M. and Tsai, C-L. (1989). Regression and time series model selection in small samples. *Biometrika* **76**, 297–307.
Jongman, R.H.G., Ter Braak, C.J.F. and Van Tongeren, O.F.R., eds (1995). *Data Analysis in Community and Landscape Ecology*. Cambridge: Cambridge University Press.
Landis, J.R. and Koch, G.C. (1977). The measurement of observer agreement for categorical data. *Biometrics* **33**, 159–174.
Macarthur, R.H. (1972). *Geographical Ecology*. New York: Harper & Row.
McPherson, J.M., Jetz, W. and Rogers, D.J. (2004). The effects of species' range sizes on the accuracy of distribution models: ecological phenomenon or statistical artefact? *Journal of Applied Ecology* **41**, 811–823.
Robinson, T., Rogers, D. and Williams, B. (1997). Mapping tsetse habitat suitability in the common fly belt of Southern Africa using multivariate analysis of climate and remotely sensed vegetation data. *Medical and Veterinary Entomology* **11**, 235–245.
Rogers, D.J. (1993). Remote sensing and the changing distribution of tsetse flies in Africa. In: *Insects in a Changing Environment* (R. Harrington and N.E. Stork, eds), pp. 167–183. Symposium Number 17. London: Royal Entomological Society.
Rogers, D.J., (1998). Satellite imagery and the prediction of tsetse distributions in East Africa. In: *Towards Livestock Disease Diagnosis and Control in the 21st Century. Proceedings of an International Conference in Vienna, Austria*, 7–11 April 1997, pp. 397–420. Vienna: International Atomic Energy Agency.
Rogers, D.J. (2000). Satellites, space, time and the African trypanosomiases. *Advances in Parasitology* **47**, 129–171.
Rogers, D.J. and Randolph, S.E. (1993). Distribution of tsetse and ticks in Africa: past, present and future. *Parasitology Today* **9**, 266–271.
Rushton, S.P., Ormerod, S.J. and Kerby, G. (2004). New paradigms for modelling species' distributions? *Journal of Applied Ecology* **41**, 193–200.
Stephens, P.A., Buskirk, S.W., Hayward, G.D. and Martinez del Rio, C. (2005). Information theory and hypothesis testing: a call for pluralism. *Journal of Applied Ecology* **42**, 4–12.

Stockwell, D. and Peters, D. (1999). The GARP modelling system: problems and solutions to automated spatial prediction. *International Journal of Geographical Information Science* **13**, 143–158.

Swain, P.H. (1978). Remote sensing: the quantitative approach. In: *Remote Sensing: The Quantitative Approach.* (P.H. Swain and S.M. Davis, eds), pp. 136–187 & 221–223, New York: McGraw-Hill.

Williams, B., Rogers, D., Staton, G., Ripley, B. and Booth, T. (1992). Statistical modelling of georeferenced data: mapping tsetse distributions in Zimbabwe using climate and vegetation data. In: *Modelling Vector-Borne and other Parasitic Diseases* (B.D. Perry and J.W. Hansen, eds), pp. 267–280. Nairobi: ILRAD.

Global Environmental Data for Mapping Infectious Disease Distribution

S.I. Hay[1,2], A.J. Tatem[1], A.J. Graham[1], S.J. Goetz[3] and D.J. Rogers[1]

[1]*TALA Research Group, Tinbergen Building, Department of Zoology, University of Oxford, South Parks Road, Oxford OX1 3PS, UK*
[2]*Malaria Public Health & Epidemiology Group, Centre for Geographic Medicine, KEMRI, P.O. Box 43640, 00100 Nairobi GPO, Kenya*
[3]*The Woods Hole Research Center, P.O. Box 296, Woods Hole, MA 02543-0296, USA*

Abstract 38
1. Introduction 38
2. The AVHRR Sensor 39
2.1. History and Overview 39
2.2. AVHRR Archives 41
2.3. From Digital to Environmental Data 41
3. Temporal Fourier Analysis (TFA) 49
3.1. History and Application 49
3.2. Fourier Data Products 50
3.3. Inter-Comparison of TFA Surfaces 51
4. Future Global Environmental Data 57
4.1. Terra, Aqua and Modis 57
4.2. Other Satellite Sensors 58
4.3. Data Continuity and NPOESS 65
4.4. Other Data Sources 67
5. Conclusions 70
Acknowledgements 70
References 71

ADVANCES IN PARASITOLOGY VOL 62
ISSN: 0065-308X $35.00
DOI: 10.1016/S0065-308X(05)62002-7

ABSTRACT

This contribution documents the satellite data archives, data processing methods and temporal Fourier analysis (TFA) techniques used to create the remotely sensed datasets on the DVD distributed with this volume. The aim is to provide a detailed reference guide to the genesis of the data, rather than a standard review. These remotely sensed data cover the entire globe at either 1 × 1 or 8 × 8 km spatial resolution. We briefly evaluate the relationships between the 1 × 1 and 8 × 8 km global TFA products to explore their inter-compatibility. The 8 × 8 km TFA surfaces are used in the mapping procedures detailed in the subsequent disease mapping reviews, since the 1 × 1 km products have been validated less widely. Details are also provided on additional, current and planned sensors that should be able to provide continuity with these environmental variable surfaces, as well as other sources of global data that may be used for mapping infectious disease.

1. INTRODUCTION

The growth in the use of remote sensing (RS) and geographic information systems (GIS) has been fuelled, in part, by scientific demands to address many environmental issues at the global scale. The focus of RS and GIS in public health has been on infectious disease mapping (Hay *et al.*, 1997, 2000; Rogers *et al.*, 2002; Tatem *et al.*, 2004; Rogers, this volume, pp. 1–35). This has rarely been attempted at global scales primarily due to a lack of readily available environmental data for epidemiologists (Hay *et al.*, 1996, 1997; Hay, 2000). The RS data provided with this volume, and described in this review, go some way towards resolving this problem. These wide-area RS and other environmental data will be of utility to a variety of applications, but our emphasis here is entirely on infectious disease mapping.

Infectious diseases that have poikilothermic arthropod intermediate hosts are very sensitive to environmental conditions (Rogers and Packer, 1993). This is often highlighted by explaining sensitivities to temperature, rainfall and humidity of components of the basic reproductive number (R_0) of vector-borne diseases (Hay *et al.*, 1997; Rogers *et al.*, 2002; Rogers and Randolph, this volume, pp. 345–381), which mathematically describes the potential for parasite/pathogen persistence through time (Anderson, 1993). The RS-based data reviewed here have been developed to obtain information or "surrogates" of the more traditional climatic variables of relevance to the transmission of vector-borne diseases and thus infectious disease mapping (Hay and Lennon, 1999; Goetz *et al.*, 2000; Hay, 2000; Green and Hay, 2002). We focus primarily on the Advanced Very High Resolution Radiometer (AVHRR) which, more than any other satellite sensor, has found considerable application in large area epidemiology (Hay, 2000; Rogers, this volume, pp. 1–35). We do not describe the various applications of these data as this is done elsewhere (Hay *et al.*, 2000; Rogers *et al.*, 2002) and is amply illustrated throughout this volume.

2. THE AVHRR SENSOR

The basic principles of RS and satellite sensor systems have been reviewed previously (Hay *et al.*, 1997; Hay, 2000; Tatem *et al.*, 2004). Here we provide a brief overview of the AVHRR sensor, a series of which have collected the RS data distributed on the accompanying DVD. It is beyond the scope of this review to extend this treatment to other sensors.

2.1. History and Overview

The National Oceanographic and Atmospheric Administration's (NOAA) series of polar-orbiting Television Infrared Observation Satellites (TIROS) has been operational since 1978 (Cracknell, 1997). TIROS-N (later renamed NOAA-6) was the first satellite to carry the AVHRR, originally for meteorological purposes, and has been followed by eleven satellites, each with an operational lifetime of 2–4 years. The definitive description of the NOAA polar-orbiting satellites, their radiometer payloads and the data they generate are given in Cracknell

(1997) and Kidwell (1998). Information on the NOAA-AVHRR satellite series is updated regularly by the NOAA Satellite and Information Services (NOAASIS) [http://noaasis.noaa.gov/].

The NOAA-AVHRR satellites complete 14.1 near-polar, Sun-synchronous orbits every 24 hours at an altitude of 833–870 km. The NOAA-AVHRR can view a 2800 km swath of the Earth and so, at this orbital frequency, daily data are recorded for the entire Earth surface. Radiation is currently measured in six wavebands (channels) of the electromagnetic spectrum (five on platforms prior to NOAA-15) so that six images are recorded for each area sensed. The visible (channel 1) and near infrared (channel 2) measure reflected solar radiation, whereas the thermal channels (4 and 5) measure emitted thermal infrared. Channel 3 (split into 3A and 3B on the latest platforms) senses the middle-infrared (MIR) and is sensitive to a combination of both reflected and emitted radiances.

The nominal 1.1×1.1 km spatial resolution data are transmitted continuously and may be received by stations near to the satellite's path, where they are referred to as High Resolution Picture Transmission (HRPT) data (Cracknell, 1997). On request to NOAA, these data may also be recorded on an on-board tape storage system and later transmitted to Earth as the satellites pass over a network of receiving stations. The data are then referred to as Local Area Coverage (LAC). These data have found application in a very wide range of disciplines that have been reviewed by Ehrlich *et al.* (1994) and Cracknell (1997).

Two processing steps reduce the spatial resolution of most of the NOAA-AVHRR data available to the research community. Since the on-board tape system is incapable of holding global coverage data at 1.1×1.1 km spatial resolution, the information from each area of five (across-track) by three (along-track) pixels is stored as a single value corresponding to the average of the first four pixels only, of the first row of the 5×3 block. The resulting imagery is referred to as Global Area Coverage (GAC) data. GAC data, with a stated nominal spatial resolution of 4×4 km, are far from ideal representations of the raw data (Justice *et al.*, 1989; Robinson, 1996). Nevertheless, the GAC data are the only form in which the NOAA-AVHRR archive was and continues to be collected. Reasonable quality global datasets are available at a variety of spatial resolutions (4 × 4 km and coarser) starting in the early 1980s (James *et al.*, 1994). When further processed, these data are often re-sampled to an 8 × 8 km spatial resolution before distribution (James *et al.*, 1994).

2.2. AVHRR Archives

A series of 1 × 1 km spatial resolution NOAA-AVHRR data are available at the "Global Land 1-KM AVHRR Project" homepage on the United States Geological Survey (USGS), EROS Data Center, website [URL: http://edcdaac.usgs.gov]. The dataset was generated to obtain a standard year of observations for the global land cover mapping project by the International Geosphere Biosphere Programme – Data and Information System (IGBP-DIS) (Eidenshink and Faundeen, 1994; Townshend *et al.*, 1994; Teillet *et al.*, 2000).

The 8 × 8 km spatial resolution NOAA-AVHRR data are available at the Global Land Biosphere Data and Information Web Site at the Goddard Space Flight Center's Distributed Active Archive Center [http://daac.gsfc.nasa.gov/]. The data were archived with the purpose of providing a long-term database for Earth observation, with a particular emphasis on tropical deforestation (Townshend, 1994). These images are referred to as the Pathfinder AVHRR Land (PAL) dataset (James *et al.*, 1994).

2.3. From Digital to Environmental Data

2.3.1. Temporal Range

The 1 × 1 km NOAA-AVHRR data are available by decad (10 day unit) from April to December 1992 (9 months, 27 decads, 162 files); January–September 1993 (9 months, 27 decads, 162 files); February–December 1995 (11 months, 33 decads, 198 files) and January–April 1996 (2 months, 6 decads, 36 files) (see Table 1).

The 8 × 8 km NOAA-AVHRR data are available from August 1981 to September 2001 inclusive; a 20-year time series. While the 1 × 1 km data are obviously more detailed spatially, the longer time series of the

Table 1 Dates of the NOAA-AVHRR 1 × 1 km data acquisition archived for IGBP-DIS. A dot indicates acquisition of all decads

Month	1992	1993	1995	1996
January		•		•
February		•	•	
March		•	•	
April	•	•	•	•
May	•	•	•	
June	•	•	•	
July	•	•	•	
August	•	•	•	
September	•	•	•	
October	•		•	
November	•		•	
December	•		•	

8 × 8 km imagery provides opportunities for examining changing land-surface patterns over a critical period of the Earth's history. These issues are explored in detail in Sections 3.2 and 3.3 respectively.

2.3.2. Data Obtained and File Sizes

For each decad of the 1 × 1 km NOAA-AVHRR the following parameters were downloaded: channels 3, 4, 5, the normalized difference vegetation index (NDVI), satellite and solar zenith angles (see Hay, 2000 for definitions). This amounts to ~680 gigabytes (Gb) of compressed data. When each global image is uncompressed it is 17 347 lines/rows by 40 031 samples/columns. For 8-bit imagery the file size is therefore (17 347 × 40 031 = 694 417 757 bytes) or ~678 Mb. For 16-bit data the file size is ~1.36 Gb.

The global 8 × 8 km data, in contrast, are considerably smaller, at 2168 lines/rows by 5004 samples/columns, and so for 8-bit data the files (2168 × 5004 = 10 848 672 bytes) are ~10.6 Mb and double that for 16-bit files. The 20-year duration of the 8 × 8 km data archive makes the total multitemporal database close to 3.5 Gb when compressed. The spatial resolution of the 1 × 1 km NOAA-AVHRR data and the temporal duration of the 8 × 8 km NOAA-AVHRR data result in such large volumes of data, that distribution on DVD in their original form is not possible, an issue addressed in Section 3.

2.3.3. Geo-Registration and Projection

All RS data represent conditions on a spherical Earth as a regular, gridded raster array of picture elements (pixels). Capturing features of a 3-dimensional object (the globe) on a 2-dimensional map involves some compromise in accurately representing direction, distance, shape and area (Snyder and Voxland, 1994; Snyder, 1997) and different map projections are used for applications that need to show one or other of these map features as accurately as possible. The projection chosen for both the 1×1 km and 8×8 km NOAA-AVHRR data at source is the Interrupted Goode's Homolosine, which is a combination of a Sinusoidal projection covering the tropical regions of the Earth (to latitudes of $40^\circ 44'$ North and South) and the Mollweide projection elsewhere. This combination of projections was considered to be the best at maintaining the shape and area of the continents (Steinwand, 1994). Despite the geographical superiorities of the Goode's projection the majority of potential non-specialist RS or GIS users, and the software with which they are familiar, use latitude and longitude data in an equi-rectangular grid technically known as the Plate-Carrée projection. After temporal Fourier processing (see Section 3) of the data in their original projection we resampled the resulting files using a bilinear interpolation algorithm (Mather, 1999) that ignored sea pixels along coastlines (i.e. sea pixels were given weights of zero in the interpolation algorithm), and distribute both 1×1 and 8×8 km NOAA-AVHRR products in the Plate-Carrée projection, at nominal resolutions of 0.01° and 0.1°, respectively. The resulting images are therefore 36000 columns $\times$ 18 000 rows or 3600 columns $\times$ 1800 rows, respectively. Table 2 gives the full details that are required for manipulating these data in RS and GIS software packages.

2.3.4. Rescaling

Values of geophysical variables are usually rescaled before storage as image files, and represented as either 8-bit (i.e. byte) or 16-bit (i.e. integer) binary numbers. This saves on hard disk storage space (a limiting component in earlier computers) and facilitates image compression, which further saves storage space. Eight-bit numbers are always positive, and span the range from 0 to 255. Sixteen-bit numbers

Table 2 Geo-referencing information for the TALA RS data

Details	1 × 1 km imagery	8 × 8 km imagery
Projection	Geographic (latitude/longitude)	Geographic (latitude/longitude)
Image size, columns	36 000	3600
Image size, rows	18 000	1800
Upper-left coordinate, pixel centre	−179.995	−179.95
Upper-left coordinate, pixel centre	89.995	89.95
Units	decimal degrees	decimal degrees
Pixel size: x	0.01	0.1
Pixel size: y	0.01	0.1
Spheroid	WGS84	WGS84
Datum	WGS84	WGS84

may be stored either in signed 16-bit format (range between −32 767 and +32 767) or unsigned 16-bit format (range between 0 and 65 535). Much confusion arises when unsigned 16-bit integers (exceeding +32767) are regarded as signed integers, and vice versa, and users must take care to specify to their software the correct format (i.e. 'signed' or 'unsigned') of 16-bit integer image data. Parts of the full digital range of stored data are often 'reserved' for mask values (indicating the sea, image quality, or some detail of the map projection used) and these should be specified by the providers of the imagery.

The original images were stored at source in either 8-bit or 16-bit format, with certain values reserved to indicate masks, sea or projection interruptions. Full details of this conversion process for the 1 × 1 km imagery are given as examples. The satellite zenith angle (SaZA) data have units of degrees and were stored in 8-bit files. In the original files, the binary min/max = 10/190 were used to represent the geophysical min/max = −90/90 and mask values were 0 = missing data over land; 1 = ocean; 2 = Goode's interrupted area. The satellite zenith angle was calculated from SaZA = (DN − 100) where DN is the original stored 8-bit number. The solar zenith angle (SoZA) also has units of degrees and was originally stored as 8-bit files. The binary min/max were 10/190 and the geophysical min/max were 0/180. The mask and other values were exactly the same as for the SaZA imagery, so SoZA = (DN − 10). The NDVI is a ratio and has no units.

It was archived as 8-bit files, the binary min/max of which were 10/210 with the geophysical min/max = −1/1 and mask and other values as before; so NDVI = (DN−110) × 0.01. Finally, the radiance data (channels 3, 4 and 5) had units of degrees Kelvin and were originally stored as unsigned 16-bit data with binary min/max = 10/1018 and the geophysical min/max = 160/340. Mask and other values were as before, so channel 3, channel 4 or channel 5 = (DN + 886.32)/5.602.

In general, before the processing steps outlined below, the imagery was stored in signed 16-bit format with the mask and other values (0, 1, 2) being stored as −999, −998 and −997, respectively. The 8 × 8 km NOAA-AVHRR data were archived similarly with further details and (different) rescaling values given by James *et al.* (1994).

2.3.5. *Quality Control and Data Pre-Processing*

Ancillary data provided with the original imagery were used to exclude unreliable pixels in both the 1 × 1 and 8 × 8 km datasets. Quality control removed pixels taken at satellite zenith angles (SaZA) greater than 40°, to reduce bi-directional effects and parallax errors due to image pixels acquired at viewing angles far from nadir (Hay and Lennon, 1999; Hay, 2000). Pixels with solar zenith angles (SoZA) exceeding 80° were also excluded, to eliminate data retrieved at low sun elevations (i.e. at dusk and dawn) that result in shadows that affect image quality and therefore interpretation (Hay and Lennon, 1999; Hay, 2000). In addition, the 8 × 8 km NOAA-AVHRR data included a data layer that could be used to mask those pixels determined as cloudy by the 'Clouds from AVHRR' (CLAVR) algorithm (Stowe *et al.*, 1991; Hay and Lennon, 1999).

The NDVI data were then maximum-value composited (Holben, 1986) into monthly files. The (incomplete) set of monthly 1 × 1 km NOAA-AVHRR data for 1992–1996 were further subjected to maximum value composition to generate a single set of 12 monthly images representing the average (i.e. synoptic) year using all the available data. The particular files contributing to each synoptic month are shown in Table 1. These data were then rescaled conditionally (if $\leqslant -996 = 0$, else = ((NDVI + 1) × 1000)) and stored as signed 16-bit integers ready for temporal Fourier analyses (TFA).

MIR (channel 3) data were also maximum-value composited in the same way (Lambin and Ehrlich, 1996). The subsequent processing procedure was identical to that applied to the NDVI. Data values indicating a temperature estimate of 350 K or greater were screened out at this stage, as they were clearly in error. These data were then conditionally rescaled (if $\leqslant -999 = 0$, else $= (\text{MIR} \times 10)$) for storage as signed 16-bit integers.

Land surface temperature (LST) data were derived using a split window algorithm (Price, 1983), using quality-controlled channel 4 and 5 data and maximum-value composited (Lambin and Ehrlich, 1996). Subsequent processing was identical to that of the channel 3 data. These data were conditionally rescaled (if$\leqslant -999 = 0$, else$=$ $(\text{LST} \times 10)$) for storage as signed 16-bit integers.

The 1×1 and 8×8 km datasets were not corrected systematically for the effects of satellite orbit drift over the lifespan of each satellite, which can affect the inter-comparability of these multi-temporal data (Gutman, 1999). This can be particularly problematic in the thermal channels because satellite orbit decays progressively delay the timing of the daily measurement, which is obviously critical for a variable such as temperature that shows strong diurnal variation (Gleason *et al.*, 2002). For these and other reasons, the AV-HRR data for 1981 and all data after 1999 were excluded from the 8×8 km NOAA-AVHRR time series before TFA (Nemani *et al.*, 2003). Many small artefacts in satellite data are smoothed by TFA (see Section 3) but we emphasize that these uncorrected data are not appropriate for the analyses of temporal change. Datasets that systematically deal with such artefacts for monitoring temporal change are becoming available (Tucker *et al.*, 2005). Finally, only NDVI, MIR and LST data products are processed and distributed here. Although it is possible to derive air temperature and vapour pressure deficit (Goetz *et al.*, 2000), we have avoided these indices as they have been tested less widely, can require further geophysical data inputs and show co-linearity with these existing products, a number of which are used in their generation. Data provided on the accompanying DVD are in the format and with the new scaling as outlined above, with further details in Table 3 and the DVD README file.

Table 3 DVD imagery contents with details of scaling factors to be applied to the data, the data units and expected values

File name	Imagery type (°)	Image	Scaling	Units	Geophysical min.	Geophysical max.
wd1003a0	TFA 0.1	mean MIR	(x/10) −273	°C	−7.8	48.9
wd1003a1	TFA 0.1	MIR annual amplitude	(x/10)	°C	0	30.8
3.1wd1003a2	TFA 0.1	MIR bi-annual amplitude	(x/10)	°C	0	1.0
wd1003a3	TFA 0.1	MIR tri-annual amplitude	(x/10)	°C	0	0.6
wd1003d1	TFA 0.1	MIR pvs described by annual cycle	No scaling	%	0	94
wd1003d2	TFA 0.1	MIR pvs described by bi-annual cycle	No scaling	%	0	45
wd1003d3	TFA 0.1	MIR pvs described by tri-annual cycle	No scaling	%	0	13
wd1003mn	TFA 0.1	Minimum MIR	(x/10) −273	°C	−26	32.4
wd1003mx	TFA 0.1	Maximum MIR	(x/10) −273	°C	−3.6	50.9
wd1003p1	TFA 0.1	MIR phase of annual cycle	(x/10)	Months	0	12
wd1003p2	TFA 0.1	MIR phase of bi-annual cycle	(x/10)	Months	0	6
wd1003p3	TFA 0.1	MIR phase of tri-annual cycle	(x/10)	Months	0	4
wd1003vr	TFA 0.1	MIR variance	(x/10)	%	0	563
wd1007a0	TFA 0.1	Mean LST	(x/10) −273	°C	−30.7	55.4
wd1007a1	TFA 0.1	LST annual amplitude	(x/10)	°C	0	3.7
wd1007a2	TFA 0.1	LST bi-annual amplitude	(x/10)	°C	0	1.0
wd1007a3	TFA 0.1	LST tri-annual amplitude	(x/10)	°C	0	0.5
wd1007d1	TFA 0.1	LST pvs described by annual cycle	No scaling	%	0	96
wd1007d2	TFA 0.1	LST pvs described by bi-annual cycle	No scaling	%	0	48
wd1007d3	TFA 0.1	LST pvs described by tri-annual cycle	No scaling	%	0	11
wd1007mn	TFA 0.1	Minimum LST	(x/10) −273	°C	−43.5	54.3
wd1007mx	TFA 0.1	Maximum LST	(x/10) −273	°C	−26.5	58.9
wd1007p1	TFA 0.1	LST phase of annual cycle	(x/10)	Months	0	12
wd1007p2	TFA 0.1	LST phase of bi-annual cycle	(x/10)	Months	0	6
wd1007p3	TFA 0.1	LST phase of tri-annual cycle	(x/10)	Months	0	4
wd1007vr	TFA 0.1	LST variance	(x/10)	%	0	758
wd1014a0	TFA 0.1	Mean NDVI	(x/1000) −1	No units[a]	−0.086	0.803
wd1014a1	TFA 0.1	NDVI annual amplitude	(x/1000)	No units[a]	0	0.445

(*Continued*)

Table 3 (*continued*)

File name	Imagery type (°)	Image	Scaling	Units	Geophysical min.	Geophysical max.
wd1014a2	TFA 0.1	NDVI bi-annual amplitude	(x/1000)	No units[a]	0	0.251
wd1014a3	TFA 0.1	NDVI tri-annual amplitude	(x/1000)	No units[a]	0	0.131
wd1014d1	TFA 0.1	NDVI pvs described by annual cycle	No scaling	%	0	94
wd1014d2	TFA 0.1	NDVI pvs described by bi-annual cycle	No scaling	%	0	80
wd1014d3	TFA 0.1	NDVI pvs described by tri-annual cycle	No scaling	%	0	39
wd1014mn	TFA 0.1	Minimum NDVI	(x/1000) −1	No units[a]	−0.12	0.764
wd1014mx	TFA 0.1	Maximum NDVI	(x/1000) −1	No units[a]	−0.066	0.865
wd1014p1	TFA 0.1	NDVI phase of annual cycle	(x/10)	Months	0	12
wd1014p2	TFA 0.1	NDVI phase of bi-annual cycle	(x/10)	Months	0	6
wd1014p3	TFA 0.1	NDVI phase of tri-annual cycle	(x/10)	Months	0	4
wd1014vr	TFA 0.1	NDVI variance	(x/100)00		0	0.116
wd1030dm	Globe D.E.M.	Globe DEM	No scaling	Meters	−295	6815
afpop00	Pop. surf.	UNEP-GRID Population for Africa	No scaling	Persons	0	1 580 304
afpopd00	Pop. d. surf.	UNEP-GRID Population for Africa	No scaling	Persons/km^2	0	84 923
gpw00	Pop. surf.	SEDAC Gridded population of the World	No scaling	Persons	0	1 467 471
gpw00d	Pop. d. surf.	SEDAC Gridded population of the World	No scaling	Persons/km^2	0	122 755
grump-2000	M. pop. surf.	SEDAC Gridded population of the World	No scaling	Persons	0	1 346 722

TFA = Temporal Fourier analysis; pvs = proportion of variance in original signal;
[a]no units because it is a ratio. Geophysical minimum and geophysical maximum values do not include masks. Globe DEM—digital elevation model. Pop. surf. = population surface, Pop. d. surf. = population density surface, M. pop. surf. = modelled population surface.

3. TEMPORAL FOURIER ANALYSIS (TFA)

3.1. History and Application

Monthly composite imagery usually shows strong serial correlations (i.e. seasons wax and wane in a relatively predictable manner) and this data redundancy may be eliminated in two different ways. Traditionally, the data are subjected to principal components analysis (PCA), and the resultant significant principal components are used in analyses (Eastman and Falk, 1993; Lillesand and Kiefer, 2000). Alternatively, the data may be subjected to TFA, which describes natural environmental cycles such as temperature and vegetation growth in terms of annual, bi-annual, tri-annual and other cycles with shorter or longer wavelengths (periods) (Rogers, 2000). The great attraction of TFA is that it produces a set of orthogonal (i.e. uncorrelated) outputs while retaining a description of seasonality (lost in PCA) that is of vital interest in vector and disease mapping (Rogers *et al.*, 1996; Rogers, 2000; Rogers and Robinson, 2004). One disadvantage of both PCA and TFA is that they both assume stationarity (i.e. constant mean and variance) of the data over time. Trends in data can first be removed by differencing the time series from a moving average spanning a number of annual cycles, and then analysing the de-trended time series. Alternatively, if the trend itself is of interest, a windowing approach can be applied, whereby the data are analysed in a series of (usually overlapping) windows in time. Within each window the data are assumed to be mathematically stationary, and the windowed TFA results can be compared to look for changes in the Fourier components (means, amplitudes, phases etc.) through time.

The origins, mathematical basis and arguments for the biological appropriateness of TFA are developed in detail elsewhere (Rogers, 2000; Rogers *et al.*, 2002; Rogers and Robinson, 2004). In brief, the trajectory through the year (the sequence of 12 monthly images) of every picture element (pixel) in the environmental time series is described by a series of orthogonal sine curves (cycles/harmonics) with different frequencies. Each cycle is described by its amplitude (the maximum variation of the cycle around the mean) and phase (its timing). A total of six cycles is required to describe a dataset of 12 months perfectly (with cycle periods of 12, 6, 4, 3, 2.4 and 2 months). We find that around 90% or more of the variation in the original images is usually captured by the first three (i.e. annual, bi-annual and tri-annual) cycles, and much of the remainder is often simply 'noise', so we normally use only these to capture important features of the seasonal variation at each site.

During TFA the raw time series of data were first examined for obvious drop-out values, that arise either from absent imagery or else from pixels masked out in the quality control algorithm of image processing. Data for such months were linearly interpolated from data for the months before and after the dropout months (with data wrap-around when dropout months occurred right at the start or right at the end of the sequence). The resulting time series were subjected to TFA and the raw data were then compared with the recomposed result (i.e. the sum of the Fourier-fitted annual, bi-annual and tri-annual cycles only). Months where the absolute difference between the raw and recomposed data exceeded user-determined threshold values were also regarded as incorrect (these threshold values were generously set so that only obvious outliers would be trapped at this stage). These months were therefore linearly interpolated as before, using data from adjacent months, and TFA carried out again on the corrected data. This process was repeated until no further outliers were identified (generally requiring no more than one or two rounds of interpolation of a few data points). Finally, no 8×8 km imagery was available for the months of September–December 1994. Images for these months were first created using the averages of the same months from the 1993 and 1995 images, and these were used in TFA.

3.2. Fourier Data Products

The TFA algorithm developed by the TALA group produces 14 different products for each input satellite channel; the overall mean (a0); the amplitude of the annual (a1), bi-annual (a2) and tri-annual (a3) cycles; the phase (i.e. peak timing) of the annual (p1), bi-annual (ph2) and tri-annual (p3) cycles in months (starting at zero in January);

the proportion of the variance in the original time series that is described by the annual (d1), bi-annual (d2), tri-annual (d3) and all three cycles combined (da); the maximum (mx) and minimum (mn) of the seasonal cycle recomposed from the first three harmonics only; and finally the variance (vr) of the original (i.e. not the fitted) time series. These products are made available in full for the 8 × 8 km imagery but only a subset is given for the 1 × 1 km products due to space constraints (see DVD README file). Specifications for each 8 × 8 km NOAA-AVHRR TFA file are given in Table 3 and examples for each band given in Figure 1a–n. Note that equivalent details for the 1 × 1 km NOAA-AVHRR TFA data are only in the DVD README file. Figure 2a and b provides a vignette of the processed 1 × 1 and 8 × 8 km imagery for the lower reaches of the Nile river in Egypt, Africa. The following section briefly discusses the inter-comparability of these data.

3.3. Inter-Comparison of TFA Surfaces

We have not assumed that 1 × 1 and 8 × 8 km TFA surfaces can be used interchangeably, because of the rather different origins, time periods and processing chains to which the data were subjected. Regression tests between 1 × 1 and 8 × 8 km surfaces were therefore performed at different latitudes to examine whether the overpass times of the various satellites, and varying solar and viewing illumination levels produced regional effects. The global area represented by the NOAA-AVHRR imagery was divided into eight latitude bands: 90–70°N, 70–50°N, 50–30°N, 30–10°N, 10°N–10°S, 10–30°S, 30–50°S, 50–90°S (see Figure 3). The 1 × 1 km NOAA-AVHRR images were resampled using a mean filter to match the spatial resolution of the 8 × 8 km imagery. A stratified random sample of 10000 points within each region was created and values of the pixels representing these points extracted from the mean, maximum and minimum MIR, LST and NDVI. Figure 4a–f shows a sample of these scatter plots of mean MIR, LST and NDVI 8 × 8 km pixel values against mean MIR, LST and NDVI 1 × 1 km pixel values for the 50°N–30°N and 10°N–10°S latitudinal segment. Regression analysis was carried out between the

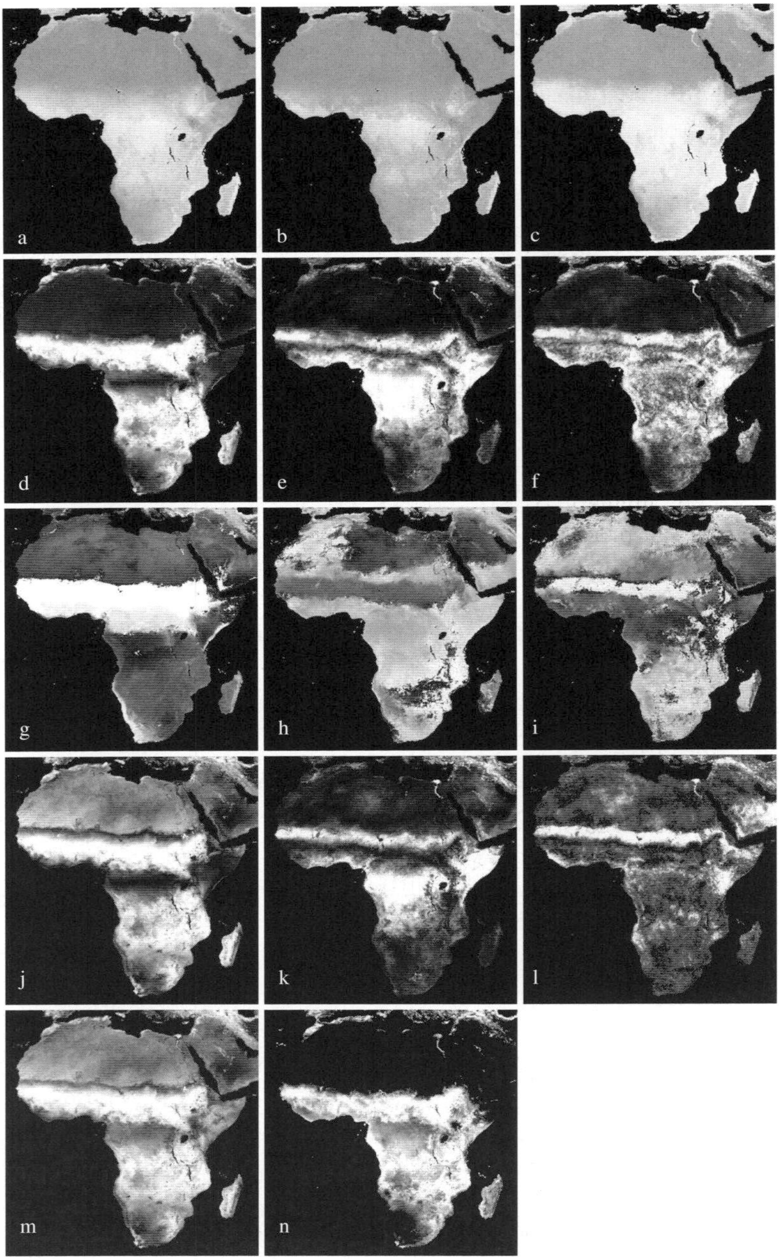
a
b
c
d
e
f
g
h
i
j
k
l
m
n

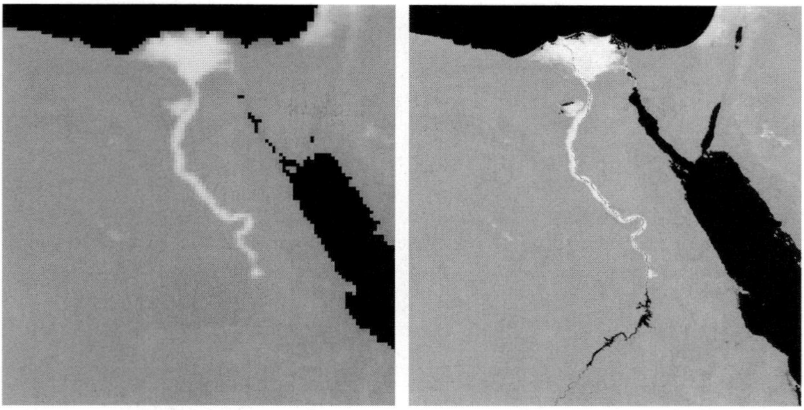

Figure 2a,b The lower reaches of the Nile river in Egypt illustrate the difference between the 8 × 8 km spatial resolution imagery (on the left) and the 1 × 1 km imagery (right). Images are of the mean NOAA-AVHRR NDVI TFA a0 product (Table 3). Data are histogram equalized stretched from the minimum data value (black, also water mask) to maximum (white) for display.

two sets of extracted pixel values. No significant differences were observed between the fit of a linear model (see Table 4) and various nonlinear alternatives.

Figure 4 highlights scatterplots for two of the most epidemiologically important latitudinal segments; the 50–30°N segment encompassing the US, Southern Europe and much of Asia, and the 10°N–10°S segment encompassing the equatorial tropical regions of

Figure 1a–n Panel of the fourteen 8 × 8 km NOAA-AVHRR NDVI TFA images (a) a0, mean, (b) mn, minimum and (c) mx, maximum signal recomposed from the first three Fourier cycles. The amplitude of the (d) a1, annual cycle, (e) a2, bi-annual and (f) a3, tri-annual are also shown, in addition to the (g) p1, phase of annual, (h) p2, bi-annual, (i) and p3, tri-annual cycle in months. The proportion of the variance in the original time series described by the (j) d1, the annual, (k) d2, the bi-annual, (l) d3, the tri-annual and (m) da, all three cycles combined is also shown with (n) vr, the variance of the original data time series. Data are histogram equalized stretched from the minimum data value (black, also water mask) to maximum (white) for display.

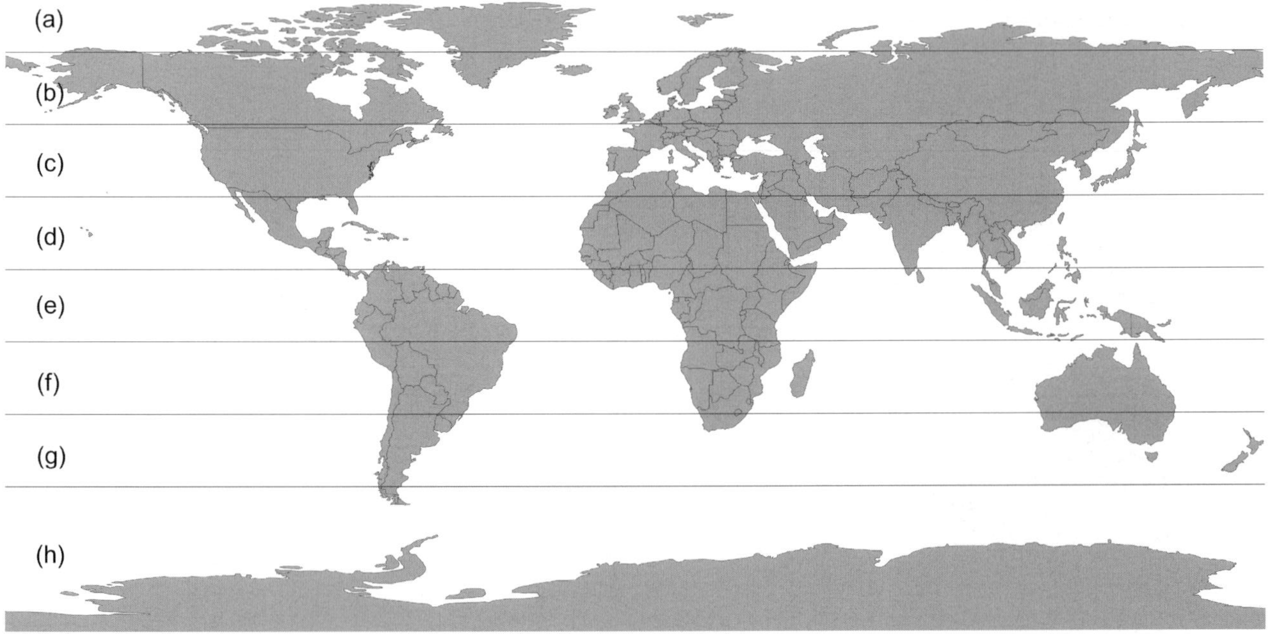

Figure 3 Latitudinal sample regions, (a) 90–70°N, (b) 70–50°N, (c) 50–30°N, (d) 30–10°N, (e) 10N–10°S, (f) 10–30°S, (g) 30–50°S, (h) 50–90°S.

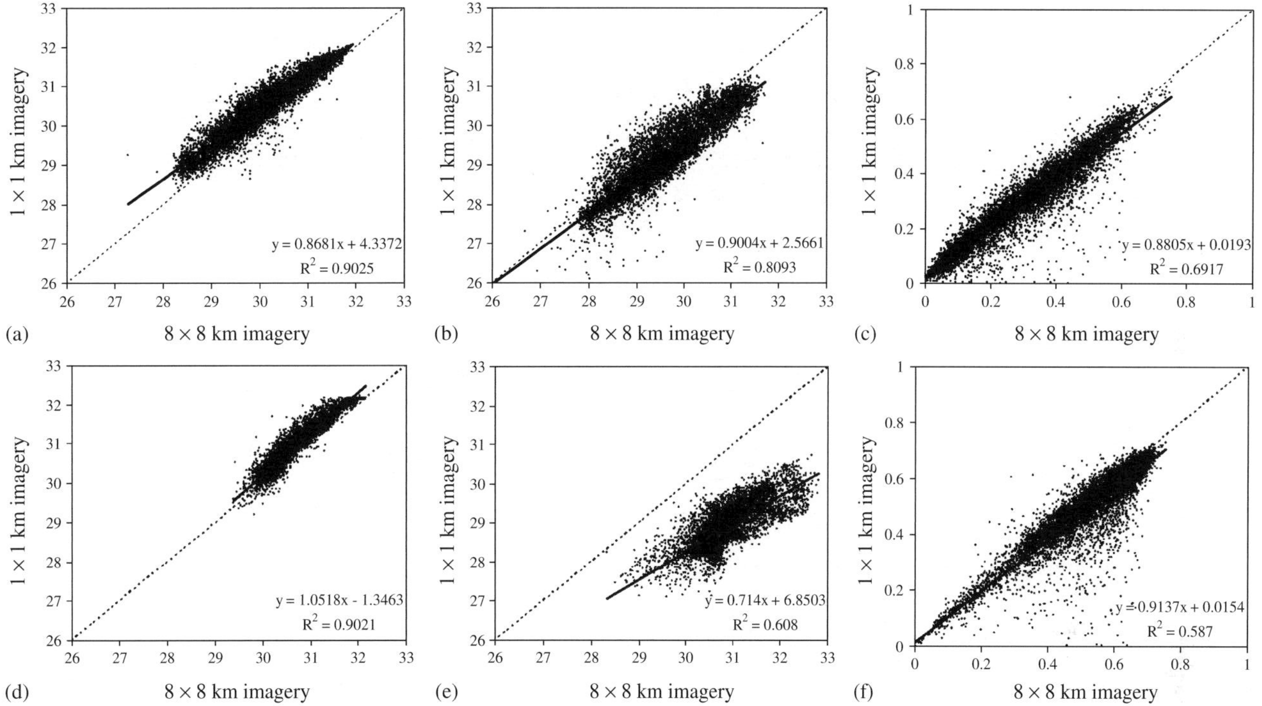

Figure 4a–f Scatterplots for latitudinal segments shown in Figure 3 of 8 × 8 km imagery sample points against 1 × 1 km imagery sample points for 50°N–30°N for MIR (a), LST (b) and NDVI (c) and for 10°N–10°S for MIR (d), LST (e) and NDVI (f). One-to-one lines (dashed) are added for ease of interpretation. Solid lines (and equations) are least squares linear fits to the data.

Table 4 Values of r^2 produced through linear regression analysis between 8 × 8 and 1 × 1 km imagery by latitudinal segment (latitudinal segments shown in Fig. 3)

Latitude	Mean MIR	Mean LST	Mean NDVI
70–90°N	0.368	0.778	0.135
50–70°N	0.705	0.755	0.523
30–50°N	0.903	0.809	0.692
10–30°N	0.950	0.510	0.822
10S–10°N	0.902	0.608	0.587
10–30°S	0.869	0.434	0.860
30–50°S	0.931	0.858	0.651
50–90°S	0.737	0.819	0.546

the World. Figure 4a shows that there is correspondence between the 1 × 1 and 8 × 8 km imagery for all three environmental variables at 50–30°N, with the vast majority of sample pixels clustered along the one-to-one lines (i.e. indicating little bias). These results are reflected in Table 4, showing r^2 values above 0.69 for all three variables. An area of concern is the clusters of rogue NDVI pixels which exhibit very low values in the 1 × 1 km NOAA-AVHRR imagery. These represent small water-bodies and coast pixels not distinguishable at the 8 × 8 km scale (but certainly affecting the values of the pixels at this scale). Figure 4b shows that, for MIR and NDVI, little bias between the 1 × 1 km and 8 × 8 km imagery exists at tropical latitudes, and that large differences again exist in a small handful of NDVI pixels. A more significant concern is that the 8 × 8 km NOAA-AVHRR LST data are showing apparently higher temperature than the corresponding 1 × 1 km imagery. This may be an artefact of the difference in time periods over which each TFA product was produced, with the 8 × 8 km imagery providing a more reliable synoptic estimate given the much greater numbers of contributing years. Table 4 shows that, despite this bias, at tropical latitudes all three mean environmental variables exhibit relatively strong correlations (all $r^2 > 0.58$). We have focused on the 8 × 8 km imagery in our studies, but we have included a smaller subset of the 1 × 1 km imagery for wider experimentation which should, nevertheless, bear these comparisons in mind.

4. FUTURE GLOBAL ENVIRONMENTAL DATA

4.1. Terra, Aqua and Modis

The Terra (EOS AM-1) and Aqua (EOS PM-2) satellites, launched in December 1999 and May 2002 respectively, are a part of (NASA) Earth Observing System (Parkinson, 2003). A range of onboard sensors capture a variety of image types, but most relevant in this context are the MODerate Resolution Imaging Spectroradiometer (MODIS) and, exclusively for Terra, the Advanced Spaceborne Thermal Emission and Reflection (ASTER) radiometer.

MODIS is particularly attractive for epidemiological applications due to; (i) a better spectral resolution than AVHRR, with 36 spectral channels with smaller waveband ranges and significantly improved signal-to-noise ratios (Justice *et al.*, 2002), (ii) a one to two-day repeat time temporal resolution at significantly higher spatial resolution (250×250 to 1000×1000 m depending on the channel) than AVHRR (Townshend and Justice, 2002) and (iii) fully processed and quality assessed data products, giving unparalleled, rapid access to contemporary and reliable data on large-area ecosystem processes. MODIS is also potentially attractive to the public health community thanks to the availability of its products at no charge to users, and its longer mission lifespan (Tatem *et al.*, 2004). A drawback of MODIS data, when compared to AVHRR, is the considerably greater computing resources needed to cope with the larger data volumes of some of its products. These constraints are likely to diminish rapidly with the exponential increases in computer power and storage capacities, but these spatial resolutions at the global scale will still present a significant challenge to the majority of users.

ASTER is an alternative data source for studies that have traditionally used SPOT-HRV or Landsat TM sensors (see below). The spatial resolution of ASTER varies with wavelength, yielding 15, 30 and 90 m resolutions at visible–near infrared (VNIR), short wave infrared (SWIR) and thermal infrared (TIR), respectively (Yamaguchi *et al.*, 1998), and the images may prove a powerful tool for studying local disease processes (Tatem *et al.*, 2004).

Spectral bands within the MODIS 36 band array are broadly similar to those of AVHRR, which may permit temporal continuity of datasets and thus the potential for extending the AVHRR time series (Friedl *et al.*, 2002), providing funding for the Terra and Aqua satellites continues (a significant consideration given the uncertainty of most USA-based Earth observation systems (Lawler, 2005)). Likewise, ASTER uses spectral channels similar in characteristics to those of Landsat TM. Significant ongoing efforts in RS are seeking to overcome issues that arise from differences in radiometric and spatial resolutions between old and new sensors (NOAA – NPOESS, 2003). We develop these issues further in Section 4.3.

4.2. Other Satellite Sensors

This section focuses on those sensors that can provide information at the global scale. Many sensors that have not proven useful in epidemiological studies and those with evident constraints, such as difficulties of image costs or data access, are not considered. A more complete account of satellite sensors is available in Campbell (2002) and Verger *et al.* (2003). There are many planned enhancements to the existing sensors which, due to the frequency of modification, are best reviewed at their relevant web sites. These, and the basic technical specifications of the sensors reviewed here, are detailed in Table 5.

4.2.1. Geostationary Satellites

The principal payload of Meteosat is the Meteosat Visible and Infra-Red Imager (MVIRI) (EUMETSAT, 2000). The radiometer operates in a broad visible waveband, a water vapour absorption infrared waveband and a thermal infrared waveband. The Meteosat satellites were designed for meteorological applications, so part of their spectral range is located in the thermal infrared area of maximal water vapour absorption, making it ideal for monitoring clouds. At nadir the spatial resolution is 2.5×2.5 km for the visible images and 5×5 km for the thermal infrared and water vapour images. Further from the equator, the spatial resolution decreases so that over northern

Table 5 Technical specifications of satellite sensors that have been used in epidemiology or that show significant potential for epidemiological applications

Satellite[a]	Sensor	Bands[b]	Bandwith (μm)	Spatial res. (m)	Temporal res.	Swath (km)	URL
Landsat-5	MSS[c]	1	0.500–0.600	80	16 days	185	http://landsat.gsfc.nasa.gov
		2	0.600–0.700	80			
		3	0.700–0.800	80			
		4	0.800–1.100	80			
	TM	1	0.450–0.520	30	16 days	185	
		2	0.520–0.600	30			
		3	0.630–0.690	30			
		4	0.760–0.900	30			
		5	1.550–1.750	30			
		6	10.40–12.50	120			
		7	2.080–2.350	30			
Landsat-7	ETM +	1	0.450–0.520	30	16 days	185	
		2	0.530–0.610	30			
		3	0.630–0.690	30			
		4	0.780–0.900	30			
		5	1.550–1.750	30			
		6	10.40–12.50	60			
		7	2.090–2.350	30			
		8 (P)	0.520–0.900	15			
NOAA-16 & 17	AVHRR	1	0.580–0.680	1090	12 hours	2800	http://edc.usgs.gov/products/satellite/avhrr.html
		2	0.725–1.000	1090			
		3A	1.580–1.640	1090			
		3B	3.550–3.930	1090			
		4	10.300–11.300	1090			
		5	11.500–12.500	1090			
SPOT-4	HRVIR	1	0.500–0.590	20	1–4 days	60	http://www.spotimage.fr/
		2	0.610–0.680	20			

(*Continued*)

Table 5 (*Continued*)

Satellite[a]	Sensor	Bands[b]	Bandwith (μm)	Spatial res. (m)	Temporal res.	Swath (km)	URL
		3	0.780–0.890	20			
		4	1.580–1.750	20			
	VGT-1	0	0.450–0.520	1000	1 day	2250	
		2	0.610–0.680	1000			
		3	0.780–0.890	1000			
		4	1.580–1.750	1000			
SPOT-5	HRG	P	0.480–0.710	5	1–4 days	60	
		1	0.500–0.590	10			
		2	0.610–0.680	10			
		3	0.780–0.890	10			
		4	1.580–1.750	20			
	VGT-2	Same as VGT-1	Same as VGT-1				
Terra	MODIS	1–2	0.620–0.876	250	1–2 days	2330	http://terra.nasa.gov/
		3–7	0.459–2.155	500			
		8–36	0.405–14.385	1000			
	ASTER	VNIR (3 bands)	0.500–0.900	15	4–16 days	60	
		SWIR (6 bands)	1.600–2.500	30			
		TIR (5 bands)	8.000–12.000	90			
GOES-12	Imager	1	0.550–0.750	1000	26 minutes	3000	http://www.goes.noaa.gov/
		2	3.800–4.000	4000			
		3	6.500–7.000	8000			
		4	10.200–11.200	4000			
		5	11.500–12.500	4000			
Meteosat-7	MVIRI	1	0.450–1.000	2500	30 minutes		http://www.eumetsat.de/
		2	5.700–7.100	5000			
		3	10.500–12.500	5000			
MSG-1	SEVIRI	1	0.560–0.710	4800	15 minutes		http://www.esa.int/msg/

(Meteosat-8)	2	0.740–0.880	4800
	3	1.500–1.780	4800
	4	3.480–4.360	4800
	5	5.350–7.150	4800
	6	6.850–7.850	4800
	7	8.300–9.100	4800
	8	9.380–9.940	4800
	9	9.800–11.800	4800
	10	11.000–13.000	4800
	11	12.400–14.400	4800
	12 (HRV)	0.400–1.100	1670

[a]Operational and/or latest satellite of the series in orbit.
[b]P = Panchromatic.
[c]Also onboard Landsat-1, -2, -3 and -4. For all acronyms refer to text. Res. = Resolution.

Europe, for example, it is 4 × 4 km in the visible wavebands and 8 × 8 km in the thermal infrared and water vapour wavebands. Each image is transmitted to the Earth in real time as each scan line is completed, and new images are generated at 30-min intervals.

Meteosat Second Generation (MSG) satellites are designed to give continuity to Meteosat missions with improved spatial, spectral and temporal resolutions (EUMETSAT-ESA, 1998; EUMETSAT, 2001; Schmetz *et al.*, 2002). MSG-1 was launched in August 2002 and, after a period of commissioning and validation, routine operations started in January 2004 (EUMETSAT, 2004). Of particular potential on board this satellite is the Spinning Enhanced Visible and Infrared Imager (SEVIRI) sensor, designed to acquire images every 15 min from 12 different bands at a spatial resolution of 1.4 × 1.4 km for the visible, and 3 × 3 km for all other bands. MSG high-quality datasets have raised expectations as powerful tools for studying temporal and spatial tropical disease patterns in Africa (Hay, 2000). These have largely not been realized due to data archival and distribution restrictions in Europe.

It is important to stress that despite the fact that Meteosat products have found wide application in malaria studies in Africa (Hay *et al.*, 1998, 2003b; Rogers *et al.*, 2002), they cannot be used in the Americas, Oceania or much of Asia as the satellite's sensors capture only that part of the Earth's disc visible from a stationary orbit positioned over the equator at the Greenwich meridian (i.e. longitude 0°). This area includes Africa, Europe and the Middle East. The satellites with equivalent capabilities for the American and Pacific region are NOAA's Geostationary Operational Environmental Satellite (GOES) (NASA, 1999). The satellite stationed at 75° W is known as GOES-E while that at 135°W is called GOES-W. GOES has been used mainly for weather observations and forecasts for the last 25 years (NASA, 1999), and its products offer great potential for epidemiological applications. The general properties of GOES-12 imagery are included in Table 5 for reference.

Two other satellite systems complete the constellation of five satellites that provide geostationary sensor coverage for the entire globe. Indian Ocean Data Coverage (IODC) is provided by Meteosat 5, which has found continued use in a different geostationary orbit position. The sensor details have already been discussed in this section. The final satellite and sensor to mention is the Geostationary Meteorological Satellite (GMS) series which sits in orbit above 140°E. GMS-1 was launched in 1977, the first in a series of satellites culminating in GMS-5 which completed its observational mission in May 2003. GOES-9 has been used as a backup to GMS-5 that had to shut down its imager due to a technical fault. The replacement for the GMS series (and GOES-9) is a new Japanese satellite called the Multi-Functional Transport Satellite (MTSAT-1R), which became operational in 2005 and is to be followed by MTSAT-2. These satellites carry sensors that image in 5 wavebands: visible, near-infrared, two infrared channels and a water vapour channel. The visible images have a spatial resolution of 1×1 km while all other channels are sensed at 4×4 km spatial resolution at nadir.

4.2.2. Landsat

The launch of Landsat-1 in 1972 heralded a new era of high resolution RS (Lauer *et al.*, 1997; Markham *et al.*, 2004). Since then, the Landsat programme has generated a continuous supply of high resolution imagery for the entire globe, from the first Multispectral Scanner (MSS) aboard Landsat-1 to the latest Enhanced Thematic Mapper (ETM+) on board Landsat-7 (Mika, 1997). During this time, there has been a substantial evolution in the quality of the radiometers (Mika, 1997), their calibration (Chander *et al.*, 2004; Thome *et al.*, 2004) and the development of multi-spectral data analysis techniques developed to process captured data (Landgrebe, 1997). The novelty and conspicuous success of the Landsat programme forced issues regarding data distribution and cost (Draeger *et al.*, 1997) and the feasibility of commercial RS (Williamson, 1997) to be considered seriously for the first time. Many countries have emulated and extended features of the Landsat programme, and other high-resolution RS data sources are becoming increasingly available (Campbell, 2002).

On May 2003, the scan line corrector (SLC) subsystem on board Landsat-7 developed an anomaly (Markham *et al.*, 2004). Later that year the problem was identified as a permanent mechanical failure and

Landsat-7 resumed its mission with the SLC turned off. The device was designed to compensate for the forward motion of the satellite. This malfunction causes data loss at the edges of images, with an effective 30% loss of information (Markham *et al.*, 2004). Simple interpolation techniques have been applied to the datasets with promising results, but further validation is needed. Since the remaining pixels, comprising 30 km^2 in the center, are unaffected, Landsat-7 is still able to yield useful imagery in the middle of each scene (USGS, 2003).

4.2.3. SPOT

The French Satellite Pour l'Observation de la Terre (SPOT) programme began in 1986 with the launch of SPOT-1, carrying the High Resolution Visible (HRV) payload. There were many similarities to Landsat-TM imagery, but, essentially, the SPOT-HRV achieved a slightly higher spatial resolution with fewer spectral channels. Data collection has continued with SPOT-4, which carries the High Resolution Visible and Infrared (HRVIR) sensor and the multi-spectral VEGETATION (VGT-1) instrument. The more recent SPOT-5 has a VGT-2, similar to its predecessor, and a High Resolution Geometric (HRG) camera that achieves spatial resolutions of up to 5 m (Campbell, 2002). Of particular relevance from an epidemiological perspective is the VGT instrument, due to the high-quality vegetation index it offers, the availability of fully corrected imagery at a constant 1 × 1 km spatial resolution, and an almost daily global coverage in four spectral bands (0.45–1.75 μm) (Campbell, 2002; Maisongrande *et al.*, 2004). The VGT sensor data can be provided as a number of different products: VGT-P products (the physical values); VGT-S1 products (daily maximum value composited syntheses), VGT-S10 products (10-day maximum value composited syntheses) and VGT–D10 products (10-day BiDirectional Composite syntheses). The VGT-S10 and VGT-D10 products are also available in degraded resolutions of 4 × 4 and 8 × 8 km.

4.2.4. ADEOS

The Japanese polar orbiting Advanced Earth Observation Satellite (ADEOS) program provides global observation datasets, albeit with limited temporal coverage due to instrument problems. The ADEOS-1 satellite, launched in August 1996, went out of service in June 1997 due to structural damage to its solar array. The ADEOS-II platform, launched in December 2002 went out of service in late October 2003, also due to loss of power and subsequent sensor failure. Both platforms carried a Global Imager (GLI) with a spectral range spanning the visible to thermal infrared portion of the spectrum. The 36 GLI spectral bands acquired data at 10:30 local time from an altitude of 800 km in a Sun-synchronous polar orbit. The additional spectral information provided by GLI data has potential for improving estimates of land surface variables, including fire disturbance mapping and "hot spot" monitoring, as well as more visible channels useful for monitoring surface waters. A swath width of 1600 km provided observational data suitable for global land surface monitoring with a repeat interval of four days. The GLI data have a spatial resolution of 250 × 250 m in six channels, comparable to MODIS, and 1 × 1 km in other spectral bands. Improved atmospheric water vapour absorption bands aided estimation of vertical humidity profiles and near-surface water content, which were being used for improved weather forecasting and have utility in disease vector mapping (Goetz *et al.*, 2000). ADEOS science teams have provided higher order datasets of environmental variables derived from the raw GLI data. Production of vegetation cover type, density, productivity and change have experienced severe set-backs from instrument failures, but refined datasets of potential utility to disease applications exist for most of 2003.

4.3. Data Continuity and NPOESS

Continuity between NOAA-AVHRR and MODIS, the current "work horses" of global RS observing systems, requires both comparability assessments between sensors and the development of future new technology sensors. The latter will have improved radiometric and spectral properties, often provided at finer spatial resolution. Inter-sensor comparability requires assessments of the ability to retrieve consistent surface environmental information from instruments with different spectral, spatial, geometric, radiometric and orbital acquisition properties. NASA has recently funded several studies to address these issues in order to ensure continuity of the primary surface reflectance, as well as spectral vegetation indices and other data products that are used to derive many of the environmental variables relevant to epidemiological research. These efforts include empirical comparisons with sensor cross calibrations and associated transformation statistics, similar to what has been done with higher resolution sensors (Goetz, 1997). More analytical assessments are planned based on modelling radiative transfer and incorporating relative sensor responses across the energy spectrum within the sensor band passes. Both approaches will lead, either separately or in combination, to releases of improved datasets compared to those distributed on the accompanying DVD, in addition to providing an extended observational record.

The primary satellite platforms that will provide data continuity with AVHRR, Terra and Aqua over the next two decades of Earth observation are the National Polar-orbiting Operational Environmental Satellite Systems (NPOESS). An NPOESS Preparatory Project (NPP) satellite is scheduled for launch in May 2006 [URL: http://jointmission.gsfc.nasa.gov/]. It will carry four primary sensors including the Visible Infrared Imager Radiometer Suite (VIIRS), the Cross-Track Infrared Sounder, the Advanced Technology Microwave Sounder, and the Ozone Mapping and Profiler Suite. The sensor of primary interest to epidemiological research is the VIIRS, which is intended to provide global observations of land, ocean and atmosphere parameters on a near-daily basis. VIIRS is a 22-band instrument, with a spectral range from the visible through the thermal infrared wavelengths (0.4–12 μm). Each image swath will extend 3000 km, from a satellite orbit of 833 km, by scanning fore and aft $\pm 56^{\circ}$ of nadir. Image data will be acquired at two spatial resolutions, 370 and 740 m. A unique feature of VIIRS is that it will reduce pixel size across track (along scan), which will therefore compensate for the pixel size expansion that typically accompanies off-nadir viewing. The ground track of NPP will mimic that of Terra, with a 16-day repeat interval consistently acquiring data at 10:30 local time in its descending node (the ascending node data are acquired at night). On the planned series of NPOESS satellites that follow NPP, data will be acquired at 09:30, 13:30 and 17:30, respectively, in order to provide measures of diurnal variability and continuity with Aqua. These satellites are currently planned for launch after 2010. Although delays are expected, the life expectancy of the instruments will ensure there are periods of overlap between VIIRS sensors. A broad range of environmental variables will be provided from VIIRS observational data by NPP and NPOESS project science teams, including vegetation indices, surface temperature, land cover type classifications, energy and heat fluxes, atmospheric water vapour and soil moisture dynamics. Much of this work will be based on algorithms already developed by the MODIS science teams. The production of VIIRS continuity datasets will also make use of a wide range of geometric, radiometric and atmospheric corrections of the raw image data.

4.4. Other Data Sources

This brief section is not intended to be a comprehensive list of the digital data available to the epidemiological community but simply selected highlights that we have found to be particularly useful. A more comprehensive listing can be found at [http://www.eden-fp6project.net/].

4.4.1. Digital Elevation

Elevation influences temperature, rainfall and humidity and has been used widely in the mapping of infectious disease (Guerra *et al.*, this volume, pp. 157–179). The 1 km Global Land One-kilometre Base Elevation (GLOBE) Digital Elevation Model (DEM) is provided on the accompanying DVD. Full documentation is provided elsewhere [URL: http://www.ngdc.noaa.gov/seg/topo/report/]. We have resampled these data to be compatible with the 1×1 km and 8×8 km RS datasets. Table 6 summarizes the important features of the data.

The Shuttle Radar Topography Mission (SRTM) obtained elevation data on a near-global scale to generate the most complete high-resolution digital topographic database of Earth. The SRTM consisted of a specially modified radar system that flew on board the space

Table 6 Geo-referencing information for the GLOBE DEM data

Details	30 arc second imagery
Projection	Geographic latitude/longitude)
Image size, columns	43200
Image size, rows	21600
Upper-left coordinate, pixel edge	−180
Upper-left coordinate, pixel edge	90
Units	Decimal degrees
Pixel size: x	0.00833333
Pixel size: y	0.00833333
Spheroid	WGS84
Datum	WGS84
Vertical (z) units	Metres above mean sea level

shuttle Endeavour during an 11-day mission in February of 2000 (Rabus *et al.*, 2003; Smith and Sandwell, 2003; Sun *et al.*, 2003). The SRTM is an international project led by the National Geospatial-Intelligence Agency (NGA) and NASA. Elevation data at 1, 3 and 30 arc-second spatial resolution (USGS, 2004) are freely available from Global Land Cover Facility [http://www.landcover.org/data/srtm/].

4.4.2. Gridded Climatologies

The Climate Research Unit at the University of East Anglia, UK, produces a range of global gridded climatologies (New *et al.*, 1999, 2000, 2002; Mitchell and Jones, 2005) derived from interpolated meteorological station data [http://www.cru.uea.ac.uk/cru/data/hrg.htm]. These include $0.5 \times 0.5°$ time series from 1901 to 2000 (New *et al.*, 1999, 2000) and 10×10 arc-second climatology of synoptic months (1961–1990) (New *et al.*, 2002). They include useful comparative information on precipitation, wet day frequency, daily mean temperature, diurnal temperature range, cloud cover, frost day frequency, vapour pressure, wind speed, relative humidity and sunshine hours. These data have found extensive use in epidemiological studies (Rogers and Randolph, 2000; Hay *et al.*, 2002; Shanks *et al.*, 2002; Small *et al.*, 2003).

4.4.3. The Africa Data Dissemination Service (ADDS)

The ADDS is operated by the US Agency for International Development (USAID) as part of its Famine Early Warning System Network (FEWS NET) [http://igskmncnwb015.cr.usgs.gov/adds/]. It provides a regularly updated archive of AVHRR-derived NDVI for Africa, as well as a suite of climatic products, including rainfall estimate data related to disease risk and food insecurity. The advantage of these data is that they are available in near real-time and hence useful for malaria epidemic monitoring (Hay *et al.*, 2003a,b).

4.4.4. Gridded Human Population Data

Demographic data are crucial if visually striking maps are to be turned into useful operational tools. The "state-of-the-art" in population mapping is detailed extensively elsewhere in this volume (Balk *et al.*, this volume, pp. 119–156) and the data can also be found on the accompanying DVD (see Table 3 and Balk *et al.*, this volume).

4.4.5. Administrative Boundary Data

One frequent difficulty in interpreting disease and other data that are mapped to local administrative boundaries, and given local administrative names, is that there is no agreed global dataset of administrative boundaries beyond admin level 1 (the major division below country level zero). Although currently incomplete, the best current prospects for a global documented standard seems to be the United Nations funded Second Level Administrative Boundaries project [http://www3.who.int/whosis/gis/salb/salb_home.htm]. Data for many countries are already available and the site is frequently updated.

4.4.6. Global Landcover Facility (GLCF)

The GLCF at the University of Maryland, USA holds a vast array of freely downloadable satellite imagery and other data sources [http://www.glcf.umiacs.umd.edu/index.shtml]. Satellite imagery available includes global coverage of MODIS and Landsat MSS, TM and ETM+, with ASTER data made available recently. A variety of vegetation, land cover, forest and burned area products, all derived from satellite imagery, are available globally. We include it here because it is an unparalleled resource.

5. CONCLUSIONS

The 8 × 8 km TALA TFA RS dataset provides the most stable synoptic surfaces our group has used for monitoring global scale environmental conditions of relevance to infectious disease mapping. The data represent a significant spatial resolution advance on synoptic climatologies and complement the suite of data these surfaces provide. The 1 × 1 km TALA TFA data streams are less comprehensive temporally, show some latitudinally dependent differences with the 8 × 8 km TFA data and are hence more experimental. They should therefore be used with greater caution. Ancillary DEM, population and human settlement data are also distributed on the DVD. These collective data represent a contemporary snapshot of environmental conditions of use to those engaged in infectious disease mapping and wider environmental studies. It is difficult to predict their useful lifespan but we hope the continuity missions outlined will make their regular updating a necessity.

ACKNOWLEDGEMENTS

We are grateful to Simon Brooker, Archie Clements, Marlies Craig, Benjamin McCormick, Bethan Purse, Victoria Sanderson and William Wint for their comments on this manuscript and testing beta versions of the DVD. SIH and AJT are funded by a Research Career Development Fellowship (to SIH) from the Wellcome Trust (#069045). SJG acknowledges support from the NASA Applied Sciences program. AJG was funded by EU grant GOCE-2003-010284 EDEN and the paper is catalogued by the EDEN Steering Committee as EDEN0009 (http://www.eden-fp6project.net/). The contents of this publication are the sole responsibility of the authors and do not reflect the views of the

European Union. The datasets discussed in this review were generated using resources from a range of grants supplied principally by the UK Department for International Development and the Wellcome Trust, for which we are grateful. The following formal acknowledgement is requested of those who use Pathfinder AVHRR Land data:

"Data used by the authors in this study include data produced through funding from the Earth Observing System Pathfinder Program of NASA's Mission to Planet Earth in co-operation with the National Oceanic and Atmospheric Administration. The data were provided by the Earth Observing System Data and Information System (EOSDIS) Distributed Active Archive Center at Goddard Space Flight Center, which archives, manages and distributes this dataset."

We provide the information on the DVD that accompanies this article without restriction, and request that when the data are used in further studies, due acknowledgement is given to the TALA Research Group, Oxford University, UK, for the TFA images and this review is referenced as the source. All reasonable efforts have been made to check the fidelity of these data but the authors can accept no liability for subsequent use and application of these datasets.

REFERENCES

Anderson, R.M. (1993). Epidemiology, 2nd ed. In: *Modern Parasitology* (F.E.G. Cox, ed.), pp. 75–116. Blackwell: Oxford.

Campbell, J.B. (2002). *Introduction to Remote Sensing*, 3rd ed. London: Taylor & Francis.

Chander, G., Helder, D.L., Markham, B.L., Dewald, J.D., Kaita, E., Thome, K.J., Micijevic, E. and Ruggles, T.A. (2004). Landsat-5 TM reflective-band absolute radiometric calibration. *IEEE Transactions on Geoscience and Remote Sensing* **42**, 2747–2760.

Cracknell, A.P. (1997). *The Advanced Very High Resolution Radiometer*. London: Taylor & Francis.

Draeger, W.C., Holm, T.M., Lauer, D.T. and Thompson, R.J. (1997). The availability of Landsat data: past, present and future. *Photogrammetric Engineering & Remote Sensing* **63**, 869–875.

Eastman, J.R. and Falk, M. (1993). Long sequence time series evaluation using standardized principal components. *Photogrammetric Engineering & Remote Sensing* **59**, 1307–1312.

Ehrlich, D., Estes, J.E. and Singh, A. (1994). Applications of NOAA-AVHRR 1 km data for environmental monitoring. *International Journal of Remote Sensing* **15**, 145–161.

Eidenshink, J.C. and Faundeen, J.L. (1994). The 1 km AVHRR global land data set - 1st stages in implementation. *International Journal of Remote Sensing* **15**, 3443–3462.

EUMETSAT (2000). *The Meteosat System*. Darmstadt: European Organisation for the Exploitation of Meteorological Satellites.

EUMETSAT (2001). *Meteosat Second Generation—MSG System Overview*. Darmstadt: European Organisation for the Exploitation of Meteorological Satellites.

EUMETSAT (2004). *Meteosat-8 (MSG-1) Dissemination Trial*. Darmstadt: European Organisation for the Exploitation of Meteorological Satellites.

EUMETSAT (1998). *The Meteosat Second Generation Programme*. Darmstadt: European Organisation for the Exploitation of Meteorological Satellites.

Friedl, M.A., McIver, D.K., Hodges, J.C.F., Zhang, X.Y., Muchoney, D., Strahler, A.H., Woodcock, C.E., Gopal, S., Schneider, A. and Cooper, A. (2002). Global land cover mapping from MODIS: algorithms and early results. *Remote Sensing of Environment* **83**, 287–302.

Gleason, A.C.R., Prince, S.D., Goetz, S.J. and Small, J. (2002). Effects of orbital drift on land surface temperature measured by AVHRR thermal sensors. *Remote Sensing of Environment* **79**, 147–165.

Goetz, S.J. (1997). Multi-sensor analysis of NDVI, surface temperature, and biophysical variables at a mixed grassland site. *International Journal of Remote Sensing* **18**, 71–94.

Goetz, S.J., Prince, S.D. and Small, J. (2000). Advances in satellite remote sensing of environmental variables for epidemiological applications. *Advances in Parasitology* **47**, 289–307.

Green, R.M. and Hay, S.I. (2002). The potential of Pathfinder AVHRR data for providing surrogate climatic variables across Africa and Europe for epidemiological applications. *Remote Sensing of Environment* **79**, 166–175.

Gutman, G.G. (1999). On the monitoring of land surface temperatures with the NOAA/AVHRR: removing the effect of satellite orbit drift. *International Journal of Remote Sensing* **20**, 3407–3413.

Hay, S.I. (2000). An overview of remote sensing and geodesy for epidemiology and public health application. *Advances in Parasitology* **47**, 1–35.

Hay, S.I., Cox, J., Rogers, D.J., Randolph, S.E., Stem, D.I., Shanks, G.D., Myers, M.F. and Snow, R.W. (2002). Climate change and the resurgence of malaria in the East African highlands. *Nature* **415**, 905–909.

Hay, S.I. and Lennon, J.J. (1999). Deriving meteorological variables across Africa for the study and control of vector-borne disease: a comparison of remote sensing and spatial interpolation of climate. *Tropical Medicine & International Health* **4**, 58–71.

Hay, S.I., Packer, M.J. and Rogers, D.J. (1997). The impact of remote sensing on the study and control of invertebrate intermediate hosts and vectors for disease. *International Journal of Remote Sensing* **18**, 2899–2930.

Hay, S.I., Randolph, S.E. and Rogers, D.J., (eds), (2000). *Remote Sensing and Geographical Information Systems in Epidemiology. Advances in Parasitology* **47**.

Hay, S.I., Renshaw, M., Ochola, S.A., Noor, A.M. and Snow, R.W. (2003a). Performance of forecasting, warning and detection of malaria epidemics in the highlands of western Kenya. *Trends in Parasitology* **19**, 394–399.

Hay, S.I., Snow, R.W. and Rogers, D.J. (1998). From predicting mosquito habitat to malaria seasons using remotely sensed data: practice, problems and perspectives. *Parasitology Today* **14**, 306–313.

Hay, S.I., Tucker, C.J., Rogers, D.J. and Packer, M.J. (1996). Remotely sensed surrogates of meteorological data for the study of the distribution and abundance of arthropod vectors of disease. *Annals of Tropical Medicine and Parasitology* **90**, 1–19.

Hay, S.I., Were, E.C., Renshaw, M., Noor, A.M., Ochola, S.A., Olusanmi, L., Alipui, N. and Snow, R.W. (2003b). Forecasting, warning, and detection of malaria epidemics: a case study. *Lancet* **361**, 1705–1706.

Holben, B.N. (1986). Characteristics of maximum-value composite images from temporal AVHRR data. *International Journal of Remote Sensing* **7**, 1417–1434.

James, M.E., Kalluri, S.N.V. and Townshend, J.R.G. (1994). The Pathfinder AVHRR land data set: an improved coarse resolution data set for terrestrial monitoring. *International Journal of Remote Sensing* **15**, 3347–3363.

Justice, C.O., Markham, B.L., Townshend, J.R.G. and Kennard, R.L. (1989). Spatial degradation of satellite data. *International Journal of Remote Sensing* **10**, 1539–1561.

Justice, C.O., Townshend, J.R.G., Vermote, E.F., Masuoka, E., Wolfe, R.E., Saleous, N., Roy, D.P. and Morisette, J.T. (2002). An overview of MODIS Land data processing and product status. *Remote Sensing of Environment* **83**, 3–15.

Kidwell, K.B. (1998). *NOAA Polar Orbiter Data User's Guide*. Washington DC: US Department of Commerce, NESDIS, NOAA, National Climate Data Center, Satellite Data Service Division.

Lambin, E.F. and Ehrlich, D. (1996). The surface temperature-vegetation index space for land cover and land-cover changes analysis. *International Journal of Remote Sensing* **17**, 463–487.

Landgrebe, D. (1997). The evolution of Landsat data analysis. *Photogrammetric Engineering & Remote Sensing* **63**, 859–867.

Lauer, D.T., Morain, S.A. and Salomonson, V.V. (1997). The Landsat Program: its origins, evolution and impacts. *Photogrammetric Engineering & Remote Sensing* **63**, 831–838.

Lawler, A. (2005). Earth observation program "at risk", academy warns. *Science* **308**, 614–615.

Lillesand, T.M. and Kiefer, R.W. (2000). *Remote Sensing and Image Interpretation*. New York: Wiley.

Maisongrande, P., Duchemin, B. and Dedieu, G. (2004). VEGETATION/SPOT: an operational mission for the Earth monitoring; presentation of new standard products. *International Journal of Remote Sensing* **25**, 9–14.

Markham, B.L., Storey, J.C., Williams, D.L. and Irons, J.R. (2004). Landsat sensor performance: history and current status. *IEEE Transactions on Geoscience and Remote Sensing* **42**, 2691–2694.

Mather, P.M. (1999). *Computer Processing of Remotely sensed Images: an Introduction*, 2nd ed. Chichester: Wiley.

Mika, A.M. (1997). Three decades of Landsat instruments. *Photogrammetric Engineering & Remote Sensing* **63**, 839–852.

Mitchell, T.D. and Jones, P.D. (2005). An improved method of constructing a database of climate observations and associated high-resolution. *International Journal of Climatology* **25**, 693–712.

NASA. (1999). *The GOES Program*. http://www.earth.nasa.gov/history/goes/goes.html. 2004.

Nemani, R.R., Keeling, C.D., Hashimoto, H., Jolly, W.M., Piper, S.C., Tucker, C.J., Myneni, R.B. and Running, S.W. (2003). Climate-driven increases in global terrestrial net primary production from 1982 to 1999. *Science* **300**, 1560–1563.

New, M., Hulme, M. and Jones, P. (1999). Representing twentieth-century space-time climate variability. Part I: Development of a mean 1961–90 monthly terrestrial climatology. *Journal of Climate* **12**, 829–857.

New, M., Hulme, M. and Jones, P. (2000). Representing twentieth-century space-time climate variability. Part II: Development of 1901–96 monthly grids of terrestrial surface climate. *Journal of Climate* **13**, 2217–2238.

New, M., Lister, D., Hulme, M. and Makin, I. (2002). A high-resolution data set of surface climate over global land areas. *Climate Research* **21**, 1–25.

NOAA – NPOESS. (2003). National Polar-orbiting Operational Environmental Satellite System.

Parkinson, C.L. (2003). Aqua: an earth-observing satellite mission to examine water and other climate variables. *IEEE Transactions on Geoscience and Remote Sensing* **41**, 173–183.

Price, J.C. (1983). Estimating surface temperatures from satellite thermal infrared data – a simple formulation for the atmospheric effect. *Remote Sensing of Environment* **13**, 353–361.

Rabus, B., Eineder, M., Roth, A. and Bamler, R. (2003). The shuttle radar topography mission – a new class of digital elevation models acquired by spaceborne radar. *ISPRS Journal of Photogrammetry and Remote Sensing* **57**, 241–262.

Robinson, T.P. (1996). Spatial and temporal accuracy of coarse resolution products of NOAA-AVHRR NDVI data. *International Journal of Remote Sensing* **17**, 2303–2321.

Rogers, D.J. (2000). Satellites, space, time and the African trypanosomiases. *Advances in Parasitology* **47**, 128–171.

Rogers, D.J., Hay, S.I. and Packer, M.J. (1996). Predicting the distribution of tsetse flies in West Africa using temporal Fourier processed meteorological satellite data. *Annals of Tropical Medicine and Parasitology* **90**, 225–241.

Rogers, D.J. and Packer, M.J. (1993). Vector-borne diseases, models and global change. *Lancet* **342**, 1282–1284.

Rogers, D.J. and Randolph, S.E. (2000). The global spread of malaria in a future, warmer world. *Science* **289**, 2283–2284.

Rogers, D.J., Randolph, S.E., Snow, R.W. and Hay, S.I. (2002). Satellite imagery in the study and forecast of malaria. *Nature* **415**, 710–715.

Rogers, D.J. and Robinson, T.P. (2004). Tsetse distribution. In: Maudlin, I., Holmes, P.H. and Miles, M.A. (eds), *The Trypanosomiases*. Wallingford: CAB International, pp. 139–179.

Schmetz, J., Pili, P., Tjemkes, S., Just, D., Kerkmann, J., Rota, S. and Ratier, A. (2002). An introduction to Meteosat Second Generation (MSG). *Bulletin of the American Meteorological Society* **83**, 977–992.

Shanks, G.D., Hay, S.I., Stern, D.I., Biomndo, K. and Snow, R.W. (2002). Meteorologic influences on *Plasmodium falciparum* malaria in the highland tea estates of Kericho, western Kenya. *Emerging Infectious Diseases* **8**, 1404–1408.

Small, J., Goetz, S.J. and Hay, S.I. (2003). Climatic suitability for malaria transmission in Africa, 1911–1995. *Proceedings of the National Academy of Sciences of the USA* **100**, 15341–15345.

Smith, B. and Sandwell, D. (2003). Accuracy and resolution of shuttle radar topography mission data. *Geophysical Research Letters* **30**, art. no.-1467.

Snyder, J.P. (1997). *Map projections: a working manual. USGS Professional Paper 1395*. Boulder, CO: United States Geological Survey (USGS), United States Government Printing.

Snyder, J.P. and Voxland, P.M. (1994). *An album of map projections. USGS Professional Paper 1453*. Boulder, CO: United States Geological Survey (USGS), United States Government Printing.

Steinwand, D.R. (1994). Mapping raster imagery to the Interrupted Goode Homolosine Projection. *International Journal of Remote Sensing* **15**, 3463–3471.

Stowe, L.L., McClain, E.P., Carey, R., Pellegrino, P., Gutmann, G.G., Davia, P., Long, C. and Hart, S. (1991). Global distribution of cloud cover derived from NOAA/AVHRR operational satellite data. *Advances in Space Research* **11**, 51–54.

Sun, G., Ranson, K.J., Khairuk, V.I. and Kovacs, K. (2003). Validation of surface height from shuttle radar topography mission using shuttle laser altimeter. *Remote Sensing of Environment* **88**, 401–411.

Tatem, A.J., Goetz, S.J. and Hay, S.I. (2004). Terra and Aqua: new data for epidemiology and public health. *International Journal of Applied Earth Observation and Geoinformation* **6**, 33–46.

Teillet, P.M., El Saleous, N., Hansen, M.C., Eidenshink, J.C., Justice, C.O. and Townshend, J.R.G. (2000). An evaluation of the global 1-km AVHRR land dataset. *International Journal of Remote Sensing* **21**, 1987–2021.

Thome, K.J., Helder, D.J., Aaron, D. and Dewald, J.D. (2004). Landsat-5 TM and Landsat-7 ETM+ absolute radiometric calibration using the reflectance-based method. *IEEE Transactions on Geoscience and Remote Sensing* **42**, 2777–2785.

Townshend, J.R.G. (1994). Global data sets for land applications from the advanced very high resolution radiometer: an introduction. *International Journal of Remote Sensing* **15**, 3319–3332.

Townshend, J.R.G. and Justice, C.O. (2002). Towards operational monitoring of terrestrial systems by moderate-resolution remote sensing. *Remote Sensing of Environment* **83**, 351–359.

Townshend, J.R.G., Justice, C.O., Skole, D., Malingreau, J.P., Cihlar, J., Teillet, P., Sadowski, F. and Ruttenberg, S. (1994). The 1 km resolution global data set: needs of the International Geosphere Biosphere Program. *International Journal of Remote Sensing* **15**, 3417–3441.

Tucker, C.J., Pinzon, J.E., Brown, M.E., Slayback, D.A., Pak, E.W., Mahoney, R., Vermote, E.F. and El Saleous, N. (2005). The global inventory mapping and monitoring 1981–1999 AVHRR 8-Km dataset. An extended AVHRR 8-Km NDVI dataset compatible with MODIS and SPOT vegetation NDVI data. *International Journal of Remote Sensing* **26**, 4485–4498.

United States Geological Survey (USGS). (2003). *Landsat 7 SLC Anomaly Investigation, Vol. 2003.*

United States Geological Survey (USGS). (2004). *(1, 3, 30) Arc second SRTM elevation, reprocessed to GeoTIFF.* College Park, MD: The Global Land Cover Facility.

Verger, F., Sourbès-Verger, I. and Ghirardi, R. (2003). *The Cambridge Encyclopaedia of Space: Missions, Applications and Exploration*. Cambridge: Cambridge University Press.

Williamson, R.A. (1997). The Landsat legacy: remote sensing policy and the development of commercial remote sensing. *Photogrammetric Engineering & Remote Sensing* **63**, 877–885.

Yamaguchi, Y., Kahle, A.B., Tsu, H., Kawakami, T. and Pniel, M. (1998). Overview of advanced spaceborne thermal emission and reflection radiometer (ASTER). *IEEE Transactions on Geoscience and Remote Sensing* **36**, 1062–1071.

Issues of Scale and Uncertainty in the Global Remote Sensing of Disease

P.M. Atkinson[1] and A.J. Graham[2]

[1]*School of Geography, University of Southampton, Highfield, Southampton SO17 1BJ, UK*
[2]*TALA Research Group, Tinbergen Building, Department of Zoology, University of Oxford, South Parks Road, Oxford OX1 3PS, UK*

Abstract . . . 80
1. Introduction . . . 80
   1.1. Indirect Relations . . . 81
   1.2. Sensor Characteristics . . . 82
   1.3. Choice of Modelling Framework . . . 84
   1.4. The Image Processing Chain . . . 85
2. Issues of Scale and Spatial Resolution . . . 86
   2.1. Preliminaries . . . 86
   2.2. Measurement Scales . . . 88
   2.3. Scale(s) of Spatial Variation . . . 90
3. Issues of Uncertainty . . . 96
   3.1. Preliminaries . . . 96
   3.2. Uncertainty in Data and Methods . . . 97
   3.3. Data-Based Accuracy Assessment . . . 99
   3.4. Model-Based Accuracy Assessment . . . 103
   3.5. Validation of Global Disease Maps: Some Considerations . . . 105
   3.6. Satellite Sensor Datasets Used in this Volume . . . 106
4. Summary . . . 107
Acknowledgements . . . 108
References . . . 108

ADVANCES IN PARASITOLOGY VOL 62
ISSN: 0065-308X $35.00
DOI: 10.1016/S0065-308X(05)62003-9

ABSTRACT

Scale and uncertainty are important issues for the global prediction of disease. Disease mapping over the entire surface of the Earth usually involves the use of remotely sensed imagery to provide environmental covariates of disease risk or disease vector density. It further implies that the spatial resolution of such imagery is relatively coarse (e.g., 8 or 1 km). Use of a coarse spatial resolution limits the information that can be extracted from imagery and has important effects on the results of epidemiological analyses. This paper discusses geostatistical models for (i) characterizing the scale(s) of spatial variation in data and (ii) changing the scale of measurement of both the data and the geostatistical model. Uncertainty is introduced, highlighting the fact that most epidemiologists are interested in accuracy, aspects of which can be estimated with measurable quantities. This paper emphasizes the distinction between data- and model-based methods of accuracy assessment and gives examples of both. The key problem of validating global maps is considered.

1. INTRODUCTION

Global epidemiology has been a focus of spatial epidemiological research for many years (Vonreyn and Mann, 1987; Elliott, 1993). A range of diseases has been studied, usually based on extensive literature searches, including asthma and related conditions (Beasley *et al.*, 1998), lymphatic filariasis (Michael and Bundy, 1997), influenza and avian influenza (Cox and Subbarao, 2000), hepatitis B virus (Custer *et al.*, 2004), and HIV/AIDS (De Cock and Weiss, 2000; Ippolito *et al.*, 2000; Morison, 2001).

Today, epidemiologists have available a growing range of increasingly large spatial datasets potentially covering very large areas of the Earth's surface. Spatial data are characterized by an attribute value z and a location (x, y), usually expressed in a two-dimensional Cartesian coordinate system (such as latitude and longitude). The widespread availability of large spatial datasets has been facilitated by rapid increases in the power of affordable personal computer hardware, the dissemination and uptake of increasingly sophisticated geographical information system (GIS) software (Nicholson and Mather, 1996; Staubach *et al.*, 2001; Kirby, 2003; Elliott and Wartenberg, 2004; Jarup, 2004) and the sudden success of the internet, supporting web-based dissemination and sharing of data (LaPorte *et al.*, 1996; Gotway and Young, 2002). Prime among the sources of large spatial datasets is *remote sensing*, defined as the use of electromagnetic radiation (EMR) sensors to record images of the environment, which can be interpreted to yield useful information (Curran, 1985). Remote sensing has the potential to provide complete cover, in the form of one or more images, synoptically at a range of spatial and temporal scales (Hay, 2000; Graham *et al.*, 2004). No other source of environmental data can provide such spatially rich information. This is especially true at the global scale.

1.1. Indirect Relations

Remotely sensed imagery and other forms of spatial data can be used to support a wide range of epidemiological analyses, although remote sensing rarely leads to direct measurement of the property of interest. Rather, the property that is remotely sensed (for our purposes EMR in optical or microwave wavelengths) is usually related indirectly to the property of interest (Lillesand and Kiefer, 2004). Curran *et al.* (2000) described this situation in the context of disease mapping using land cover as the indirect link between radiation and vector or parasite prevalence. The amount of EMR reflected or emitted in a specific waveband from the Earth's surface is a direct function of the land cover or, more generally, Earth surface properties. Radiation may also be reflected by the atmosphere.

Reflected or emitted EMR is only indirectly related to disease vectors (Rogers and Randolph, 1991; Cross *et al.*, 1996; Hay *et al.*, 1997; Crombie *et al.*, 1999). For example, the distribution of malaria-carrying *Anopheles* mosquitoes is related to the signal recorded by a sensor through properties such as mosquito habitat, proximity to water and land surface temperature (LST) (all of which are functions of land cover). It is also related to proxies for rainfall lagged by one or two months [such as the normalized difference vegetation index (NDVI) and cold cloud duration (CCD)] (Hay *et al.*, 1996; Rogers *et al.*, 1996; Thomson *et al.*, 1996, 1999). Such information has been used to predict the seasonality of malaria transmission (Hay *et al.*, 1998a, b). Further, reflected radiation is indirectly related to parasite prevalence (Linthicum *et al.*, 1999). For example, the malaria parasite is related to the signal recorded by a sensor through climate and land surface properties, as well as the malaria vector and human settlement distributions (Curran *et al.*, 2000). This means that if remotely sensed images are to be useful in epidemiological studies then it is necessary to model the indirect relations between reflected (or emitted) radiation and vector or parasite prevalence through (often non-linear) functions of land surface and atmospheric properties.

1.2. Sensor Characteristics

Remotely sensed images may be acquired in a range of different wavelengths of the electromagnetic spectrum (EMS). Most remotely sensed imagery used commonly in the study of disease is recorded in visible to thermal-infrared wavelengths. For example, the National Aeronautics and Space Administration (NASA) Landsat Enhanced Thematic Mapper (ETM +) satellite sensor acquires images in seven discrete wavebands or portions of the EMS, while the National Oceanographic and Atmospheric Administration (NOAA) Advanced Very High Resolution Radiometer (AVHRR) satellite sensor acquires imagery in five wavebands (Cracknell, 1997) (Hay *et al.*, this volume, pp. 37–77). Such multi-band optical imagery can be invaluable for mapping classes of land cover and continua such as vegetation biophysical (Goel *et al.*, 2003) and biochemical properties (Curran *et al.*, 1998). Many modern imaging sensors record radiation in several hundred spectral wavebands, referred to as hyperspectral sensing (Curran, 1994).

Microwave remote sensing measures radiation in microwave wavelengths either emitted (passive sensing) or backscattered (active sensing) from the Earth's surface. Most active microwave remote sensing is achieved using synthetic aperture radar (SAR) imagery, which can be useful for predicting a range of structural properties (e.g., of vegetation) and has been used to predict the flooding status of Kenyan Rift Valley fever vector habitats (Pope *et al.*, 1992). Interferometric SAR (e.g., from the highly successful Shuttle Radar Topography Mission; Kellndorfer *et al.*, 2004; Eineder and Adam, 2005) is a comparatively recent technique that uses SAR to predict height, often with very high accuracy. Many other parts of the EMS and methods of remote sensing (e.g., LiDAR to sense surface height directly; Naesset *et al.*, 2005) may be useful in an epidemiological context.

The number, size and position of wavebands are just three of the several sensor characteristics that are important in determining the suitability of sensor imagery for a given epidemiological purpose and the quality of the resulting predictions (e.g., disease map). A further important sensor characteristic in the context of global disease prediction is the spatial resolution. While the number, size and position of each waveband in the EMS determines the spectral (attribute) information in an image, the spatial resolution of the imagery has equally important and predictable effects on the resulting information and its utility for particular applications. The spatial resolution of satellite sensors can vary from several kilometres (e.g., NOAA AVHRR imagery has two spatial resolutions of 1.1 and 8 km) to less than 1 m (e.g., Quickbird panchromatic imagery has a spatial resolution of 0.6 m, although the multiband imagery has a spatial resolution of 2.4 m) (Danson *et al.*, 2003).

For the global mapping of disease it is impractical to cover the entire surface of the Earth with fine spatial resolution (e.g., 1–100 m) imagery. This is partly because of the enormous numbers of data involved at fine spatial resolutions and the amount of effort involved in "stitching" together large numbers of images. An early example of a global remote sensing product is provided by the DISCover project which used NOAA AVHRR imagery to map land cover over the entire globe with a spatial resolution of 1 km (Loveland *et al.*, 2000). This has been superseded by the Global Land Cover (GLC) 2000 project (Latifovic *et al.*, 2004; Giri *et al.*, 2005), components of which were produced using Moderate Resolution Imaging Spectrometer (MODIS) imagery (Friedl *et al.*, 2002). The requirement to adopt relatively coarse spatial resolution data (e.g., 500 m–8 km) when mapping over large areas such as continents (e.g., Africa) or the entire globe has important implications for subsequent analyses, which are explored further in this paper.

As already described, remotely sensed imagery is rarely the object of scientific interest itself. To be useful, the imagery must usually be combined, via a model, with primary data acquired on the ground of the property of interest. Such "ground data" may represent areas (e.g., ground surveys of tsetse flies, Hendrickx *et al.*, 2001; number of clinical cases of a disease per health facility, each of which serves an unknown, but definable catchment population, Gething *et al.*, 2004) or points (e.g., individuals in an intensive community-based survey). Even in the former case, the geographical catchments are likely to be smaller than, or at least of different geometry to, the image pixels, such that both cases present a serious problem for modelling: specifically, the problem of incompatible spatial units (Hay *et al.*, 2001; Dungan *et al.*, 2002; Gotway and Young, 2002). This problem is discussed in Section 2.

1.3. Choice of Modelling Framework

In remote sensing, it is common to use the multivariate (multi-band) information to predict some property of interest (e.g., land cover, NDVI, LST) using one of several classes of model (e.g., physical model, statistical model). Physical models are usually analytical and are fitted to empirical data in the forward sense (Strahler *et al.*, 1986). The fitted model is then inverted to predict the property of interest. Statistical models, also referred to as empirical models, are usually fitted to empirical data directly and rarely require inversion (Schowengerdt, 1997). Semi-empirical models (Graham and Harris, 2003) lie between these two.

In a statistical sense, remotely sensed imagery are most commonly associated with the random field or random function (RF) model in which each point in space is treated as being characterized by a random variable (RV) and the relation between RVs at different locations is modelled using a spatial covariance or equivalent "structure" function (Atkinson, 1999). Several different models are applicable to such data including geostatistical models (Matheron, 1965, 1971; Chilès and Delfiner, 1999) and spatially autoregressive models (e.g., Augustin *et al.*, 1996, developed the autologistic model which is popular in ecology). An alternative to the RF model is the object-based model or view (Hay *et al.*, 2003), as might be applicable to point data on health facilities in a given country (Gething *et al.*, 2004). Statistical techniques such as point pattern analysis and cluster modelling (Diggle and Rowlingson, 1994; Cockings *et al.*, 2004; Dragicevic, 2003a; Kulldorff *et al.*, 2003) may aid epidemiological investigation of such data.

Although not the primary interest of this paper, process models should be considered alongside the static RF and object-based views of the world. Dynamic models of disease processes are becoming ever more common based on increased availability of suitable data and high-performance computer power. Such models are being fitted at the local or even individual level, replacing previous population-based models (Anderson *et al.*, 1991) and introducing new insights into the factors influencing the space–time dynamics of disease (Bian, 2004; Cisternas *et al.*, 2004; Eubank *et al.*, 2004). Such models ultimately rely on high-quality data both on the disease and the environment in which it is transmitted, and the latter can often be provided by remote sensing. Where this is the case, the efficacy of the model at representing real processes will depend on the characteristics of the model and the input data (such as spatial resolution).

1.4. The Image Processing Chain

Whichever modelling approach is chosen, fitting a model can be seen as one of several components in what is known as the processing chain (Schott, 1987). Common components in the processing chain are image acquisition, radiometric calibration (e.g., Guyot and Gu, 1994), atmospheric correction (e.g., Chavez, 1996), geometric correction, model fitting, prediction and accuracy assessment. Each of these components can have an effect on the information content and accuracy of the final output.

The purpose of this paper is to evaluate the effect of two components of the processing chain, scale and uncertainty, on epidemiological predictions made based on remotely sensed imagery. Issues of spatial scale and spatial resolution are considered in Section 2, and issues of uncertainty are considered in Section 3. Section 4 provides a conclusion.

2. ISSUES OF SCALE AND SPATIAL RESOLUTION

2.1. Preliminaries

Scale and the scaling of data are central subjects in remote sensing and GIS (Woodcock and Strahler, 1987; Quattrochi and Goodchild, 1997). A fundamental concept that underpins the understanding of scale and scaling is that data $z_{\mathbf{v}}(\mathbf{x})$ on property z defined on a support $\mathbf{v}$ (the size, geometry and orientation of the space on which each measurement is made) at locations $\mathbf{x}$ are a function of both reality (i.e., the underlying property of interest $z(\mathbf{y})$, where $\mathbf{y}$ denotes any point location within the target space) and the sampling framework $s(\mathbf{x}_s, \boldsymbol{\psi})$ with parameters $\boldsymbol{\psi}$ fixed at location $\mathbf{x}_s$:

$$z_{\mathbf{v}}(\mathbf{x}) = f(z(\mathbf{y}), s(\mathbf{x}_s, \boldsymbol{\psi})) \tag{1}$$

Equation (1) ignores the temporal dimension, and the uncertainty in measurement which is developed in Section 3. The key point is that mathematical operations are applied to the data, not reality, and so the outcome of any operation is partially a function of the sampling framework. Only by understanding and modelling the effects of the sampling framework is it possible to characterize adequately and predict optimally the underlying property of interest.

The sampling framework comprises both spatial and temporal parameters with the spatial set of parameters comprised of the sample size n (e.g., 30), sampling scheme ϕ (e.g., random, stratified random), sampling density d (e.g., 30 observations per km^2) and support $\mathbf{v}$ (e.g., a point, a pixel). The support itself is defined by a further set of parameters: geometry $\rho_{\mathbf{v}}$ (e.g., square wave response over a square pixel, 2-D Gaussian function), orientation $\theta_{\mathbf{v}}$ (e.g., north-to-south) and size $|\mathbf{v}|$ (e.g., 1 km on a side). These six parameters $\boldsymbol{\psi}_{\mathbf{x}}$ define the spatial sampling framework completely:

$$s(\boldsymbol{\psi}_{\mathbf{x}}) = f\left(n, \phi, d, \rho_{\mathbf{v}}, \theta_{\mathbf{v}}, |\mathbf{v}|\right) \tag{2}$$

The temporal sampling framework $s(\boldsymbol{\psi}_t)$ is defined by an equivalent five parameters $\boldsymbol{\psi}_t$ (support orientation is meaningless in 1-D). Together, these 11 generic parameters $\boldsymbol{\psi}_{\mathbf{x},t}$ define the sampling framework $s(\boldsymbol{\psi}_{\mathbf{x},t})$. Once defined, the sampling framework can be applied to any region of interest to obtain data by fixing location $\mathbf{x}$ and time t; conceptually involving definition of a further three parameters. Equation (1) can now be updated to include the temporal dimension:

$$z_{\mathbf{v}}(\mathbf{x}, t) = f\left(z(\mathbf{y}), s\left(\mathbf{x}_S, t, \boldsymbol{\psi}_{\mathbf{x},t}\right)\right) \tag{3}$$

The sampling framework has an important influence on the measured data. From this point onwards, the sampling framework $s\left(\mathbf{x}_S, t, \boldsymbol{\psi}_{\mathbf{x},t}\right)$ is represented by $s(\mathbf{x}, t)$, where t represents time.

The word scale has numerous different meanings (Curran and Atkinson, 1998; Dungan *et al.*, 2002). The most common meaning in the present context is the cartographic definition of scale: the ratio between a distance in a representation, model or map of reality to the actual distance in reality (e.g., 1:10 000). This definition can be confusing for many reasons that have been explained previously (Atkinson and Tate, 2000; Dungan *et al.*, 2002) and so, in this paper, the everyday usage is adopted in which scale simply means size (e.g., large-scale investigation or process simply means large investigation or process). This is the definition commonly adopted in ecology and physics. This definition is required because it is necessary to define the scale(s) of measurement in the sampling framework and the scale(s) of spatial variation in the data, neither of which are ratios.

One of the most important effects of $s(\mathbf{x}, t)$ on observed data is to determine the scales of spatial variation present in the data. More fully, the scales of spatial variation present in spatial data are a function of the scales of spatial variation present in reality and the sampling framework. Certain parameters of $s(\mathbf{x}, t)$ defined above, for example the support, have a greater effect than others. Knowledge of which scale(s) of spatial variation are revealed (and which are omitted) is important. The filter on reality induced by $s(\mathbf{x}, t)$ determines, fundamentally, the information content of data, and the outcome of subsequent analyses on those data. For example, Belward and Lambin (1990) describe limits to characterizing spatial structure using NOAA AVHRR imagery. It is common in remote sensing that pixels are compared directly to ground data and the correlation coefficient and predictive power of a fitted regression model will depend on the image and ground supports, both in relative and absolute terms. The choice of spatial resolution for remote sensing investigations has been the subject of extensive research (e.g., Woodcock and Strahler, 1987; Atkinson and Curran, 1995, 1997; Curran and Atkinson, 1999).

The advantages of modelling scale in the above context are that (i) the scale(s) of spatial variation in data may be characterized and (ii) by combining such information with a model of the convolution process implicit in sampling it is possible to change the scales of variation in data as a function of hypothetical changes in $s(\mathbf{x}, t)$. This provides a very powerful means for addressing directly some of the most important issues of scale inherent in the global prediction of disease.

2.2. Measurement Scales

2.2.1. The Support

Of all the parameters of $s(\mathbf{x}, t)$, the support is the most important in that it provides a fundamental limit on the scales of spatial variation that can be exhibited in data. The size of support $|\mathbf{v}|$ has the greatest influence, but geometry ρ_{v} also has an effect and orientation θ_{v} can be important where variation is anisotropic. These support effects are best modelled as a convolution of the underlying signal by the support:

$$Z_{\mathrm{v}}(\mathbf{x}) = \int_{\mathbf{v}^*} Z(\mathbf{y}) h_{\mathrm{v}}(\mathbf{y})\, dy \tag{4}$$

where $\mathbf{v}$ is the support, $\mathbf{y}$ a point and $h_{\mathrm{v}}(\mathbf{y})$ represents the support. Thus, the support changes the basic character of the dataset, obscuring certain fine scales of spatial variation from the investigator.

Further effects are induced where the supports of individual data vary, for example, with census data. Such variation leads to the modifiable areal unit problem (MAUP) (Openshaw, 1984; Unwin, 1996). Then, the so-called aggregation effect of $|\mathbf{v}|$ is intermingled with a zonation effect. Wakefield (2003, 2004) provides methods for assessing the effects of area-based (i.e., cell-based) data used in epidemiology by comparing the predictions to point statistics using a random effects model. Krieger *et al.* (2002) provide an example in which the relation between socioeconomic status and cancer incidence depends on support size.

2.2.2. Spatial Resolution

In remote sensing, the terms support and spatial resolution are often used interchangeably. Whereas the support is defined above as a first-order parameter of $s(\mathbf{x}, t)$ (because it relates to a single observation), the spatial resolution is a second-order property (because it is a function of more than one observation). Spatial resolution provides an important lower limit to the scales of spatial variation that can be captured in a given dataset. Spatial resolution is a function of the set of smallest distances between observations as well as the support. In remote sensing the spatial resolution is equal to the pixel size, which is an approximate representation of the support. In ecology, the term "grain" is often used interchangeably to mean spatial resolution (Dungan *et al.*, 2002).

2.2.3. Spatial Extent

The spatial extent is defined here as a second-order property that provides an upper limit on the scales of spatial variation present in data. It is a function of the set of largest distances between observations.

Although it is helpful conceptually to distinguish between the spatial resolution and spatial extent as lower and upper limits imposed by $s(\mathbf{x}, t)$, in reality, the set of all possible pairs of observations implicit in $s(\mathbf{x}, t)$ determines the full sampling of the scales of spatial variation existing in the convolved variable of interest $Z_{\mathbf{v}}(\mathbf{x})$ (i.e., the property of interest measured on the given support).

2.3. Scale(s) of Spatial Variation

2.3.1. Characterizing Spatial Variation

There exist many different methods for characterizing spatial variation. These can be organized according to the basic data model adopted. For the RF model, various forms of regression analysis (e.g., Kelsall and Diggle, 1998) and geostatistical approaches (Chilès and Delfiner, 1999) are popular. For geostatistics, basic structure functions such as the spatial covariance, autocorrelation function and variogram are used for characterization (Curran and Atkinson, 1998; Atkinson, 1999). Geostatistics has been applied to non-communicable disease (e.g., Oliver *et al.*, 1992, 1998; Webster *et al.*, 1994; Minozzo and Fruttini, 2004), and, less commonly, to communicable disease (e.g., Diggle *et al.*, 2002).

For the object-based model, point-pattern analysis provides a wide range of functions (e.g., the *K*-, *L*- and *D*-function) where the objects are points (Cressie, 1991). Point pattern analysis and cluster analysis have been applied widely in exploring spatial variation in disease risk (Diggle and Rowlingson, 1994; Gatrell *et al.*, 1996; Lawson, 2000; Dragicevic, 2003b). Hand and Bolton (2004) provide a review of pattern discovery techniques from a statistical perspective. For other types of object (i.e., non-point objects) the tools are not as well founded statistically, mostly relying on technical developments in GIS and landscape ecology (e.g., Hay *et al.*, 2003).

Much spatial variation in disease vector distributions and predictors of disease (e.g., environmental controls on vector distribution) at the *global* scale may be modelled appropriately using RF. Therefore, the geostatistical approach provides the focus in this section. Of the available functions, the variogram is usually preferred because it exists in circumstances where the covariance and autocorrelation function do not. Specifically, the variogram relies on a weaker model of stationarity (referred to as intrinsic stationarity) than the covariance and autocorrelation (second-order stationarity) (Myers, 1989) and consequently this discussion is restricted to variography.

For continuous variables, such as NDVI, LST and CCD, the sample semivariance is defined as half the average squared difference between values separated by a given lag **h**. The sample variogram $\hat{\gamma}(\mathbf{h})$ may be estimated using

$$\hat{\gamma}(\mathbf{h}) = \frac{1}{2P(\mathbf{h})} \sum_{\alpha=1}^{P(\mathbf{h})} [z(\mathbf{x}_\alpha) - z(\mathbf{x}_\alpha + \mathbf{h})]^2 \tag{5}$$

where $P(\mathbf{h})$ is the number of paired comparisons at a specific lag **h** and $Z(\mathbf{x}_\alpha)$ an observation (treated as a realization of $Z(\mathbf{x})$ at location $\mathbf{x}_\alpha$). The sample variogram may be estimated for all directions lumped together (omnidirectional) or for several orientations where variation is thought to be anisotropic.

The sample variogram is a set of semivariances at a set of discrete lags. To allow statistical inference, it is usually necessary to fit a continuous mathematical model to the estimated variogram. To ensure that all linear combinations of the RF result in non-negative variances, the variogram model must be conditional negative semi-definite (CNSD). It is common practice to select from a pre-defined set of CNSD models (McBratney and Webster, 1986; Webster and Oliver, 1990). Two commonly selected models are given below.

(i) the nugget effect model:

$$\gamma(h) = c_0 \begin{cases} 0 & \text{if } h = 0 \\ 1 & \text{otherwise} \end{cases} \tag{6}$$

where c_0 is the nugget variance parameter; and

(ii) the spherical model:

$$\gamma(h) = c_1 \begin{cases} 1.5\frac{h}{a} - 0.5\left(\frac{h}{a}\right)^3 & \text{if} h \leq a \\ 1 & \text{otherwise} \end{cases} \tag{7}$$

where c_1 is the sill variance parameter and a the range parameter. Webster and Oliver (1990) provide further examples. The model is usually fitted to the sample variogram by some automatic process such as weighted least squares (e.g., Cressie, 1985, 1991; Goovaerts, 1997).

It is important to understand the meaning of the information that the variogram conveys. For example, the sill c_1 of the spherical model provides information on the *amount* of variation present in V. More precisely, the sill estimates the *a priori* variance $D^2(v, \infty)$ of Z (that is the variance obtained on a support v within an infinitely sized region).

The model form and the range a provide information on the *scale(s)* of spatial variation. The variogram models described may be used singly or in a positive linear combination, in which case the resolved variation is said to be 'nested' (see Webster and Oliver, 1990). The nugget model is often fitted together with a structured component (such as a spherical model) to represent a discontinuity at the origin. The nugget variance c_0 represents unresolved variation that exists on a micro-scale *and* measurement error, but it can also arise from uncertainty both in estimating the variogram and in fitting the model at short lags (Atkinson, 1997). Where observations are abutting or overlapping, as for remotely sensed imagery, then the amount of micro-scale variation can be assumed to be small and the nugget variance can be assumed to be due primarily to measurement error (e.g., Curran and Dungan, 1989). While fractals are an important mathematical model of scaling in the natural world because they are scale-invariant, they are beyond the scope of this discussion. However, it is interesting to note that fractal dimension can be estimated from the modelled variogram (Klinkenberg and Goodchild, 1992).

The modelled variogram provides a basis for statistical inference via geostatistical spatial prediction, known as kriging. Geostatistical kriging has been applied to a variety of disease prediction problems. For example, Oliver *et al.* (1992) and Webster *et al.* (1994) were one amongst the first to apply geostatistics to characterize and map disease pattern. Kelsall and Wakefield (2002) used kriging to map colorectal cancer in Birmingham, UK. Geostatistical cokriging has been applied to map the risk of childhood cancer (Oliver *et al.*, 1998) and tick habitats from NOAA AVHRR imagery (Estrada-Pena, 1998). Bayesian statistical approaches have become popular in epidemiology (e.g., Wakefield and Morris, 2001). Model-based geostatistics, in which a Bayesian framework is used to incorporate the uncertainty in variogram estimation into spatial predictions, has been applied to mapping malaria in the Gambia (Diggle *et al.*, 2002).

Generally, the focus of interest in disease mapping is in characterizing and mapping spatial variation at a particular point in time, but recent advances have been made in space–time modelling of disease processes (e.g., Baker, 2004; Grenfell *et al.*, 2001; Mugglin *et al.*, 2002).

2.3.2. Upscaling and Downscaling the Data

Whereas several papers have appeared in which geostatistical and related spatial statistical methods are applied to epidemiological problems, including characterizing the scales of spatial variation in disease, little research has considered the potential for scaling offered by the geostatistical framework. Since data are often acquired or assembled with different supports, and varying sampling strategies more generally, researchers should give serious consideration to the potential solutions provided for handling such diverse datasets (Gotway and Young, 2002).

Upscaling refers to an increase and downscaling to a decrease, in the size of support. Upscaling and downscaling are recognized as important operations in ecology and related subjects (Bierkens *et al.*, 2000). The importance of these operations is now being realized by epidemiologists who have shifted their focus to national, continental and now global scales, through increased utilization of remotely sensed imagery. Upscaling is usually relatively easy to achieve. For example, where sufficient data exist at the finer spatial scale (e.g., an image), data upscaling is achieved readily by averaging over a kernel defined by the new larger support. The averaging is usually through a linear combination of the smaller cells, but can be non-linear (Bierkens *et al.*, 2000). Where data are sparse spatially and interpolation is required, data upscaling may be achieved by geostatistical block kriging (Burgess and Webster, 1980) or block cokriging (Atkinson *et al.*, 1992) (although care is needed to distinguish desirable regularization over the support from unwanted convolution due to smoothing; Atkinson and Kelly, 1997; Atkinson and Tate, 2000). Further, upscaling can be achieved using block conditional simulation, which removes the smoothing of kriging, but retains the uncertainty of prediction (Deutsch and Journel, 1992; Journel, 1996).

Downscaling is more difficult to achieve because it implies an increase in information content above that provided by the original data. For continua, a solution may be achieved using the geostatistical methods of area-to-point kriging for the univariate case (Kyriakidis, 2004) and downscaling cokriging for the multivariate case (Pardo-Iguzquiza *et al.*, 2005). For classification, a solution may be achieved using super-resolution mapping (Tatem *et al.*, 2001, 2002; Atkinson, 2005). The latter has already been applied in an epidemiological context to map settlement distribution in Kenya (Tatem *et al.*, 2004). These techniques for downscaling allow an increase in the spatial resolution of the predicted map.

There are many different situations in epidemiology and remote sensing in which it may be desirable to manipulate one or more variables to provide data on matching supports. As the pixel is usually larger than the support of the ground data, the mis-match may mean that certain (fine) scales of variation expected to be present in the ground data will automatically be missing from the imagery. By default, this reduces the correlation between image and ground data. The appropriate action depends on whether the fine-scale variation is of interest or not. If of interest, the imagery should be downscaled to match the support of the ground data through linear methods (area-to-point kriging and cokriging) or non-linear constrained optimization (super-resolution mapping). If this is not done, the fine-scale variation will be missed in predictions based on the imagery. The fine-scale variation may be of interest where other datasets are defined on the smaller support (i.e., resolving the problem of incompatible supports between variables). Further, downscaling may provide a means by which to conduct validation at the scale of ground observation. If the fine scale of spatial variation is not of interest, correlation and regression analyses may be applied directly, with the caveat that the additional fine-scale variation at the ground will increase the uncertainty of the prediction.

More complicated scenarios exist. For example, several different variables to be used in an epidemiological study may have been captured at fundamentally different measurement scales. Such scales may include different levels of cartographic generalization in digitized lines and choropleth maps, different spatial resolutions in scanned aerial photographs and digital sensor imagery, and different enumeration areas in population censuses. The same basic principles of convolution and scaling apply to all of these datasets. The fitting of appropriate models and application of proper scaling relations is non-trivial given such a diversity of datasets. Nevertheless, the scenario is realistic, and in many ways typical of epidemiological analysis at the global scale where data are rarely collected for the purpose and investigators must use that which is available. While dealing with such incompatible spatial measurement scales is difficult, researchers should be aware of the problems, the effects of such scales on the data and results of analyses, and where possible take action to model such effects as part of their analyses.

2.3.3. Upscaling and Downscaling the Model

Upscaling (downscaling) can refer to the variable of interest, but it can also refer to the model (e.g., the variogram). Journel and Huijbregts (1978) give a series of equations for the geostatistical operation of regularization that are helpful in understanding the effect of the support on the variogram. The relation between the punctual or 'point' semivariance and the regularized (or convolved) semivariance at a lag $\mathbf{h}$ is given by (Journel and Huijbregts, 1978):

$$\gamma_{\mathbf{v}}(\mathbf{h}) = \bar{\gamma}(\mathbf{v}, \mathbf{v_h}) - \bar{\gamma}(\mathbf{v}, \mathbf{v}) \tag{8}$$

where $\bar{\gamma}(\mathbf{v}, \mathbf{v_h})$ is the integral punctual semivariance between two supports of size v whose centroids are separated by $\mathbf{h}$, given formally by

$$\bar{\gamma}(\mathbf{v}, \mathbf{v_h}) = \frac{1}{\mathbf{v}^2} \int_{\mathbf{v}} \int_{\mathbf{v(h)}} \gamma(\mathbf{y}, \mathbf{y}')\, d\mathbf{y}\, d\mathbf{y}' \tag{9}$$

where $\mathbf{y}$ describes an observation of size v and $\mathbf{y}'$ describes independently another observation of equal size and shape at a lag $\mathbf{h}$ away. The quantity $\bar{\gamma}(\mathbf{v}, \mathbf{v})$ is the integral punctual semivariance within an observation of size v:

$$\bar{\gamma}(\mathbf{v}, \mathbf{v}) = \frac{1}{\mathbf{v}^2} \iint \gamma(\mathbf{y}, \mathbf{y}')\, d\mathbf{y}\, d\mathbf{y}' \tag{10}$$

where $\mathbf{y}$ and $\mathbf{y}'$ now cover the same pixel independently.

The fact that observed data are always a function of $s(\mathbf{x}, t)$, and particularly the support, means that in order to convolve the variogram model, it is first necessary to deconvolve. Downscaling can be applied to the model (e.g., variogram) representing the character of spatial variation (Pardo-Iguzquiza *et al.*, 2005). Such a deconvolved or deregularized model must be estimated for area-to-point kriging and cokriging above, and it must be estimated empirically from data for super-resolution mapping of fine-scale spatial pattern (Tatem *et al.*, 2002).

3. ISSUES OF UNCERTAINTY

3.1. Preliminaries

Observed spatial data $z_{\mathbf{v}}^{*}(\mathbf{x})$ (e.g., of spectral response z on support $\mathbf{v}$ at location $\mathbf{x}$) can be viewed as being comprised of the true values $z_{\mathbf{v}}(\mathbf{x})$ plus some variable $e_{\mathbf{v}}(\mathbf{x})$ representing measurement error:

$$z_{\mathbf{v}}^{*}(\mathbf{x}) = z_{\mathbf{v}}(\mathbf{x}) + e_{\mathbf{v}}(\mathbf{x}) \tag{11}$$

It has been demonstrated above that the "true" value $z_{\mathbf{v}}(\mathbf{x})$ is a convolved version of the unobservable point property of interest $z(\mathbf{x})$ [Equation (3)]. Thus, combining Eqs. (3) and (11):

$$z_{\mathbf{v}}^{*}(\mathbf{x}, t) = f\left(z(\mathbf{y}), s\left(\mathbf{x}_{s}, t, \psi_{\mathbf{x},t}\right), e_{\mathbf{v}}(\mathbf{x})\right) \tag{12}$$

The effect of $e_{\mathbf{v}}(\mathbf{x})$ is to add uncertainty to the already convolved observable variable.

Uncertainty is defined here as a general concept, with meaning similar to that in its everyday usage (Atkinson and Foody, 2002). There are many forms of uncertainty, including ambiguity (expressed using a probabilistic model) and vagueness (expressed using fuzzy set theory; Klir, 2001). Here we are concerned primarily with ambiguity. Fundamentally, whether the interest is in measurement uncertainty or prediction uncertainty, the focus of attention in most epidemiological studies is accuracy. Accuracy is defined as the combination of unbiasedness and precision. Bias arises from systematic error $e_{s}(\mathbf{x})$ and imprecision from random error $e_{r}(\mathbf{x})$:

$$e_{\mathbf{v}}(\mathbf{x}) = e_{s}(\mathbf{x}) + e_{r}(\mathbf{x}) \tag{13}$$

Accuracy, bias and precision can be estimated via measurable quantities. Such estimates of bias and (im)precision amount to estimates of the expectation of the systematic $\mathrm{E}\left(\left|e_{s}(\mathbf{x})\right|\right)$ and random $\mathrm{E}\left(\left|e_{r}(\mathbf{x})\right|\right)$ components of the error in Eq. (13).

It is rarely possible to estimate the actual error $e_{\mathbf{v}}(\mathbf{x}_{0})$. If the actual error were known, the observed value could be adjusted to the true value and the uncertainty would disappear. Rather, it is common to estimate the expected error. To provide an expectation, a statistical model (no matter how trivial) is required, and the model is fitted to an ensemble of data (no matter how small). The statistical model is usually stationary in the parameter being estimated.

3.2. Uncertainty in Data and Methods

Uncertainty manifests itself in all stages of the processing chain (Schott, 1987). Measurement error can have deleterious effects on modelling and prediction based on the data, and prediction inaccuracy can, in turn, lead to inappropriate public health decisions.

The remotely sensed data used in this volume have inherent measurement errors (Nemani *et al.*, 2003), in addition to those related to georectification and atmospheric correction. Problems such as striping and missing pixels introduce serious, albeit obvious and detectable, errors into the remotely sensed data (Figure 1). Clearly, it is important to identify such errors in imagery so that they can be handled appropriately.

In addition to the satellite sensor data, the epidemiological data used in the modelling process can be, and often are, replete with errors. Such errors can include incorrect georeferencing of a location, usually as a result of user error; confusion regarding which location reference in a gazetteer is correct; incorrect diagnosis of a specific disease leading to presence or absence records where the opposite is correct; issues over the reliability of the record source, especially for historical sources or less often recorded diseases; confusion over the aggregation methods applied to the data, which may have implications for the final location and data values used; and the use of randomized input points (for presence and/or absence) taken from within areal records.

Further sources of error and uncertainty are introduced by the modelling methods themselves. Generally, no method or model is perfect and prediction always involves uncertainty, usually expressed through the predictive error distribution.

If the investigator processes the satellite sensor data, collects the field epidemiological data and codes the computer programs then much of

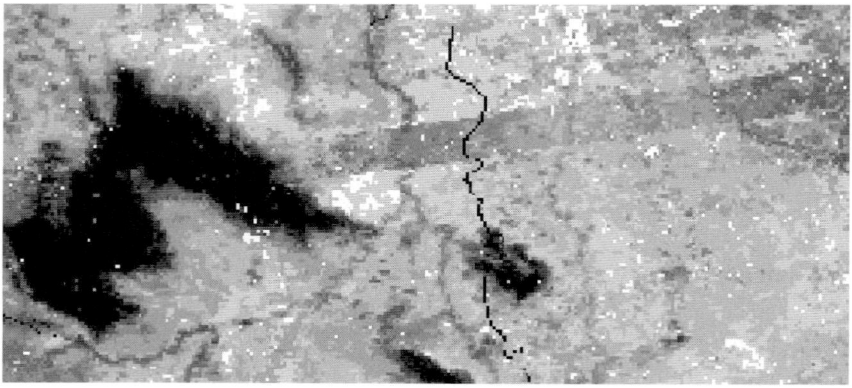

Figure 1 A subset image of the 1 × 1 km data demonstrating the types of error in the 1 km spatial resolution imagery. Notice the gap in the coverage (black area to the left of the image) due to masking out of cloud, the offset river (black line in the centre of the image) due to misregistration and the striping (dark grey feature running diagonally across the image) linked to data collection problems with the sensor.

the uncertainty and error can be controlled, or at least accounted for. When using data or programs from other sources it is important that the investigator is aware of the quality of the data, and the methods employed to produce them, via appropriate metadata. A lack of metadata creates uncertainty about (data and model) uncertainty.

Given the above discussion, it should be clear that information on uncertainty, in the form of estimates of accuracy (precision, and where possible bias) is important in epidemiological analysis. Information on uncertainty should be provided as an average globally (e.g., precision for an entire map), but should, where possible, also be provided at the local level (e.g., per-country, -region, -local area or even per-pixel). No map of disease, or vector distribution, is complete without some assessment of uncertainty.

Accuracy assessment can be achieved in two fundamentally different ways. One approach is to use unseen data (e.g., jack-knifing) or in some cases the data used in prediction (e.g., cross-validation) to evaluate the predictions through direct comparison of known and predicted values. This approach is data based. Another approach is to use the fitted model, together with assumptions or knowledge of the predictive error distribution to estimate uncertainty in the predictions. This approach is model based.

3.3. Data-Based Accuracy Assessment

Data-based assessment of accuracy involves comparison of model-based predictions with data that have not been used in prediction and, depending on the method, may not have been used in fitting the model. Such a comparison results in direct estimation of the actual error. Although an extremely valuable spatially located source of information, the data are rarely used directly. Usually, a spatially stationary statistical model is adopted and the errors lumped to provide a global estimate of accuracy.

3.3.1. Categorical Variables

Accuracy assessment for categorical variables usually centres on the confusion matrix or contingency table (Foody, 2002). Many statistics can be derived from the confusion matrix, including the overall accuracy (the number of correct predictions divided by the total number of predictions), the users' and producers' accuracies, and sensitivity and specificity (Cohen, 1960; Rosenfield and Fitzpatrick-Lins, 1986; Story and Congalton, 1986; Congalton, 1991; Foody, 2002). Since several of these statistics are adopted in this volume they are given brief attention in this section (also see Table 1 in Rogers, this volume, pp. 1–35).

An example of a contingency table for a binary classification is given in Table 1. A contingency table requires two or more variables from which the frequency of agreement can be calculated. This is undertaken by listing all categories of the classification (e.g., presence/absence or class) in rows and the concomitant reference class in columns. The cell frequency is computed for each cell and summed (for rows and columns). Dividing a correctly modelled class cell value by the row total results in a value known as the 'user's accuracy'. Dividing the same cell value by the column total results in the 'producer's accuracy'. The user's accuracy is a measure of commission error, indicating the probability of a correct prediction, while the

Table 1 Example of a contingency table for a hypothetical presence/absence system denoting the true-/false-positive values, true-/false-negative values, users/producers and total accuracies and κ

	Actual presence	Actual absence	Row total	Users accuracy
Predicted Presence	15 True positive (TP)	3 False positive (FP)	18	83%
Predicted absence	1 False negative (FN)	12 True negative (TN)	13	92%
Column total	16	15	31	
Producers Accuracy	93%	80%		87% Total accuracy
		$\kappa = 0.741$ *95% confidence interval: 0.503–0.978*		

producer's accuracy indicates how well the training data have been predicted (Lillesand and Kiefer, 2004). The overall accuracy of the prediction is calculated by dividing the sum of the diagonal (true positive values) by the sum of all values in the table (Table 1).

Kappa (κ) is a widely used statistic for estimating the accuracy of prediction of a categorical variable. The κ statistic is a "discrete multivariate technique ... for determining statistically if one error matrix is significantly different from another" (Zhan *et al.*, 2002) while accounting for chance. κ is calculated from the contingency table using the following conceptual equation:

$$\kappa = \frac{\textit{observed accuracy} - \textit{chance agreement}}{1 - \textit{chance agreement}} \tag{14}$$

The full equation requires the number of rows in the contingency table (r), the total observations in a row (x_{i+}), the total observations in a column (x_{+i}), the number of observations in row i and column i (the diagonal) (x_{ii}) and the total number of observations in the matrix (N):

$$\kappa = \frac{N\sum_{i=1}^{r} x_{ii} - \sum_{i=1}^{r}(x_{i+} \cdot x_{+i})}{N^2 - \sum_{i=1}^{r}(x_{i+} \cdot x_{+i})} \tag{15}$$

Despite the utility of κ in the context of disease distribution modelling, as a summary of the contingency table, some caveats do exist. A major criticism is that κ does not distinguish between types or causes of disagreement and that the statistic may be sensitive to sample size (Fielding and Bell, 1997; Forbes, 1995), failing when one class is substantially larger than the other. Another concern is that comparison between κ values for classifications based on the same input data may be statistically flawed (Foody, 2004). The Tau coefficient may also be derived from the confusion matrix (Foody, 2002). Kendall's Tau is a distribution-free correlation coefficient, which can be applied to ordinal (rank) data (Shaw and Wheeler, 1994) and can be applied readily to abundance distribution mapping. This statistic is linked closely to κ but relies on *a priori* probabilities (as opposed to the *a posteriori* probabilities used when calculating κ) (Fielding and Bell, 1997).

The percentage of pixels correctly assigned to either presence or absence classes are summary statistics, providing initial indications of the accuracy of a classification (although unlike κ these do not account for chance occurrence). Two further tests linked to the contingency table are useful for assessing slightly different aspects of accuracy. These are sensitivity, which identifies true positives as a proportion of all positive predictions; and specificity which identifies true negatives as a proportion of all negative predictions. Table 2 shows the links between sensitivity and specificity and the contingency table.

Receiver operating characteristic (ROC) curves were devised to define thresholds in early radar backscatter returns from equipment operated by different technicians, but the method has now been directed at epidemiological studies (Goddard and Hinberg, 1990; Thompson and Zucchini, 1989) to define cut-off points in diagnostic tests. The ROC curve (Brooker *et al.*, this volume, pp. 221–262) is created by plotting sensitivity against 1-specificity, and the area under the curve (AUC) is frequently used to discriminate between different curves. A completely non-effective model would produce a flat line, indicating that for every true positive, the procedure also generated a false positive resulting in an AUC value of 50% (i.e., half the plot is under the line).

Where it is available, the "true" image subtracted from the observed image provides a more basic primitive than the confusion

Table 2 The inter-relationships between sensitivity and specificity. A test with high values of specificity and sensitivity has few false positives or false negatives and demonstrates that the predicted output is reasonably accurate

Result	Name
True-positive result	Sensitivity
True-negative result	Specificity
False-negative result	1-Sensitivity
False-positive result	1-Specificity

matrix because it retains important spatial information in the error-per-pixel. Given such an error or difference image, the spatial character of the error may be explored as well as its magnitude.

3.3.2. Continuous Variables

For continua, several common statistics may be used to describe the different aspects of accuracy of interest. The scatterplot of the predicted against the observed variable may provide a useful visual description, in some ways equivalent to the confusion matrix. Given an appropriate (i.e., Gaussian) error distribution, the correlation coefficient provides a measure of precision, while the mean error provides a measure of bias. The root-mean-squared error provides information on overall accuracy, as does the mean absolute error. Many similar and related statistics exist. Such statistics are well known and so are not described further here.

As for categorical variables, a difference image, if available, provides a useful tool for exploring the spatial character of the error. For continua, variography of the error is a useful approach that can reveal scales of variation in the autocorrelated error.

3.3.3. Cross-Validation

Cross-validation is a method for assessing accuracy based on the data that were used to fit the model (Chilès and Delfiner, 1999). The basic principle is that a datum is omitted, and all other data are used to predict at that location. The observed value is subtracted from the predicted value to estimate the actual error. The datum is replaced. The process is repeated until all observed data have been predicted, allowing construction of a full error distribution. Cross-validation provides an important check on accuracy where external data are not available for accuracy assessment.

3.4. Model-Based Accuracy Assessment

3.4.1. Continuous Variables

Model-based assessment of accuracy refers to the use of the model to generate estimates of precision (it is often not possible to estimate bias without external data and moreover, if it were estimated the model should be adjusted to remove it). At the most basic level, the model might be used to estimate the error variance, per-prediction. Regression-type predictors can be used to provide such an estimate, based on the linear model involved and the assumption of a Gaussian error distribution. For example, the estimation of confidence intervals around predictions made using a regression model has been standard practice for many years (De Groot and Schervish, 2001; Jarner *et al.*, 2002).

A good example of a regression-based estimate of prediction variance is the popular kriging variance. The kriging variance is provided for every spatial prediction made. The distinction between actual error variance and expectation of error variance (i.e., kriging variance) is important: the kriging variance has many, well-documented shortcomings (Goovaerts, 1997). An estimate of the expected error variance is only as good as the statistical model used to provide it. Nevertheless, where it is possible to make model-based estimates of precision, they should be given. They cost little in terms of additional effort and are sometimes the only information on uncertainty available.

The error variance is only one parameter of an assumed Gaussian error distribution, and so is of limited value, particularly where interest is in some non-linear function of the data (e.g., the probability $p(z(\mathbf{x}_0) > k | z(\mathbf{x}))$ of exceeding some threshold k). Researchers have recently turned their attention to the problem of estimating the entire conditional cumulative distribution function (ccdf). The predicted value is then seen as the mean of the ccdf, with alternatives less likely, but nevertheless possible. Given the entire ccdf, it is possible to consider non-linear functions of the data (e.g., $p(z(\mathbf{x}_0) > k | z(\mathbf{x})))$. It should again be remembered that an estimate of the ccdf depends on the goodness of the model.

Proper treatment of uncertainty is the core concern of statistics. Several researchers have developed Bayesian approaches that are applicable to disease data (e.g., Cressie, 1995; Stern and Cressie, 2000; Diggle *et al.*, 2002). Markov chain Monte Carlo (MCMC) has gained popularity in statistics in recent years as a method for solving integration problems that are analytically intractable (Lawson, 2000). It allows fitting of statistical models, while estimating the predictive error distribution as well as the distributions for each of the parameters in the model. MCMC is likely to appear as an important tool in the epidemiologists' toolkit in time.

3.4.2. Categorical Variables

Many classification algorithms can be used to provide information on the uncertainty of prediction, although this is sadly rarely the case. For example, the Mahalanobis distances used in discriminant analysis and maximum likelihood classification can be used to provide information on the discriminatory power of individual input variables. Various measures such as the sum of the Mahalanobis distances between all classes and the sum of the distances between non-zero classes (i.e., classes of prevalence) can be useful in determining the variables that are selected for inclusion in a stepwise discriminant analysis.

Similarly, the maximum likelihood classifier predicts the most likely class for a given case by selecting from the set of posterior probabilities for all classes. Although rarely used, this set of probabilities provides information on the degree of uncertainty of each individual allocation. For example, from a set of six classes, a maximum likelihood of only 0.4 indicates an uncertain allocation (Foody *et al.*, 1992). Entropy estimated for all six probabilities encapsulates the uncertainty in allocation more completely (Maselli *et al.*, 1994). Similar measures can be estimated for other classifiers including, for example, artificial neural networks where the strength of activation of output neurons indicates much the same information as for the maximum likelihood classifier.

3.5. Validation of Global Disease Maps: Some Considerations

A key problem in the prediction of disease globally is validation. Predictions are often made on the support of the remotely sensed covariates (e.g., 8×8 km) and it may be very difficult to find suitable methods of acquiring "known" or high-accuracy data on the property of interest with which to assess the accuracy of such predictions. A major problem is in matching the supports of the remotely sensed and ground variables.

Suppose that the image pixel size is nominally 8 km on a side and complete coverage of several variables (e.g., NDVI, LST, CCD) for Africa has been used to predict malaria parasite rate (PR) by month. Let us suppose that PR data are available for 50 health facilities in Kenya and that these data have been suggested as a possible source for checking the accuracy of the map. There are multiple problems with the use of such data including:

(i) Sometimes imperfect data are the only source of information available for validation: the PR data may represent certain age groups or be conditioned in other ways. The accuracy assessment will depend on the suitability of PR as a measure of what has *actually* been predicted.

(ii) Only the area of Kenya is covered and this limits the range of values that will be included in the correlation analysis used for accuracy assessment. Strictly the accuracy assessment applies to Kenya only: extrapolation of the accuracy estimate to the rest of Africa requires assumptions of stationarity.

(iii) If the PR data are averages per facility then the support is unknown. It may be approximated by the catchment of the facility, but that also is most likely unknown.

(iv) The sample size at the ground will influence the precision of the resulting accuracy assessment. The sampling scheme used will have an influence: it may be necessary to deal with autocorrelation in the residuals of a regression.

(v) It may be extremely difficult to match the PR data with the correct pixel from the imagery, particularly if the imagery are not geometrically corrected, but also because the large pixel makes geometric correction and identification of precise location within the image difficult.
(vi) The support of the imagery is not a square pixel but a centre-weighted function due to the point spread function (PSF) of the sensor.
(vii) Where the support of the PR data is smaller than the pixel, the PR data will include variance that is not included in the predicted values leading to a smaller correlation.

The frequency of spatial variation in disease risk (or PR in the example) can be very fine relative to a support of 8 km. For example, while malaria risk is often demarcated into spatially homogeneous zones based on seasonality, in reality malaria risk can vary over distances of a few hundred metres, particularly in a country such as Kenya which encompasses a diverse range of habitats, changes in elevation and local climates. Where the frequency of spatial variation in disease risk is fine relative to the support, researchers need to be especially careful to construct their ground data on the same support as that of the imagery (predictions). This implies the use of geostatistical methods to predict the ground variable on the same support as the imagery. Potential methods include kriging and cokriging, but researchers need to be aware that such linear regression-based methods involve smoothing (conditional bias), and that such smoothing may affect the correlation coefficient (Atkinson and Tate, 2000). Geostatistical conditional simulation provides a potential partial solution (Journel, 1996; Chilès and Delfiner, 1999).

3.6. Satellite Sensor Datasets Used in this Volume

Two satellite sensor-derived data sets are provided on the DVD with this volume, each with a different spatial resolution. Both datasets are temporal Fourier processed, the first having an original spatial resolution of $\sim 8 \times 8$ km and the second having a finer spatial resolution of 1×1 km (Hay *et al.*, this volume, pp. 37–77). Modelled predictions of disease and vector distributions presented in subsequent reviews are based on the reprojected (to 0.10 degree spatial resolution) 8×8 km data due to issues of computing performance, data storage and effective viewing at the global scale. Almost 20 years of monthly composites of 10-day averaged AVHRR data were used as input to the $\sim 8 \times 8$ km Fourier dataset, the methods of creation of which are detailed in Chapter 2 (Hay *et al.*, this volume, pp. 37–77). The $\sim 8 \times 8$ km data files are of a manageable size for analysis and distribution and have an acceptable spatial resolution for the global scale studies discussed in this volume. These are important considerations, as the spatial prediction models need to be adaptable to a variety of computer specifications and the output interpretable at the scale of the entire globe. A balance needs to be achieved between the desire for high-spatial resolution data and the increasing file size or data volume associated with increasing spatial resolution. For example, the spatial resolution of the 1 km data is arguably more useful to the field epidemiologist than the 8 km data, but the data volume is in excess of 1.4 GB per variable (compared to 10.8 MB for 8 km data). Such data volumes not only create storage and backup problems but also severely affect both the initial processing time and the computational speed when modelling disease distributions. Moving such a large data volume also creates logistical problems.

4. SUMMARY

Scale and uncertainty are important issues in the prediction of global disease distributions. Disease mapping over the entire surface of the Earth increasingly involves the use of remotely sensed imagery. Such data often provide complete coverage of the entire surface of the Earth and can be used to predict environmental covariates of disease risk or disease vector density. Global coverage implies a relatively coarse spatial resolution (e.g., 8 or 1 km) and this will impose limits on the information content of the imagery and the information that can be extracted on disease risk or disease vector distribution. This paper introduced geostatistical models for (i) characterizing the scale(s) of spatial variation in data (i.e., variogram) and (ii) changing the scale of measurement of both the data (i.e., upscaling and downscaling) and the geostatistical model (i.e., convolution and deconvolution). Such models provide a means of handling the effects of scale in epidemiological analyses. A discussion of uncertainty highlighted the fact that uncertainty is a vague concept and that most epidemiologists are interested in measurable quantities that estimate accuracy, or components of accuracy such as bias and precision. The distinction between data- and model-based methods of accuracy assessment was emphasized and examples of both were given. The key problem of validating global maps was discussed briefly.

This paper has highlighted some of the problems involved with the use of coarse spatial resolution data and uncertain methods of prediction in the global mapping of disease. The epidemiological community is encouraged to pay close attention to the effects of scale and uncertainty as discussed in this paper and adopt appropriate methods to handle such effects.

ACKNOWLEDGEMENTS

The authors are grateful to Peter Gething, Prof. David Rogers and the three referees for their constructive comments. AJG was funded by European Union grant GOCE-2003-010284 EDEN and the paper is catalogued by the EDEN Steering Committee as EDEN0012 (http://www.eden-fp6project.net/). The contents of this publication are the sole responsibility of the authors and do not reflect the views of the European Union.

REFERENCES

Anderson, R.M., May, R.M., Boily, M.C., Garnett, G.P. and Rowley, J.T. (1991). The spread of HIV-1 in Africa—sexual contact patterns and the predicted demographic-impact of AIDS. *Nature* **352**, 581–589.

Atkinson, P.M. (1997). On estimating measurement error in remotely-sensed images with the variogram. *International Journal of Remote Sensing* **18**, 3075–3084.

Atkinson, P.M. (1999). Spatial statistics. In: *Spatial Statistics for Remote Sensing* (A. Stein, F. van der Meer and B. Gorte, eds), pp. 57–81. Dordrecht: Kluwer.

Atkinson, P.M. (2005). Super-resolution target mapping from soft classified remotely sensed imagery. *Photogrammetric Engineering and Remote Sensing* **71**, 839–846.

Atkinson, P.M. and Curran, P.J. (1995). Defining an optimal size of support for remote-sensing investigations. *IEEE Transactions on Geoscience and Remote Sensing* **33**, 768–776.

Atkinson, P.M. and Curran, P.J. (1997). Choosing an appropriate spatial resolution for remote sensing investigations. *Photogrammetric Engineering and Remote Sensing* **63**, 1345–1351.

Atkinson, P.M. and Foody, G.M. (2002). Uncertainty in remote sensing and GIS: fundamentals. In: *Uncertainty in Remote Sensing and GIS* (G.M. Foody and P.M. Atkinson, eds), pp. 1–18. Chichester: Wiley.

Atkinson, P.M. and Kelly, R.E.J. (1997). Scaling-up point snow depth data in the UK for comparison with SSM/I imagery. *International Journal of Remote Sensing* **18**, 437–443.

Atkinson, P.M. and Tate, N.J. (2000). Spatial scale problems and geostatistical solutions: a review. *Professional Geographer* **52**, 607–623.

Atkinson, P.M., Webster, R. and Curran, P.J. (1992). Cokriging with ground-based radiometry. *Remote Sensing of Environment* **41**, 45–60.

Augustin, N.H., Mugglestone, M.A. and Buckland, S.T. (1996). An autologistic model for the spatial distribution of wildlife. *Journal of Applied Ecology* **33**, 339–347.

Baker, R.D. (2004). Identifying space–time disease clusters. *Acta Tropica* **91**, 291–299.

Beasley, R., Keil, U., von Mutius, E., Pearce, N., Ait-Khaled, N., Anabwani, G., Anderson, H.R., Asher, M.I., Bjorkstein, B., Burr, M.L., Clayton, T.O., Crane, J., Ellwood, P., Lai, C.K.W., Mallol, J., Martinez, F.D., Mitchell, E.A., Montefort, S., Robertson, C.F., Shah, J.R., Sibbald, B., Stewart, A.W., Strachan, D.P., Weiland, S.K. and Williams, H.C. (1998). Worldwide variation in prevalence of symptoms of asthma, allergic rhinoconjunctivitis, and atopic eczema: ISAAC. *Lancet* **351**, 1225–1232.

Belward, A.S. and Lambin, E. (1990). Limitations to the identification of spatial structures from AVHRR data. *International Journal of Remote Sensing* **11**, 921–927.

Bian, L. (2004). A conceptual framework for an individual-based spatially explicit epidemiological model. *Environment and Planning B-Planning & Design* **31**, 381–395.

Bierkens, M.F.P., Finke, P.A. and De Willigen, P. (2000). *Upscaling and Downscaling Methods for Environmental Research.* Springer: Amsterdam.

Burgess, T.M. and Webster, R. (1980). Optimal interpolation and isarithmic mapping of soil properties II. Block kriging. *Journal of Soil Science* **31**, 333–341.

Chavez, P.S. (1996). Image-based atmospheric corrections revisited and improved. *Photogrammetric Engineering and Remote Sensing* **62**, 1025–1036.

Chilès, P. and Delfiner, P. (1999). *Geostatistics: Modelling Spatial Uncertainty*. New York: Wiley.

Cisternas, J., Gear, C.W., Levin, S. and Kisvrekidis, I.G. (2004). Equation-free modelling of evolving diseases: coarse-grained computations with individual-based models. *Proceedings of the Royal Society of London Series A- Mathematical, Physical and Engineering Sciences* **460**, 2761–2779.

Cockings, S., Dunn, C.E., Bhopal, R.S. and Walker, D.R. (2004). Users' perspectives on epidemiological, GIS and point pattern approaches to analysing environment and health data. *Health & Place* **10**, 169–182.

Cohen, J. (1960). A coefficient of agreement for nominal scales. *Educational and Psychological Measurement* **20**, 37–46.

Congalton, R.G. (1991). A review of assessing the accuracy of classifications of remotely sensed data. *Remote Sensing of Environment* **37**, 35–46.

Cox, N.J. and Subbarao, K. (2000). Global epidemiology of influenza: past and present. *Annual Review of Medicine* **51**, 407–421.

Cracknell, A.P. (1997). *The Advanced Very High Resolution Radiometer (AVHRR)*. London: Taylor & Francis.

Cressie, N. (1985). Fitting variogram models by weighted least-squares. *Journal of the International Association for Mathematical Geology* **17**, 563–586.

Cressie, N. (1991). *Statistics for Spatial Data*. New York: Wiley.

Cressie, N. (1995). Bayesian smoothing of rates in geographic areas. *Journal of Regional Science* **35**, 659–673.

Crombie, M.K., Gillies, R.R., Arvidson, R.E., Brookmeyer, P., Weil, G.J., Sultan, M. and Harb, M. (1999). An application of remotely derived climatological fields for risk assessment of vector-borne diseases: a spatial study of filariasis prevalence in the Nile delta, Egypt. *Photogrammetric Engineering and Remote Sensing* **65**, 1401–1409.

Cross, E.R., Newcomb, W.W. and Tucker, C.J. (1996). Use of weather data and remote sensing to predict the geographic and seasonal distribution of *Phlebotomus papatasi* in southwest Asia. *American Journal of Tropical Medicine and Hygiene* **54**, 530–536.

Curran, P.J. (1994). Imaging spectrometry. *Progress in Physical Geography* **18**, 247–266.

Curran, P.J. and Atkinson, P.M. (1998). Geostatistics and remote sensing. *Progress in Physical Geography* **22**, 61–78.

Curran, P.J. and Atkinson, P.M. (1999). Issues of scale and optimal pixel size. In: *Spatial Statistics in Remote Sensing* (A. Stein and F. van der Meer, eds), pp. 115–133. Dordrecht: Kluwer.

Curran, P.J., Atkinson, P.M., Foody, G.M. and Milton, E.J. (2000). Linking remote sensing, land cover and disease. *Advances in Parasitology* **47**, 37–80.
Curran, P.J. and Dungan, J.L. (1989). Estimation of signal-to-noise—a new procedure applied to AVIRIS data. *IEEE Transactions on Geoscience and Remote Sensing* **27**, 620–628.
Curran, P.J., Kupiec, J.A. and Smith, G.M. (1998). Remote sensing the biochemical composition of a slash pine canopy (vol. 35, p. 415, 1997). *IEEE Transactions on Geoscience and Remote Sensing* **36**, 351.
Custer, B., Sullivan, S.D., Hazlet, T.K., Lloje, U., Veenstra, D.L. and Kowdley, K.V. (2004). Global epidemiology of hepatitis B virus. *Journal of Clinical Gastroenterology* **38**, S158–S168.
Danson, F.M., Graham, A.J., Pleydell, D.R.J., Campos-Ponce, M., Giraudoux, P. and Craig, P.S. (2003). Multi-scale spatial analysis of human alveolar echinococcosis risk in China. *Parasitology* **127**, S133–S141.
De Cock, K.M. and Weiss, H.A. (2000). The global epidemiology of HIV/AIDS. *Tropical Medicine & International Health* **5**, A3–A9.
De Groot, M.H. and Schervish, M.J. (2001). *Probability and Statistics*, 3rd ed. Reading, Massachusetts: Addison-Wesley.
Deutsch, C.V. and Journel, A.G. (1992). *GSLIB Geostatistical Software Library Users Guide*. Oxford: Oxford University Press.
Diggle, P., Moyeed, R., Rowlingson, B. and Thomson, M. (2002). Childhood malaria in the Gambia: a case-study in model-based geostatistics. *Journal of the Royal Statistical Society Series C* **51**, 493–506.
Diggle, P.J. and Rowlingson, B.S. (1994). A conditional approach to point process modeling of elevated risk. *Journal of the Royal Statistical Society Series A-Statistics in Society* **157**, 433–440.
Dragicevic, S. (2003a). Spatial cluster modelling. *Annals of the Association of American Geographers* **93**, 950–952.
Dragicevic, S. (2003b). Statistical methods in spatial epidemiology. *Annals of the Association of American Geographers* **93**, 950–952.
Dungan, J.L., Perry, J.N., Dale, M.R.T., Legendre, P., Citronpousty, S., Fortin, M.J., Jakomulska, A., Miriti, M. and Rosenberg, M.S. (2002). A balanced view of scale in spatial statistical analysis. *Ecography* **25**, 626–640.
Eineder, M. and Adam, N. (2005). A maximum-likelihood estimator to simultaneously unwrap, geocode, and fuse SAR interferograms from different viewing geometries into one digital elevation model. *IEEE Transactions on Geoscience and Remote Sensing* **43**, 24–36.
Elliott, P. (1993). Global epidemiology. *CIBA Foundation Symposia* **175**, 219–233.
Elliott, P. and Wartenberg, D. (2004). Spatial epidemiology: current approaches and future challenges. *Environmental Health Perspectives* **112**, 998–1006.
Estrada-Pena, A. (1998). Geostatistics and remote sensing as predictive tools of tick distribution: a cokriging system to estimate *Ixodes scapularis*

(Acari: Ixodidae) habitat suitability in the United States and Canada from advanced very high resolution radiometer satellite imagery. *Journal of Medical Entomology* **35**, 989–995.

Eubank, S., Guclu, H., Kumar, V.S.A., Marathe, M.V., Srinivasan, A., Toroczkai, Z. and Wang, N. (2004). Modelling disease outbreaks in realistic urban social networks. *Nature* **429**, 180–184.

Fielding, A.H. and Bell, J.F. (1997). A review of methods for the assessment of prediction errors in conservation presence/absence models. *Environmental Conservation* **24**, 38–49.

Foody, G.M. (2002). Status of land cover classification accuracy assessment. *Remote Sensing of Environment* **80**, 185–201.

Foody, G.M. (2004). Thematic map comparison: evaluating the statistical significance of differences in classification accuracy. *Photogrammetric Engineering and Remote Sensing* **70**, 627–633.

Foody, G.M., Campbell, N.A., Trodd, N.M. and Wood, T.F. (1992). Derivation and applications of probabilistic measures of class membership from the maximum-likelihood classification. *Photogrammetric Engineering and Remote Sensing* **58**, 1335–1341.

Forbes, A.D. (1995). Classification algorithm evaluation: five performance measures based on confusion matrices. *Journal of Clinical Monitoring* **11**, 189–206.

Friedl, M.A., Mciver, D.K., Hodges, J.C.F., Zang, X.Y., Muchoney, D., Strahler, A.H., Woodcock, C.E., Gopal, S., Schneider, A., Cooper, A., Baccini, A., Gao, F. and Schaaf, C. (2002). Global land cover mapping from MODIS: algorithms and early results. *Remote Sensing of Environment* **83**, 287–302.

Gatrell, A.C., Bailey, T.C., Diggle, P.J. and Rowlingson, B.S. (1996). Spatial point pattern analysis and its application in geographical epidemiology. *Transactions of the Institute of British Geographers* **21**, 256–274.

Gething, P.W., Noor, A.M., Zurovac, D., Atkinson, P.M., Hay, S.J., Nixon, M.S. and Snow, R.W. (2004). Empirical modelling of government health service use by children with fevers in Kenya. *Acta Tropica* **91**, 227–237.

Giri, C., Zhu, Z.L. and Reed, B. (2005). A comparative analysis of the Global Land Cover 2000 and MODIS land cover data sets. *Remote Sensing of Environment* **94**, 123–132.

Goddard, M.J. and Hinberg, I. (1990). Receiver operating characteristic (ROC) curves and non-normal data: an empirical study. *Statistical Medicine* **9**, 325–337.

Goel, P.K., Prasher, S.O., Landry, J.A., Bonnell, R.B. and Viau, A.A. (2003). Estimation of crop biophysical parameters through airborne and field hyperspectral remote sensing. *Transactions of the ASAE* **46**, 1235–1246.

Goovaerts, P. (1997). *Geostatistics for Natural Resources Evaluation*. New York: Oxford University Press.

Gotway, C.A. and Young, L.J. (2002). Combining incompatible spatial data. *Journal of the American Statistical Association* **97**, 632–648.

Graham, A.J., Atkinson, P.M. and Danson, F.M. (2004). Spatial analysis for epidemiology. *Acta Tropica* **91**, 219–225.

Graham, A.J. and Harris, R. (2003). Extracting biophysical parameters from remotely sensed radar data: a review of the water cloud model. *Progress in Physical Geography* **27**, 217–229.

Grenfell, B.T., Bjørnstad, O.N. and Kappey, J. (2001). Travelling waves and spatial hierarchies in measles epidemics. *Nature* **414**, 716–721.

Guyot, G. and Gu, X.F. (1994). Effect of radiometric corrections on NDVI-determined from Spot-HRV and Landsat-TM data. *Remote Sensing of Environment* **49**, 169–180.

Hand, D.J. and Bolton, R.J. (2004). Pattern discovery and detection: a unified statistical methodology. *Journal of Applied Statistics* **31**, 885–924.

Hay, G.J., Blaschke, T., Marceau, D.J. and Bouchard, A. (2003). A comparison of three image-object methods for the multiscale analysis of landscape structure. *ISPRS Journal of Photogrammetry and Remote Sensing* **57**, 327–345.

Hay, G.J., Marceau, D.J., Dube, P. and Bouchard, A. (2001). A multiscale framework for landscape analysis: object-specific analysis and upscaling. *Landscape Ecology* **16**, 471–490.

Hay, S.I. (2000). An overview of remote sensing and geodesy for epidemiology and public health application. *Advances in Parasitology* **47**, 1–35.

Hay, S.I., Packer, M.J. and Rogers, D.J. (1997). The impact of remote sensing on the study and control of invertebrate intermediate hosts and vectors for disease. *International Journal of Remote Sensing* **18**, 2899–2930.

Hay, S.I., Snow, R.W. and Rogers, D.J. (1998a). From predicting mosquito habitat to malaria seasons using remotely sensed data: practice, problems and perspectives. *Parasitology Today* **14**, 306–313.

Hay, S.I., Snow, R.W. and Rogers, D.J. (1998b). Predicting malaria seasons in Kenya using multitemporal meteorological satellite sensor data. *Transactions of the Royal Society of Tropical Medicine and Hygiene* **92**, 12–20.

Hay, S.I., Tucker, C.J., Rogers, D.J. and Packer, M.J. (1996). Remotely sensed surrogates of meteorological data for the study of the distribution and abundance of arthropod vectors of disease. *Annals of Tropical Medicine and Parasitology* **90**, 1–19.

Hendrickx, G., Napala, A., Slingenbergh, J.H.W., De Deken, R. and Rogers, D.J. (2001). A contribution towards simplifying area-wide tsetse surveys using medium resolution meteorological satellite data. *Bulletin of Entomological Research* **91**, 333–346.

Ippolito, G., Rezza, G., Girardi, E., Malkovsky, M. and Cairns, J.D. (2000). HIV infections: the global epidemiology and goals for vaccine research. *Molecular Medicine* **6**, 69–85.

Jarner, M.F., Diggle, P.J. and Chetwynd, A.G. (2002). Estimation of spatial variation in risk using matched case-control data. *Biometrical Journal* **44**, 936–945.

Jarup, L. (2004). Health and environment information systems for exposure and disease mapping, and risk assessment. *Environmental Health Perspectives* **112**, 995–997.

Journel, A.G. (1996). Modelling uncertainty and spatial dependence: stochastic imaging. *International Journal of Geographical Information Systems* **10**, 517–522.

Journel, A.G. and Huijbregts, C.J. (1978). *Mining Geostatistics*. London: Academic Press.

Kellndorfer, J., Walker, W., Pierce, L., Dobson, C., Fites, J.A., Hunsaker, C., Vona, J. and Clutter, M. (2004). Vegetation height estimation from shuttle radar topography mission and national elevation datasets. *Remote Sensing of Environment* **93**, 339–358.

Kelsall, J. and Wakefield, J. (2002). Modeling spatial variation in disease risk: a geostatistical approach. *Journal of the American Statistical Association* **97**, 692–701.

Kelsall, J.E. and Diggle, P.J. (1998). Spatial variation in risk of disease: a nonparametric binary regression approach. *Journal of the Royal Statistical Society Series C-Applied Statistics* **47**, 559–573.

Kirby, R.S. (2003). GIS and public health. *Annals of the Association of American Geographers* **93**, 261–263.

Klinkenberg, B. and Goodchild, M.F. (1992). The fractal properties of topography—a comparison of methods. *Earth Surface Processes and Landforms* **17**, 217–234.

Klir, G.J. (2001). Foundations of fuzzy set theory and fuzzy logic: a historical overview. *International Journal of General Systems* **30**, 91–132.

Krieger, N., Chen, J.T., Waterman, P.D., Soobader, M.J., Subramanian, S.V. and Carson, R. (2002). Geocoding and monitoring of US socioeconomic inequalities in mortality and cancer incidence: does the choice of area-based measure and geographic level matter? The Public Health Disparities Geocoding Project. *American Journal of Epidemiology* **156**, 471–482.

Kulldorff, M., Tango, T. and Park, P.J. (2003). Power comparisons for disease clustering tests. *Computational Statistics & Data Analysis* **42**, 665–684.

Kyriakidis, P.C. (2004). A geostatistical framework for area-to-point spatial interpolation. *Geographical Analysis* **36**, 259–289.

LaPorte, R.E., Barinas, E., Chang, Y.F. and Libman, I. (1996). Global epidemiology and public health in the 21st century—applications of new technology. *Annals of Epidemiology* **6**, 162–167.

Latifovic, R., Zhu, Z.L., Cihlar, J., Giri, C. and Olthof, I. (2004). Land cover mapping of north and central America—Global Land Cover 2000. *Remote Sensing of Environment* **89**, 116–127.

Lawson, A.B. (2000). Cluster modelling of disease incidence via RJMCMC methods: a comparative evaluation. *Statistics in Medicine* **19**, 2361–2375.

Lillesand, T.M. and Kiefer, R.W. (2004). *Remote Sensing and Image Interpretation*. New York: Wiley.

Linthicum, K.J., Anyamba, A., Tucker, C.J., Kelly, P.W., Myers, M.F. and Peters, C.J. (1999). Climate and satellite indicators to forecast Rift Valley fever epidemics in Kenya. *Science* **285**, 397–400.

Loveland, T.R., Reed, B.C., Brown, J.F., Ohlen, D.O., Zhu, Z., Yang, L. and Merchant, J.W. (2000). Development of a global land cover characteristics database and IGBP DISCover from 1 km AVHRR data. *International Journal of Remote Sensing* **21**, 1303–1330.

Maselli, F., Conese, C. and Petkov, L. (1994). Use of probability entropy for the estimation and graphical representation of the accuracy of maximum-likelihood classifications. *ISPRS Journal of Photogrammetry and Remote Sensing* **49**, 13–20.

Matheron, G. (1965). *Les Variables Régionalisées et Leur Estimation*. Paris: Masson.

Matheron, G. (1971). *The Theory of Regionalized Variables and its Applications*. Fontainebleau: Centre de Morphologie Mathématique de Fontainebleau.

McBratney, A.B. and Webster, R. (1986). Choosing functions for semi-variograms of soil properties and fitting them to sampling estimates. *Journal of Soil Science* **37**, 617–639.

Michael, E. and Bundy, D.A.P. (1997). Global mapping of lymphatic filariasis. *Parasitology Today* **13**, 472–476.

Minozzo, M. and Fruttini, D. (2004). Loglinear spatial factor analysis: an application to diabetes mellitus complications. *Environmetrics* **15**, 423–434.

Morison, L. (2001). The global epidemiology of HIV/AIDS. *British Medical Bulletin* **58**, 7–18.

Mugglin, A.S., Cressie, N. and Gemmell, I. (2002). Hierarchical statistical modelling of influenza epidemic dynamics in space and time. *Statistics in Medicine* **21**, 2703–2721.

Myers, D.E. (1989). To be or not to be .. stationary—that is the question. *Mathematical Geology* **21**, 347–362.

Naesset, E., Bollandsas, O.M. and Gobakken, T. (2005). Comparing regression methods in estimation of biophysical properties of forest stands from two different inventories using laser scanner data. *Remote Sensing of Environment* **94**, 541–553.

Nemani, R.R., Keeling, C.D., Hashimoto, H., Jolly, W.M., Piper, S.C., Tucker, C.J., Myneni, R.B. and Running, S.W. (2003). Climate-driven

increases in global terrestrial net primary production from 1982 to 1999. *Science* **300**, 1560–1563.
Nicholson, M.C. and Mather, T.N. (1996). Methods for evaluating lyme disease risks using geographic information systems and geospatial analysis. *Journal of Medical Entomology* **33**, 711–720.
Oliver, M.A., Muir, K.R., Webster, R., Parkes, S.E., Cameron, A.H., Stevens, M.C.G. and Mann, J.R. (1992). A geostatistical approach to the analysis of pattern in rare disease. *Journal of Public Health Medicine* **14**, 280–289.
Oliver, M.A., Webster, R., Lajaunie, C., Muir, K.R., Parkes, S.E., Cameron, A.H., Stevens, M.C.G. and Mann, J.R. (1998). Binomial cokriging for estimating and mapping the risk of childhood cancer. *IMA Journal of Mathematics Applied in Medicine and Biology* **15**, 279–297.
Openshaw, S. (1984). The modifiable areal unit problem. *Concepts and Techniques in Modern Geography* **38**, 41.
Pardo-Iguzquiza, E., Chico-Olmo, M. and Atkinson, P. M. (2005). Increasing the spatial resolution of remote sensing images using linear systems theory and cokriging. *Remote Sensing of Environment* (submitted).
Pope, K.O., Sheffner, E.J., Linthicum, K.J., Bailey, C.L., Logan, T.M., Kasischke, E.S., Birney, K., Njogu, A.R. and Roberts, C.R. (1992). Identification of Central Kenyan Rift-Valley Fever virus vector habitats with Landsat TM and evaluation of their flooding status with airborne imaging radar. *Remote Sensing of Environment* **40**, 185–196.
Quattrochi, D.A. and Goodchild, M.F., eds. (1997). *Scale in Remote Sensing and GIS*. New York: CRC Press.
Rogers, D.J., Hay, S.I. and Packer, M.J. (1996). Predicting the distribution of tsetse flies in West Africa using temporal Fourier processed meteorological satellite data. *Annals of Tropical Medicine and Parasitology* **90**, 225–241.
Rogers, D.J. and Randolph, S.E. (1991). Mortality-rates and population-density of tsetse-flies correlated with satellite imagery. *Nature* **351**, 739–741.
Rosenfield, G.H. and Fitzpatrick-Lins, K. (1986). A coefficient of agreement as a measure of thematic classification accuracy. *Photogrammetric Engineering and Remote Sensing* **52**, 223–227.
Schott, J.R. (1987). *Remote Sensing: The Image Chain Approach*, 2nd ed. San Diego: Academic Press.
Schowengerdt, R.A. (1997). *Remote Sensing Models and Methods for Image Processing*, 2nd ed. San Diego: Academic Press.
Shaw, G. and Wheeler, D. (1994). *Statistical Techniques in Geographical Analysis*, 2nd ed. London: David Fulton Publishers.
Staubach, C., Thulke, H.H., Tackmann, K., Hugh-Jones, M. and Conraths, F.J. (2001). Geographic information system-aided analysis of factors associated with the spatial distribution of *Echinococcus multilocularis*

infections of foxes. *American Journal of Tropical Medicine and Hygiene* **65**, 943–948.

Stern, H.S. and Cressie, N. (2000). Posterior predictive model checks for disease mapping models. *Statistics in Medicine* **19**, 2377–2397.

Story, M. and Congalton, R.G. (1986). Accuracy assessment—a users perspective. *Photogrammetric Engineering and Remote Sensing* **52**, 397–399.

Strahler, A.H., Woodcock, C.E. and Smith, J.A. (1986). On the nature of models in remote-sensing. *Remote Sensing of Environment* **20**, 121–139.

Tatem, A.J., Lewis, H.G., Atkinson, P.M. and Nixon, M.S. (2001). Superresolution target identification from remotely sensed images using a Hopfield neural network. *IEEE Transactions on Geoscience and Remote Sensing* **39**, 781–796.

Tatem, A.J., Lewis, H.G., Atkinson, P.M. and Nixon, M.S. (2002). Superresolution land cover pattern prediction using a Hopfield neural network. *Remote Sensing of Environment* **79**, 1–14.

Tatem, A.J., Noor, A.M. and Haya, S.I. (2004). Defining approaches to settlement mapping for public health management in Kenya using medium spatial resolution satellite imagery. *Remote Sensing of Environment* **93**, 42–52.

Thompson, M.L. and Zucchini, W. (1989). On the statistical analysis of ROC curves. *Statistical Medicine* **8**, 1277–1290.

Thomson, M.C., Connor, S.J., D'Alessandro, U., Rowlingson, B., Diggle, P., Cresswell, M. and Greenwood, B. (1999). Predicting malaria infection in Gambian children from satellite data and bed net use surveys: the importance of spatial correlation in the interpretation of results. *American Journal of Tropical Medicine and Hygiene* **61**, 2–8.

Thomson, M.C., Connor, S.J., Milligan, P.J.M. and Flasse, S.P. (1996). The ecology of malaria—as seen from Earth-observation satellites. *Annals of Tropical Medicine and Parasitology* **90**, 243–264.

Unwin, D.J. (1996). GIS, spatial analysis and spatial statistics. *Progress in Human Geography* **20**, 540–551.

Vonreyn, C.F. and Mann, J.M. (1987). Global epidemiology. *Western Journal of Medicine* **147**, 694–701.

Wakefield, J.C. (2003). Sensitivity analyses for ecological regression. *Biometrics* **59**, 9–17.

Wakefield, J.C. (2004). A critique of statistical aspects of ecological studies in spatial epidemiology. *Environmental and Ecological Statistics* **11**, 31–54.

Wakefield, J.C. and Morris, S.E. (2001). The Bayesian modeling of disease risk in relation to a point source. *Journal of the American Statistical Association* **96**, 77–91.

Webster, R. and Oliver, M.A. (1990). *Statistical Methods for Soil and Land Resources Survey*. Oxford: Oxford University Press.

Webster, R., Oliver, M.A., Muir, K.R. and Mann, J.R. (1994). Kriging the local risk of a rare disease from a register of diagnoses. *Geographical Analysis* **26**, 168–185.

Woodcock, C.E. and Strahler, A.H. (1987). The factor of scale in remote-sensing. *Remote Sensing of Environment* **21**, 311–332.

Zhan, Q., Molenaar, M. and Lucieer, A. (2002). Pixel unmixing at the sub-pixel scale based on land cover class probabilities: application to urban areas. In: *Uncertainty in Remote Sensing and GIS* (G.M. Foody and P.M. Atkinson, eds), pp. 59–76. Wiley: Chichester.

Determining Global Population Distribution: Methods, Applications and Data

D.L. Balk[1], U. Deichmann[2], G. Yetman[1], F. Pozzi[1,3], S.I. Hay[4,5] and A. Nelson[6,7]

[1]*Center for International Earth Science Information Network (CIESIN), Columbia University, PO Box 1000, Palisades, NY 10964, USA*
[2]*Development Research Group, World Bank, 1818 H Street, Washington, DC 20433, USA*
[3]*Università Cattolica del Sacro Cuore, sede di Brescia, via dei Musei 41, 25121, Brescia, Italy*
[4]*TALA Research Group, Department of Zoology, University of Oxford, South Parks Road, Oxford, OX1 3PS, UK*
[5]*Malaria Public Health & Epidemiology Group, Centre for Geographic Medicine, KEMRI, PO Box 43640, 00100 Nairobi, Kenya*
[6]*School of Geography, University of Leeds, Woodhouse Lane, West Yorkshire, Leeds LS2 9JT, UK*
[7]*JRC (Joint Research Centre of the European Commission), Global Environment Monitoring Unit, TP 440, Via Enrico Fermi 1, I-21020 Ispra (VA), Italy*

Abstract 120
1. Introduction 120
1.1. Rendering Population on a Global Grid 121
1.2. Institutional Stewardship 122
2. Data 124
2.1. Gridded Population of the World 127
2.2. Global Rural Urban Mapping Project 128
2.3. Accessibility Modeling 130
2.4. Highly Modeled Surfaces 132
3. Methodology 133
3.1. Adjusting Population Estimates to Target Years 133
3.2. Limitations of the Ancillary Data 136

ADVANCES IN PARASITOLOGY VOL 62
ISSN: 0065-308X $35.00
DOI: 10.1016/S0065-308X(05)62004-0

4. Health Applications 138
4.1. General Health Studies 139
4.2. Specific Diseases 140
5. Discussion. 144
5.1. Ideal Spatial Resolution 145
5.2. Conclusion 147
6. Data Dissemination 148
6.1. Data Selection 149
6.2. Methods and Issues in Analysis 150
Acknowledgements 151
References 151

ABSTRACT

Evaluating the total numbers of people at risk from infectious disease in the world requires not just tabular population data, but data that are spatially explicit and global in extent at a moderate resolution. This review describes the basic methods for constructing estimates of global population distribution with attention to recent advances in improving both spatial and temporal resolution. To evaluate the optimal resolution for the study of disease, the native resolution of the data inputs as well as that of the resulting outputs are discussed. Assumptions used to produce different population data sets are also described, with their implications for the study of infectious disease. Lastly, the application of these population data sets in studies to assess disease distribution and health impacts is reviewed. The data described in this review are distributed in the accompanying DVD.

1. INTRODUCTION

Deriving population at risk estimates as a basis for evaluation of disease burdens requires spatially explicit, moderate-resolution population data at the global scale. In this contribution, methods for constructing estimates of global population distribution that are suitable for geographic analysis are described. Though the basic approach has been used widely for more than a decade, particular attention is given to recent advances to increase both spatial and temporal resolution. As global data products are dependent on a diverse set of inputs, issues related to input and output data resolution have an immediate bearing on the suitability of the resulting datasets for a given task. This paper also reviews applications of these population databases in the health sector, in particular, for the study of infectious disease. Finally, the population and associated data files that accompany this volume are briefly described.

1.1. Rendering Population on a Global Grid

Global or broad-scale inquiry on the relationship between population and environmental factors such as disease vectors or habitats is intrinsically spatial. While notable exceptions exist, especially at the local scale, two key barriers have contributed to the paucity of spatially oriented analysis: (1) the methods of analysis require some knowledge of geographic data and tools for analysis; and (2) population data, at regional and global scales, have tended to be recorded in national units that do not permit cross-national, subnational, or cross-habitat analysis. These barriers have been slowly eroding. One trend that has contributed to this is that the collectors and custodians of demographic data—the national census and statistics offices—increasingly compile and distribute data for small administrative or statistical units. While data from population censuses conducted before the 1990 round of population censuses were often published only for the country and major divisions such as provinces or states, more recent census output often includes digital census databases with detailed demographic data for districts, subdistricts, or even "enumeration" areas (EAs), the smallest geographical unit in most census operations.

Great progress has been made in harmonizing subnational data released for different dates so that they are comparable across international borders. First, since census years are not synchronized across the world, this involves interpolation or extrapolation of population estimates to a common base year. Second, subnational reference units can be vastly different in size and shape across countries. For spatial analysis, it is often preferable to instead record population estimates on a set of standardized reporting units, such as regular grid cells. Grids are more commonly used to collect or compile data describing natural phenomena. In interdisciplinary work, conversion to a regular grid imposes consistency that would be more difficult to achieve with irregularly shaped census or administrative units. Methods that transform population data from native census units (which correspond to vector format) to a regular raster grid are the main focus of this paper. A third harmonization issue arises for other demographic variables where, despite efforts by the United Nations and others to promote common definitions, indicators are often not entirely comparable. This is a major reason why global, georeferenced demographic databases have so far focused on the simplest of all demographic variables: total population.

Efforts to estimate population distribution for a regular raster grid predate the computerization of geography that started in the 1980s. Early examples such as the map by Adams (1968) for West Africa served largely cartographic purposes. Census offices, most notably those of Japan and Sweden, also produced national population grids for inclusion in national atlases (e.g., Tufte, 1990, on Japan). Computerized population maps for individual countries were produced by the US Census Bureau using rectangular grid cells superimposed with circles for major urban areas (Leddy, 1994). Deichmann and Eklundh (1991) presented a continental, gridded population database for Africa used to investigate interactions between population and land degradation. Others, such as Martin and Bracken (1991), developed techniques for producing local-level population grids (see Clark & Rhind, 1992; Deichmann 1996a, for reviews).

1.2. Institutional Stewardship

While national statistical offices produce population estimates that are sometimes linked to spatial data, few agencies render their population estimates on a common grid. The first efforts to place population data on a global-scale latitude–longitude grid were completed in the mid-1990s at the National Center for Geographic Information and Analysis at the University of California, Santa Barbara (Tobler *et al*., 1997). This initial dataset was itself an outgrowth of prior work on regional and continental databases. The Global Demography Workshop held in 1994 at CIESIN (the Center for International

Earth Science Information Network, now part of the Earth Institute at Columbia University) brought together experts in the field and helped advance methodological development and database creation for the first global grid. CIESIN is the locus of current global efforts, though it works closely with partnering institutions. Like many complex global data products, the Gridded Population of the World (GPW) database has evolved with numerous partners. Subsequent versions have included different collaborators, inputs, and outputs, but the guiding principle is to achieve the best possible suite of data products representing the distribution of human population, sometimes heuristically (i.e., without modeling) and sometimes with light modeling (Deichmann, 1996a). The fewer the assumptions and inputs that are used in the construction of the databases, the fewer the restrictions that have to be imposed on the appropriateness of use in a wide variety of applications. For example, if land cover were used to predict population densities, one could not predict expected changes in land cover from a resulting population distribution(s) that included land cover as a reallocation factor, as it would be endogenous.

Since the first version of GPW, several key advances have been made: the spatial resolution of administrative boundary data is improving; national statistical offices and spatial data providers and related institutions are adopting more open-data policies; population and spatial data providers are increasingly aware of, and increasingly collaborate with one another; and the computing capacity to manage, manipulate, and process increasingly large datasets is continually expanding (Balk and Yetman, 2005). As a result of these advances, some countries now produce and disseminate high-resolution spatially explicit population data. In local studies, nationally produced data are typically superior (i.e., of higher resolution, with more variables, and so on) to globally rendered data. Researchers asking highly place-based questions should begin with locally available data, if possible. Nevertheless, many questions are regional in scale, or at least span across more than one country, or require data that have been transformed to a common grid. For those problems, the data in this paper are highly suitable.

The basic global database to arise out of these efforts is the GPW, now in its third revision, with large gains to resolution having been made with each revision. In addition to the key advances described above, advances in ancillary data to allow for light modeling, especially valuable where input data are of suboptimal spatial resolution, have allowed for more sophisticated but still simple modeling. Thus, GPW and related population data products are the main focus of this review. The resulting datasets are also included in the accompanying DVD. Details on the variations in these databases, their methods, assumptions, and limitations follow.

2. DATA

The georeferenced population data sets that are the focus of this paper share as a critical common characteristic: the fact that they are constructed with an emphasis on the highest-resolution input data, rather than focusing on statistical or heuristic prediction of population distribution from coarse input data. That is, they attempt to measure the distribution of the population of the world, as measured at one's usual place of residence. The basic premise is that no amount of further processing or modeling can substitute for obtaining population counts for the smallest geographic reporting units available. Censuses in many countries are far from perfect and reliable civil registration systems exist only in a small number of countries. These sources provide the only complete enumeration of a country's population and by definition, provide the only geographically complete count of residents. By making additional assumptions about regularities in population distribution, it is possible to further disaggregate the reported district or subdistrict totals, but usually one cannot then reliably assess how accurate the resulting distributions are because there is no basis for sound validation. Population distribution modeling should therefore be considered a last resort in the absence of EA population maps, rather than as a goal in itself. When modeling is undertaken, the inputs of that model and the means for the redistribution should be made as transparent as possible.

The differences in these evolving data products are reviewed in Table 1 and are discussed in subsequent sections. Fundamental modifications include an increase in input resolution by over 20 times from the first to the current version of GPW (Balk and Yetman,

Table 1 Comparison of GPW versions and related databases

Data set	GPW				Accessibility model	GRUMP v1
	GPW v1	GPW v2	GPW v3	GPW 2015		
Publication year	1995	2000	2004	2004	2004	2004
Years of estimation	1994	1990, 1995	1990, 1995, 2000	2015	1960–2000	1990, 1995, 2000
Number of input units	19 000	127 000	376 500	376 500	Varies by continent	c. 1 000 000
Modeled inputs	None	None	None	None	Infrastructure, urban areas	Urban areas
Spatial extent	Global	Global	Global	Global	Africa, Asia, Latin America	Global
Authors	Tobler *et al.*	CIESIN, IFPRI, and WRI	CIESIN & CIAT	CIESIN, FAO, and CIAT	Deichmann; WRI; CIAT, UNEP and CIESIN	CIESIN, IFPRI, World Bank, and CIAT
Gridded surface resolution[a]	5′	2.5′	2.5′	2.5′	2.5′	30″
Population density	•	•	•	•	•	•
Population counts	•	•	•	•	•	•
Land area	•	•	•	•	•	•
Population-weighted admin. units			•	•		•
Urban extent mask						•
Settlement points (xls, csv, shp formats)						•

Note: A dot indicates the data set is publicly available. ′ and ″ represent arc-minutes and arc-seconds, respectively.
[a]Gridded surfaces are available in these formats: 00, bil, ascii.

2005), and nearly a tripling again for the modeled data products of the Global Rural Urban Mapping Project (GRUMP) (Balk *et al.*, 2005a). Increases in the input data enabled a corresponding increase in output resolution from 5 arc-minutes of GPW version 1–2.5 arc-minutes for later versions of GPW and related products. The data products from the GRUMP effort utilize higher resolution inputs, and thus outputs have been rendered at a 30 arc-second resolution.

The basic method by which population counts are transformed from census units to a grid, developed for the first version of GPW (Tobler *et al.*, 1997) and modified slightly for GPW v2 (Deichmann *et al.*, 2001), remain the same in the third version; related databases with light modeling use additional methods, but the basic method underlies all of these databases. Population data are transformed from their native spatial units that are usually administrative division of irregular shape and resolutions (see Figure 1) to a global grid of square latitude–longitude cells at a resolution of 2.5 arc minutes (i.e., approximately 4.6 km at the equator). The main inputs consist of geographically referenced boundaries of administrative or statistical reporting units at the highest available resolution—ideally the EA, but more typically at district or subdistrict level. The methods used to distribute the reporting unit total population numbers across the raster grid cells that fall into that unit differ slightly between the different versions of GPW and closely related data. These will be discussed below. Temporal adjustments are discussed in Section 3.

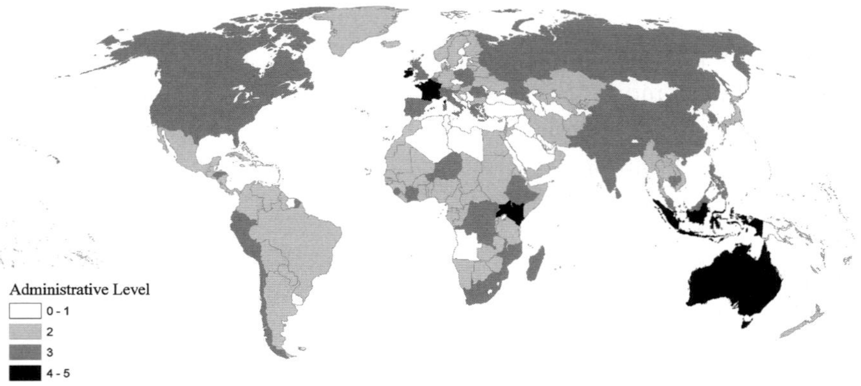

Figure 1 Administrative level used per country.

2.1. Gridded Population of the World

The GPW database uses two basic inputs: non-spatial population estimates (i.e., tables of population counts listed by administrative area names) and spatially explicit administrative boundary data. These are collected from hundreds of different data providers (often differing for the population and boundary data). The first part of the process is to match the population estimates with the administrative boundaries into what is known as polygon (or vector) format, ensure that the resulting data are geospatially consistent (e.g., that all internal boundaries match, leaving no unaccounted polygons or that island chains which might share a single population figure should they belong to the same administrative unit), and sum to the national-level population, as estimated by the data provider. These basic consistency checks mirror census principles of not leaving any resident out and not counting anyone more than once. To construct the GPW database, the administrative unit data in polygon format are converted to raster grids. In version 1, built-in geographic information systems (GIS) software functions were used to accomplish this conversion: grid cells that fall onto the boundary of two or more units were assigned to only one reporting unit based on a simple majority rule. The total unit population was then proportionally allocated over all grid cells assigned to that unit. A second product from this effort used these grids as a starting point for a re-distribution algorithm called *smooth pycnophylactic* (mass-preserving) *interpolation* (Tobler, 1979). The assumption underlying this approach is that those areas within a given administrative unit that neighbor regions with higher population densities are likely to house more people than areas that neighbor low-population density regions. The previously homogeneous population figures in grid cells within each administrative unit are thus re-distributed taking grid cells in neighboring units into account. By iteratively adjusting grid cell populations on this basis, the method results in a maximally smooth surface while preserving total population within each reporting unit.

The second and third versions of GPW retained most of the characteristics of the "unsmoothed" version of GPW v1, while significantly increasing the number of reporting units that served as input to the gridding routine. While version 1 relied on about 19 000 administrative units, version 2 used 120 000, and version 3 used 375 000 units with much of the increased precision achieved in developing countries (see Table 1). The main difference in processing in these newer versions lies in the way boundary areas between administrative units are treated. While version 1 allocated grid cells to only one unit even if it was shared by two or more (i.e., majority rule), GPW v2 and v3 use a proportional allocation so grid cells are assigned population in proportion to the area of overlap of grid cell and administrative units. Figure 2 (detail) and Table 2 illustrate this for a grid cell in the Dominican Republic. Proportional allocation is often referred to as an areal weighting scheme (e.g., Goodchild *et al.*, 1993).

2.2. Global Rural Urban Mapping Project

The allocation mechanism for the GRUMP (Balk *et al.*, 2005a) builds on the GPW approach but explicitly considers population of urban areas. In addition to data for statistical reporting units, the project

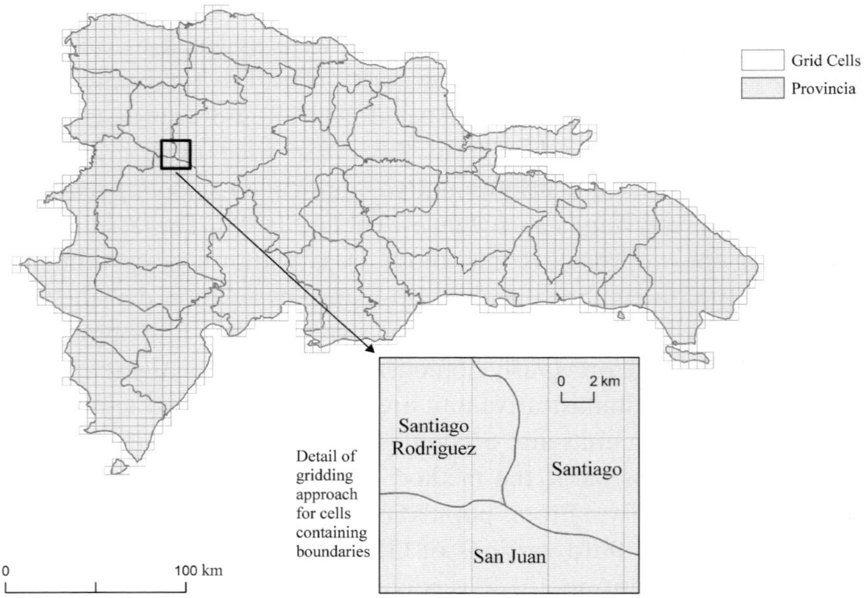

Figure 2 Grid cell size in relationship to administrative boundaries, Dominican Republic.

Table 2 Areal weighting scheme to allocation of population whose boundaries cross grid cells

Administrative unit name	Administrative unit density (persons/sq km)	Area of overlap (sq km)	Population estimate for grid cell
Santiago Rodriguez	64.2	5.3	340
Santiago	246.5	2.2	542
San Juan	75.9	12.8	972
Total for cell	91.3	20.3	1854

collected population estimates, point location, and the approximate footprint for urban centers in each country. The objective is to disaggregate the urban area populations from the total population of the administrative unit into which the urban area falls. This allows us to allocate urban and rural population separately, which effectively increases the number of input units and thus the effective resolution of the population grid.

In contrast to GPW, estimates of population for urban centers were needed in addition to population estimates associated with their census boundaries. Much less investment has been required from national statistical agencies to collect and publish population estimates for urban areas, unless these are entirely consistent with the census information for administrative units (which is rarely the case). Nevertheless, city population figures are published in a variety of sources. These data were collected and then matched with the urban footprint. That matching also occurs through a series of steps starting with simply a name-match of the populated places with geographic locations (i.e., latitude and longitude of the presumed center of the urban area). The geographic coordinates were found in national or international gazetteers, such as that of the US National Geospatial Intelligence Agency (see Balk *et al.*, 2005a, for details).

A more challenging problem was to determine the footprint of major city areas. The most important source are nighttime satellite images that show areas lit by streetlights and other permanent light sources that are concentrated in urban settlements (Elvidge *et al.*, 1999). In cases where statistical sources indicated a city that could not be detected on nighttime satellite images—a common occurrence in

Africa—urban areas were delineated from other sources (e.g., Tactical Pilotage Charts) or approximated by circles whose sizes were given by population–area relationships calibrated (through a regression analysis) on existing data. It is acknowledged that a circle is not an accurate form for any city, but this assumption was the most practical one to implement and the basic shape from lights for small extents tends toward circular. Circle-generated extents in each country were cross validated with other locations of near population size to confirm that the sizes were on the same order of magnitude. Conversely, footprints that could not be matched with populated place information were not assumed to have population and were discarded from the data. The population estimates, matched with geographic point locations were summarized for each footprint, producing an urban extent data set with population estimates.

The final step was to use these many pieces of information—which are summarized as administrative regions with population estimates and urban extents with population estimates (shown as panels 1A and 1B, respectively, in Figure 3)—and generate a population grid (panel 2B, Figure 3). (Figure 3 is Plate 4.3 in the Separate Color Plate Section.).

Because these come from different sources, it is important to make sure that the urban area population totals do not exceed those of the administrative areas in which the urban areas are located. Thus, a model is used to re-allocate population of the administrative areas given the population of the urban areas, the total population of the administrative area, and minimum and maximum criteria about each country's urbanization trends (details are given in Balk *et al.*, 2005a). The output resolution for this grid is 30 arc-seconds, similar to that of the nighttime lights' data. The GRUMP population grid also uses a proportional allocation rule in gridding.

2.3. Accessibility Modeling

The final set of gridded population datasets reviewed here are based on an additional set of assumptions about population distribution: the basic premise is that people tend to live in or close to cities and tend to move toward areas that are well connected with urban centers. Even in rural areas, it is expected that densely populated areas are closer to transport links than more isolated areas, and higher densities are nearer cities than the hinterland. These stylized facts concerning the distribution of people across space are implemented using the concept of accessibility—a measure of the ease by which destinations such as markets or service centers can be reached from a given location. In practice, these measures are adapted from the well-known gravity model of spatial interaction (Haynes & Fotheringham, 1984). They represent the sum of an indicator of size or mass at destinations (such as population of surrounding cities) inversely weighted for some function of distance. The ideal measure here is an estimate of travel time using the shortest route on a geographically referenced transportation network of roads, rivers, rails, and so forth. The resulting access estimates for each grid cell are then used to proportionally distribute each administrative unit population total across the grid cells that fall into it. This approach has been implemented for continental-scale databases for Africa, Asia, and Latin America, with support from the United Nations Environment Programme, the International Center for Tropical Agriculture (CIAT), and others. Nelson and Deichmann (2004) describe the latest version for Africa and document the modeling approach in detail.

The most important input into the model is information about the transportation network consisting of roads, railroads, and navigable rivers and their associated speeds of travel (i.e. 60 km per hour for 2-lane paved roads, 30 km per hour for railroads, etc.). The second main component is information on the location and population of urban centers, which are then linked to the transport network. These inputs are used to compute a measure of accessibility (V_i) for each node (intersection) in the network, which is based on the sum of the population of towns (P_k) in the vicinity of the current node weighted by a function of travel time across the network between the node and the towns $f(d_{ik})$. Figure 4 illustrates the computation of the accessibility index for a single node based on the weighted sum of the population of four towns that are within a given travel time threshold.

The accessibility values at each node were interpolated into a raster surface to create an accessibility index for each grid cell. Raster data on inland water bodies (lakes and glaciers), protected areas, and altitude were then used heuristically to reduce the accessibility

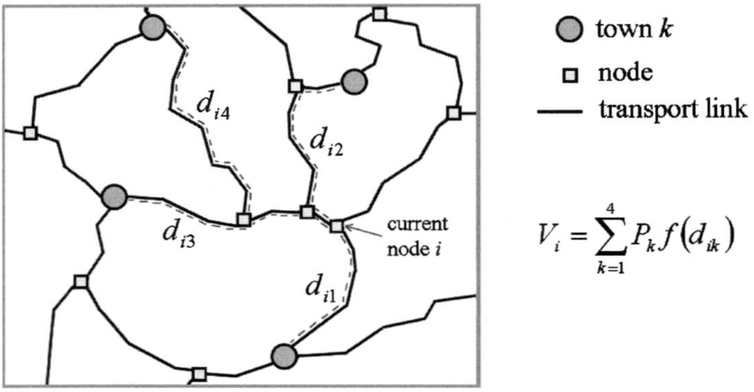

Figure 4 The computation of accessibility potential for a single node on the transport network where four towns are within the chosen travel time threshold.

potential in areas where there is little or no population. Accessibility values in water bodies and areas of extremely high altitude were set to zero. Accessibility values in protected areas and forest reserves were reduced by 80% and 50%, respectively. Both adjustments were heuristically chosen in the absence of empirical data.

The accessibility values estimated for each grid cell serve as weights to distribute population proportionately. The grid cells in the accessibility index were summed within each administrative unit. Each value was then divided by the corresponding administrative unit sum such that the resulting weights sum to one within each administrative unit. Multiplying each cell value by the total population yields the estimated number of people residing in each grid cell. The standardization of the accessibility index implies that the absolute magnitudes of the predicted access values are unimportant—only the variation within the administrative unit determines population densities within each district (Deichmann, 1997; Nelson & Deichmann, 2004)—but that, similar to GRUMP, the sum of grid cell population values for each unit cannot exceed the value for the administrative unit in which they fall.

2.4. Highly Modeled Surfaces

Another recently developed dataset, LandScan, takes a highly modeled approach, whereby much less investment is made in using the highest possible resolution population data (ORNL, 2003). This data set is categorically different from those described above, in that it does not attempt to represent nighttime, census residence, or usual population but rather it aims to measure an "ambient" population—i.e., the average location of an individual across seasons, days of the week, and times of day. Instead, effort is spent on getting annual updates to relatively coarse-level population inputs, and to ancillary data (including roads, nighttime lights, elevation, slope, and land cover) to be fitted to a complex model (Dobson *et al.*, 2000). The specific model parameters or their calibration are not published and, thus, it is difficult to assess the appropriateness or accuracy of this approach. LandScan receives less attention here, but is briefly discussed where it has been applied in the studies below.

3. METHODOLOGY

Though the basic method for re-distributing population from census and other units to a grid has been discussed, there are additional methodological requirements. For each reporting unit, a consistent population estimate for a baseline year is obtained. Where no census data or official estimates are available for the target year, a population figure is estimated using census year population and intercensal growth rates.

3.1. Adjusting Population Estimates to Target Years

Key inputs in all population databases reviewed in the previous section are subnational population totals typically available for small administrative or statistical reporting units. The standard source for such data is a national population and housing census, or, in some instances, a large demographic survey. Population censuses are undertaken periodically in many countries, once a decade. Exceptions are countries in which well-functioning civil registration systems make periodic census-taking unnecessary. Many countries take their censuses on the decadal year (1980, 1990, 2000), others take them on the first year thereafter (1991, 2001). (The US Census Bureau maintains an inventory of past and future census dates for each country at www.census.gov/ipc/www/cendates/.) Some countries produce inter-censal estimates. Other countries, particularly those experiencing civil unrest, with few resources, or where census information may be deemed to be politically threatening, tend to have less regular censuses taken at intervals wider than once per decade.

Given that the population data are collected in different years, the small area population totals need to be reconciled by estimating population for the target years of interest. In GPW v3, these are 1990, 1995, and 2000 as well as a projection for 2015. GRUMP is similarly produced for 1990, 1995, and 2000. The regional Africa and Latin America data sets that are based on the accessibility model include population estimates for 1960, 1970, 1980, 1990, and 2000. For most countries, where two native population estimates were available from the national statistical offices, an average annual population growth rate was computed, as follows:

$$r = \frac{\ln\left(\frac{P_2}{P_1}\right)}{t} \tag{1}$$

where r is the average rate of growth, P_1 and P_2 the population totals for the first and second reference years, respectively, and t the number of years between the two census enumerations. This rate was then applied to the census figures to interpolate or extrapolate population totals to the target years. For example, the 1995 estimate is calculated (where t is now the difference between P_1 and 1995):

$$P_{1995} = P_1 e^{rt} \tag{2}$$

Some countries had only one population estimate. This includes newly formed states (e.g., Croatia) as well as countries that for either economic or political reasons have not conducted a census or released census results since 1990 (e.g., Angola). Others have conducted a recent census (e.g., Afghanistan) but administrative areas have changed to an extent that it cannot be matched with prior censuses. Additionally, many small islands have infrequent censuses and do not have subnational data. In these instances, national-level growth rates from the United Nations were used in lieu of intrinsically calculated growth rates (United Nations, 2001).

3.1.1. Boundary Matching Over Time

The GPW population surfaces use only population and boundary information and the other datasets use these data in combination with other sources. These pieces of information are linked. Where boundaries have changed over time, as they often do, considerable effort is made to reconcile the differences. For example, if a district in 1990 were split into two districts in 2000, the population for the two districts in 2000 would be summed so as to represent the same areal distribution as given in 1990. (It is usually impossible to adequately divide the population for the given district of 1990 in the absence of information provided by the census office to this effect.) As higher resolution data are collected, the need for reconciling boundary changes becomes greater, because lower level units such as districts are modified more frequently than provinces or states. Fitrani *et al.* (2005) describe how decentralization in Indonesia led to a sharp increase in the number of local governments and associated boundaries (from 292 in 1998 to 434 in 2004). In many countries, changes are less dramatic, but reconciling boundaries and reporting unit identifiers nevertheless poses one of the most challenging problems in compiling detailed, cross-national population databases. Interpolating or extrapolating population figures to a common base year often requires the use of a hybrid method, whereby growth rates are calculated at a level where boundaries have not changed (e.g., provinces), and applied to higher resolution subunits such as districts.

3.1.2. Temporal Aspects of Ancillary Data for Modeled Population Grids

Unlike the GPW databases, GRUMP and the Accessibility Model also use other datasets, which represent phenomena that change over time: changes in urbanization and infrastructure. Unfortunately, the current versions of these databases are limited to a single snapshot. The urban extents are derived primarily from a stable city-lights' database from a 1994–1995 composite and the roads' data are approximately as of the year 2004. Users of these databases, interested in changes over time, should be well aware of this limitation.

Future versions of this database will be able to incorporate improved temporal coverage, since the nighttime lights' data are being processed for additional time periods. Additional research will be required to confirm that changes in nighttime satellite-derived urban extents truly reflect land-use changes surrounding major urban areas rather than changes in sensor characteristics or processing. Should time series of road networks become available, they too could be incorporated. Alternatively, historical transport networks can be approximated by altering the speed of travel over particular surfaces to represent the poorer condition of the transport network in the past and envisaged better conditions in the future.

3.2. Limitations of the Ancillary Data

GRUMP and the Accessibility Model rely on ancillary data because in all instances the best possible data are not available. For this reason, it is important to understand the strengths and weaknesses of those data sources before applying them. While some of the issues associated with the temporal shortcomings have been mentioned, there are other caveats unrelated to temporal concerns.

There have been many uses of the nighttime lights' data as a proxy for urban areas (Elvidge *et al.*, 1997; Sutton *et al.*, 2001; Pozzi *et al.*, 2003; Schneider *et al.*, 2003) and these data are the only globally consistent and repeated sources of likely urban areas. Nevertheless, they have a few key limitations: they are known to over-represent built-up area, an effect called "blooming." The blooming effect depends on intrinsic characteristics of the sensor and on geolocation errors in the compositing process (Elvidge *et al.*, 2004). Studies have shown that it is not possible to find a unique threshold to reduce the blooming effect that would work globally (Small *et al.*, 2005). In fact, a 10% threshold could reduce the blooming effect without significantly affecting many individual small settlements for the 1994/1995 dataset. But this threshold does not provide a globally consistent basis for relating lighted areas to urban extent, since the characteristics of the blooming effect are, to some extent, city and country specific. Thus, heuristic or *ad hoc* adjustments of this nature would make data analysis questionable. A second shortcoming of these data is that they under-represent small settlements that are either poorly or infrequently lit due to insufficient detection by the sensor. This is a particular problem in Africa or rural Asia, where population data are also often sparse.

Given the limitations with the nighttime lights' data, GRUMP protects against overestimation of urban extents that are false positives—i.e., lights at industrial sites which may not be (or are sparsely) populated—by requiring additional information for validation (i.e., a name, location, and population estimate corresponding to the light). GRUMP also uses additional sources and indirect techniques to estimate extents for known populations that fall below the sensor's detection threshold as discussed above (see Balk *et al.*, 2005a).

For small-scale or even regional applications, the urban mask associated with the GRUMP data may produce areal extents that are larger than expected. In these instances, use of the urban extent mask if used with the GRUMP population grid may provide sub-urban population detail that might assist in further delineating the more- and less-densely populated areas within these enlarged—or agglomerated—urban areas and thus indicating features (density) that are associated with urban gradients. Reliance on the extent mask in and of itself may lead to overestimation of urban areas. For example, Tatem *et al.* (2005) found that the GRUMP urban mask overestimates urban extents for Kenya when compared with data derived from higher-resolution satellite imagery.

Future versions may be able to use improved night-lights products, both in their ability to reduce the blooming (though that work is just underway) and to make use of lights detected at more than a single time point. As mentioned, GRUMP was developed when only the 1994–1995 product was available, but subsequent to that, 1992–1993 and 2000 releases have become available. These are not fully analogous datasets, so additional work to determine their utility for urban detection would first be required.

Similarly, for use in the Accessibility Model, there are few data sources that provide consistent, geographically referenced transportation network data for large areas such as an entire continent. The combination of the Vector Map Level 0 (VMap0) spatial data (NIMA, 1997) with the improved attribute data and the transport data used for the African Accessibility Model should be viewed as currently the best available for the given constraints. The spatial data for the African transport network is derived from the Vector Map th VMap layers for roads, rivers, and railroads (NIMA, 1997). VMap0 is an updated version of the Digital Chart of the World (DCW) and is suitable for applications at a scale of 1:1 million. While this provides a consistent level of spatial detail for Africa, the transport links in the database do not contain sufficient information about their characteristics (road quality, road type), which is essential for computing the travel times in the accessibility model. For most of Africa, roads are the most important means of transport, and so the attributes of the road links were substantially improved through the use of continental-scale paper maps of Africa at a scale of 1:4 million (Michelin Travel Publications, 2004). These maps were used to identify 132 000 km of major roads and 282 000 km of secondary roads (11% and 22% of all roads in the VMap0 layer, respectively).

There are many uncertainties in the spatial and attribute data for the transportation network. There is often no easy way to determine the original data source. It is also likely that the original scale of the data varies from country to country. It is often hard to determine how current the data are and how data from different sources were reconciled at country boundaries. Indeed, it is quite possible that the final transportation network does not represent consistently the state of the road network for any one year and it needs to be used with great caution in applications that require data at scales greater than 1:1 million or that require data for the state of the transport network for Africa pre-1990 or post-2005. Future improvements in the quality of continental-scale transport networks will most likely depend on the public release of VMap Level 1 data at 1:250 000 or concerted regional efforts to publish consistent key data layers (such as SERVIR for Central America http://servir.nsstc.nasa.gov/home.html).

4. HEALTH APPLICATIONS

Since the earliest version of GPW and the Accessibility Models in the mid-1990s, health researchers have been using the data to better understand population exposure, vector-habitat, disease distribution, mortality, and related factors (from habitat change to livestock distribution to the distribution of underweight children). These data have been used effectively at the regional and global scale, and in some instance (large areas or countries), in fairly specific local areas. Gridded population data have been used to assist in sampling for a health survey in Chad (Brooker *et al.*, 2002; Beasley *et al.*, 2002) to estimate the geographic distribution of underweight children (Balk *et al.*, 2005b), to determine changing habitat (for example, Reid *et al.*, 2000), and to estimate population at risk of a specific infectious disease. Measures of population counts and density distributions have broad-scale health applications. Although the bulk of this section addresses the latter, a brief review of the former is also included, in part, because gridded population data act as a proxy for a host of other health-related data.

4.1. General Health Studies

Regional studies of mortality and malnutrition have focused largely on understanding biological and socioeconomic factors associated with those outcomes. Spatially explicit data on those outcomes is typically not available. When survey or clinic data are georeferenced, as is increasingly the case, it becomes possible to consider a range of spatially explicit factors, including population density. Density relates to disease transmission—and ultimately health status—in a variety of ways. For example, person-to-person transmission is likely to be high in densely populated urban areas, though such areas may reduce the potential for particular vector habitats. Population density estimates also provide continuous measures of the degree of urbanness (such as high-density core urban areas or less dense semi-urban areas). In the absence of explicit data on the mode of disease transmission, or the vector habitat, and with careful use, population density may be a useful proxy for an urban continuum.

In a study of West African mortality, Balk and colleagues (2004) confirm the complexities associated with measuring and interpreting population density: in urban areas, increases in population density reduced the risk of infant deaths, and the further away from an urban area, the greater the likelihood of infant death. In this study, density (GPW v3) and the GRUMP urban extent mask (alpha version) are used as proxy variables for clinic or health services density (which were not directly measured). In a study of underweight status in African children, Balk and colleagues (2005b) find that population density (GPW v3, CIESIN and CIAT, 2004)—again acting as an urban proxy—decreases the likelihood of children being underweight. Similarly, Sachs and colleagues (2001) and Gallup and Sachs (2001) use GPW v2 to explain differences in the spatial pattern of poverty and disease burden in Africa. These studies find that coastal dwellers—in large part due to their access to ports, urban areas, and infrastructure—experience less poverty and a lower economic burden associated with malaria than inland populations.

4.2. Specific Diseases

Population grids have become a key tool to understanding the populations at risk of contracting various infectious diseases. Infectious diseases have vectors or other transmission routes that are generally highly location based or geographic in nature. The means to understanding the impact of specific disease burdens depends in part on the ability to identify spatially the areas at risk as well as understanding the population in those places. Matching these spatial units—disease numerators with population denominators—is a large part of the contribution that gridded population data make toward understanding specific infectious diseases.

In many low-income countries, lack of resources and capacity in the health system prevent the development of reliable records of malaria morbidity and mortality. A large body of work has attempted to triangulate malaria risk and human population distribution to define population at risk. This work was pioneered in Africa with the development of the MARA/ARMA model of climate suitability for *Plasmodium falciparum* transmission (Craig *et al.*, 1999). Combinations of this map and the African population database (Deichmann, 1996b) were used to define age-specific populations at risk in 1995. These estimates were derived using national-level age distribution data from the UN Population Division applied to subnational population totals. In combination with empirical epidemiological data from local studies, Snow and colleagues (1999a, b) produced estimates of morbidity and mortality for the total and under five-year-old populations of Africa (see also Hay *et al.*, 2000). This work was updated and augmented (Snow *et al.*, 2003) to the year 2000 using the African population database (Deichmann, 1996b) to determine the proportion of the population in transmission risk categories and applying these to year 2000 national population estimates from the United Nations (2001). The most accurate revision of these mortality and morbidity figures for Africa has been done by using new extractions for the year 2000 using GPWv3 (CIESIN and CIAT, 2004) and the MARA model (Hay *et al.*, 2005a). This work is also incorporating the location of urban populations in Africa to discount morbidity and mortality estimates for the significantly lower malaria transmission rates in these urban areas.

Recently, these "population at risk" assessments have been conducted using historical maps of malaria endemicity and its transmission extent to evaluate the changing population at risk between 1900 and modern times at the global scale (Hay *et al.*, 2004). Using a similar approach to MARA/ARMA morbidity, estimates for *P. falciparum* have now been conducted globally (Snow *et al.*, 2005). In addition, some (Rogers and Randolph, 2000; Van Leishout *et al.*, 2004) have used GPWv2.0 (CIESIN *et al.*, 2000) to estimate population at risk under coupled scenarios of population and climate change. There are many issues involved with the choice of population surfaces and their derivation and these have been evaluated with respect to population at risk of malaria in Kenya (Hay *et al.*, 2005b). Hay and colleagues show the paramount importance of the average spatial resolution of the input census data by comparing five population surfaces including GRUMP v1, GPW v2, and v3; the Accessibility Model (version 3 not the most current); and LandScan. Figure 5 compares the error associated with each dataset at varying levels of spatial aggregation: they all estimate about the same population at the most aggregated level (the first administrative level) but two stand apart, GPW v3 and GRUMP v1, providing notably superior estimates at the highest spatial resolution. (Note that this publication was not undertaken on the most recent versions of the Accessibility Model, in

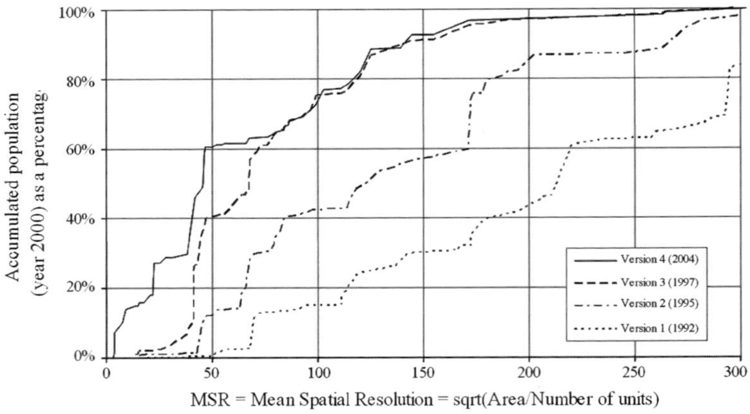

Figure 5 Cumulative percent of the African population represented by mean spatial resolution (MSR) (i.e., for version 4, 60% of the population is represented by an MSR of 50 or better).

which the underlying inputs have been improved, or of LandScan.) The results also highlight the issues involved and accuracy that can be obtained using simple interpolation techniques at different administrative levels, where these might be locally available. Although the interpolation methods differ, the best-fit datasets are those with inputs of the highest mean spatial resolution (MSR).

Given the absence of reliable data on the total number of parasitic infections in a country, estimates have often been based on prevalence data from a few limited studies and extrapolated to the country as a whole. In order to make these extrapolations more accurate, global georeferenced population datasets have been used increasingly. In particular, population totals and distribution from the Africa Population database (Deichmann, 1996b) and the first version of GPW (Tobler *et al.*, 1995), along with district-level census data when available, have been used to estimate population at risk of parasitic diseases or to estimate the number of people infected. For example, different statistical models have been developed to estimate the number of individuals to be treated based on the prevalence of infection of a given disease and population structure and distribution (Brooker *et al.*, 2000; Lindsay & Thomas, 2000; Noma *et al.*, 2002). Lindsay and Thomas (2000) use climate data to predict the distribution of lymphatic filariasis and overlay the resulting risk maps with a continental population grid (Deichmann, 1994) to estimate the number of people potentially exposed to the infection in Africa.

The issue of identifying population at risk and priority areas for treatment has been addressed by combining gridded population data with remotely sensed data. For instance, a recent methodology was developed to combine ecological zones defined using satellite-derived data (land-surface temperature and photosynthetic activity averages) with population density and prevalence data to map population at risk of parasitic infections in different countries in Africa (Brooker *et al.*, 2001a, 2002; Kabatereine *et al.*, 2004) and Asia (Brooker *et al.*, 2003). The results provide a targeted sampling frame of schools to guide valid epidemiological surveys and the identification of priority areas for national school initiatives and mass treatment. Noma *et al.* (2002) use GIS to identify bioclimatic zones of potential for onchocerciasis and to select which communities should be surveyed. The results were used to define areas of varying transmission risk to guide the implementation of control strategies. Similarly, Brooker and colleagues (2001b) used an early version of the African Population grid (Deichmann, 1996b) to determine populations at risk in particular locations resulting in observation of a significant relationship between the prevalence of *Schistosoma mansoni* and the distance of the schools from the lakeshore; as a matter of health policy, "distance to lakeshore" can now be used as a means to screen schools in East Africa.

A related application is one where global population data were used to study the relationship between population distribution changes and associated habitat changes. For example, Reid and colleagues (2000) predict that population distributional changes will, in effect, reduce the cattle population habitat leading to the reduction of the tsetse fly population and sleeping sickness prevalence in the human population.

Several uses of gridded population surfaces have demonstrated patterns in the distribution of human population *vis-à-vis* physiographic, climatic, and other environmental parameters that may be closely linked to health and disease burdens. For example, Small and Cohen (2004) use GPW v2 to show that people tend to live at low altitude (with Mexico City being an important exception) and near permanent water sources (rivers and coasts), but that population is not nearly as localized with respect to climatic variables such as precipitation or temperature. Disease vectors may be influenced by all of these factors, thus demonstrating the need for moderate-resolution population surfaces that allow for these factors to be disentangled in any given region of interest. In another study, Astrom and colleagues (2003), using GPW v2, find that populations residing above a certain altitude—due to the relationship with the physiological processing of oxygen at high altitude—experience lower tumor incidence.

In the wake of the Indian Ocean tsunami of December 26, 2004, the GRUMP population grid was used in combination with coastal buffer distances and elevation to estimate the population exposed to the great wave (Balk *et al.*, 2005c): roughly four million persons were estimated to live within a 2 km buffer in the most-affected regions. These estimates were then used to calculate death rates in some of the affected regions. National and moderate-resolution subnational population estimates could not be used rapidly, and without considerable assumptions, to generate estimates of exposure to natural hazards. (Even if some countries had high-resolution subnational data, they would need to be gridded to make such calculations.) Further, since this tragedy occurred across many national borders, it highlighted the utility of having a global population grid that is agnostic about independent of country boundaries. A global study of natural disaster hotspots has used GPW to estimate the risk of mortality and economic loss from six major natural hazards (Dilley *et al.*, 2005).

Lastly, an exploratory study considers the relationship of population density to the location of newly emerging or re-emerging infectious disease (Patel *et al.*, 2006). While the evidence is preliminary and complex, it suggests that disease emergence may be causally related to population dynamics, travel and trade routes.

5. DISCUSSION

Population input data are inevitably highly variable in terms of quality, resolution, and accuracy, in ways that are not quantifiable. In part, that is the nature of demographic data, which represent social processes, but treating them as if they were an easily measurable physical variable (on a grid). Administrative units will always be larger in sparsely populated areas, and perhaps will have more detail than may be needed for some applications in high-density places. Users should bear this constraint in mind when using these data.

5.1. Ideal Spatial Resolution

The ideal resolution for the study of infectious diseases and health will vary. Localized disease outbreaks might require information on village location, boundaries, and associated population characteristics. Emergency response studies, such as the recent tsunami in the Indian Ocean (Balk *et al.*, 2005c) require high-resolution administrative boundaries, population, and other demographic data associated with those boundaries as well as infrastructure (e.g., health clinics) at risk. Where the emergency is brought on by a geophysical phenomenon that is best estimated with physical data (such as coastal distance or elevation) gridded data are a prerequisite for establishing baseline population exposure. For broad synoptic analysis of health–environment issues, medium-resolution data would likely be sufficient.

The databases discussed herein have been constructed with enough information to incorporate uncertainty into the analysis. A simple measure for each pixel is the resolution—in this case, the size of geographic area—of the administrative unit from which the pixel population was derived or modeled. A grid of this indicator is available for version 3 of GPW. In practice, few people take the trouble to do serious uncertainty or sensitivity analyses. The responsibility of data producers is to provide all relevant information about input data, document modeling, and processing and leave it to the user to take this information into account.

In the development of the aforementioned data products, it has been useful to construct a measure of effective resolution. Measured as the country-specific average resolution, it can be thought of as the "cell size" if all units in a country were square and of equal size, which of course they are not. It is calculated as follows:

$$\text{Mean resolution(km)} = \sqrt{(\textit{country area})/(\textit{number of units})} \qquad (3)$$

A closer look at the varying resolution (or area) of the administrative units reveals other key improvements in the database in the GPW efforts. The average resolution of all countries went from 60 to 46, with improvements of 10 times or more for particular countries. Figure 6 shows the resolution improvements in Africa, for four versions of the Accessibility Model, by cumulative population. In the current version of the accessibility model, as with GPW v3, more than 60% of Africa's population is represented by a mean resolution of 50 km or better. This represents a significant improvement over previous models, including version 2 of the Accessibility Model and GPW v1, where 60% of the population was represented by much coarser resolution, more than three times coarser than the current resolution (about 170 km).

Though GPW has always sought to be based on inputs of the best-available resolution at the time, efforts to improve version 3 of GPW included acquisition of even higher resolution data for countries with

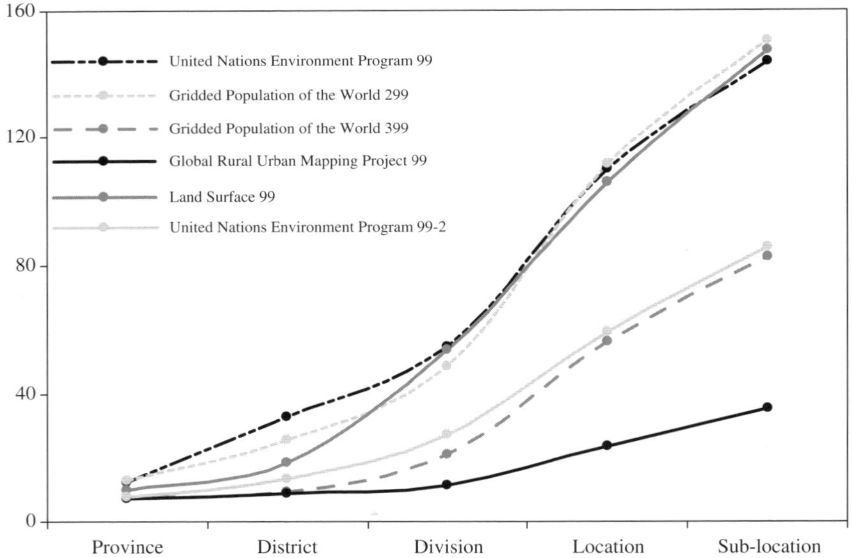

Figure 6 Graph of error structure by administrative level for the five large area public-domain human population distribution surfaces (see Hay *et al.*, 2005b). Left axis is the root mean square error expressed as a percentage of the mean population size of the administrative level.

coarse-resolution inputs and islands some of which required labor inputs to compile the basic data (such as digitizing). Earlier versions of GPW had less motivation (and resources) to do this, because the output resolution of 2.5 arc-minutes rendered finer input resolution redundant. The inputs for the third version of GPW were also used as an input to the GRUMP population surface that includes re-allocations toward urban areas and whose output resolution is 30 arc-seconds. Given the small footprint of many urban areas, the considerable investments in obtaining the highest available resolution population data were necessary to achieve the best-possible match between input and output resolution for each country. Often, these new inputs had to be digitized from imperfect source materials, since digital versions of these data were not available. For countries that are island chains, the improvements consisted of collecting island-level population data, and then assigning population to existing spatial inputs. GPW v2 had 41 countries with country-level (administrative level 0) data only, 31 of which were islands, which had an average resolution of 46. In version 3, fewer than half of these countries remain (with a slightly smaller share of them being islands) with an average resolution of 22.

5.2. Conclusion

As capabilities in refining the estimates of population distribution, urban areas, and associated infrastructure networks have increased, the more evident the localized nature of the distribution of human population has become. Improved estimates show that less, not more, land area tends to be occupied by moderate and densely populated settlements, as shown in Figure 7 (Figure 7 is Plate 4.7 in the Separate Color Plate Section), for the case of Ecuador. These spatial Lorenz curves show the cumulative fraction of the population as a function of cumulative fraction of land area, where units are ordered by increasing population density. Forty percent of Ecuador's population lives on 15% of its land area according to GPW v2. The improved resolution of GRUMP revise estimates substantially, reducing it by more than half, to only 6% of the land area in this example. People live locally, are burdened by disease locally, and receive their health services locally. Gains in the improved resolution of human population distribution will continue to lead to a better understanding of disease and health, but these gains must also be matched with improvements in information on health clinics, health catchments, and infrastructure.

In the future, more high-resolution data should become available so that modeling will be less and less necessary for most health analyses. While there may still be a need for modeled population data—for example, to understand seasonal flows—the basic improvement would be to the baseline population distribution. Hence what is important is to ensure long-term funding for maintaining and updating these data, and to ensure open-data dissemination policies so that data are made easily available for science and policy. For health studies, priority next steps, apart from continuing to increase resolution, would be more consistent global time series (e.g., going back several decades to assess recent trends), and further demographic variables such as age distribution and other variables required to make rigorous spatial projections.

6. DATA DISSEMINATION

The following data are available in the accompanying DVD: the Gridded Population of the World version 3 (beta) at 2.5 arc-minutes: population counts, land area, and population density; version 1 (alpha) of the GRUMP 30' population surface; and the Accessibility Model for Africa. All grids are available in GeoTIFF format. Users are strongly encouraged to visit the respective websites for updates and final versions. For GPW and GRUMP, see http://sedac.ciesin.columbia.edu/gpw, where users can also download the grids for 2015, the GRUMP settlement points (alpha), and the urban extent mask (alpha) as well as ancillary data products associated with GPW (e.g., national coastlines to match the population grid and a grid of national identifiers). The website for Accessibility Model for Africa is http://na.unep.net/globalpop/africa/Africa_index.html. An updated version of the Accessibility Model for Latin America and the

Caribbean is underway and users should visit CIAT's website http://gisweb.ciat.cgiar.org/population/ for updates. Users are strongly encouraged to supply feedback, and their publications that make use of these data, to gpw@ciesin.columbia.edu.

6.1. Data Selection

Before using the population surfaces in the companion DVD for analysis, a population model and spatial resolution must be chosen and the data evaluated to ensure that its precision meets the study requirements. Population surface and accessibility models should not be used as 'independent' tests of the reliability of nighttime light imagery or transportation network data (nor should they be used to modify such data in attempts to eliminate errors) when they have been derived from these original datasets in the first place. It is essential to avoid circularity in 'improvements' of both original and derived datasets. The choice of appropriate resolution—a 30 arc-second or 2.5 arc-minutes—depends on the scale of the study. In general, the 2.5 arc-minute data are most appropriate for continental and large-region studies; the 30 arc-second data are most appropriate for smaller regions and national studies. In some cases, subnational studies are possible with the 30 arc-second data, but it is not possible to derive meaningful results for small-area studies such as those for a single city.

For the GPW and GRUMP data, the administrative unit area grid (available from the GPW web site) may be used to determine the approximate locational precision of the population surfaces on a cell-by-cell basis. The administrative unit area grid indicates the area of the administrative unit from which the population value was derived. Where multiple units contributed to a cell, the value is the weighted mean of the input administrative unit sizes. A cutoff mean administrative unit area value can be approximated by calculating the area based on a given radius. For example, to identify the cells with a locational accuracy of approximately 10 km or greater, a cutoff value of 314 would be used, because cells with a value greater than this are derived from an administrative unit that cannot be enclosed by a circle with a radius of 10 km. In reality, a larger value should be used, as very few administrative units are circular in shape.

6.2. Methods and Issues in Analysis

Using the population data surfaces requires a software package capable of dealing with raster data, such as ArcGIS™ (with the Spatial Analyst extension), Erdas Imagine®, Idrisi, GRASS, MatLab®, or any number of others. GeoTIFF is a well-known format supported in most packages that handle raster data. If translation is necessary, the open source Geospatial Data Abstraction Library (GDAL), available at: http://www.remotesensing.org/gdal/, can be used to convert files to a number of other formats.

The most common form of analysis is to aggregate population totals in the surfaces by some other unit of analysis (such as ecological regions or habitats, buffers around points of interest such as health clinics, and so on) using a zonal statistics function. Population density grids may be used in a similar manner to characterize the variability of population within different zones; the minimum, maximum, mean, and standard deviation of density values within a given zone are often more useful for inter-zone comparison than just for the total populations of the zones.

Regression analysis with population counts or density as an explanatory variable or as a per capita denominator for explanatory variables other than population is another tool used commonly with these data. While there are many legitimate uses of these raster population surfaces in quantitative analysis of this type, care must be taken as raster data can invalidate the assumptions in classic regression. This occurs simply as a function of the self-replicating feature of the gridded nature of the data. A raster layer comparison is useful for explanation but cannot be relied on for rejecting the null hypothesis at a given probability level (Openshaw, 1991) because these data may be biased. That is, the original administrative area data would have had a single value that was distributed across far more grid cells. While the approximate value of each grid cell would be accurate, each observed grid cell is not independent of (i.e., they are spatially dependent, being from the same original administrative area polygon; and they inflate the number of observations). Geostatistical approaches based on point observations (GPW and GRUMP make centroids of the units used in gridding available for this purpose), or using the data to first construct variables based on zonal statistics, may be better. The examples given herein have paid attention to this caveat. These approaches can be accomplished with geostatistical extensions to any other GIS software or stand-alone software packages for working with spatial data (e.g., the ArcGIS® Geostatistical Analyst extension or the free GeoDa software package).

ACKNOWLEDGEMENTS

The authors thank Christopher Small for Figure 7. DLB, FP, and GY were funded, and primary support for the production of the Gridded Population of the World Data set was given, by National Aeronautics and Space Administration (Contract no. NAS5-03117) for the Continued Operation of the Socioeconomic Data and Applications Center (SEDAC) at CIESIN at Columbia University. SIH is funded by a Research Career Development Fellowship from the Wellcome Trust (no. #069045).

REFERENCES

Adams, J. (1968). *A population map of West Africa*. London: Graduate School of Geography, Discussion paper no. 26, London School of Economics.

Astrom, K., Cohen, J.E., Willett-Brozick, J.E., Aston, C.E. and Baysal, B.E. (2003). Altitude is a phenotypic modifier in hereditary *paraganglioma* type 1: evidence for an oxygen-sensing defect. *Human Genetics* **113**, 228–237.

Balk, D., Pullum, T., Storeygard, A., Greenwell, F. and Neuman, M. (2004b). A spatial analysis of childhood mortality in West Africa. *Population, Space and Place* **10**, 175–216.

Balk, D., Pozzi, F., Yetman, G., Deichmann, U. and Nelson, A. (2005a). The distribution of people and the dimension of place: methodologies to improve the global estimation of urban extents. *Proceedings of the Urban Remote Sensing Conference (of the International Society for the Photogrammetry and Remote Sensing)*, Tempe, Arizona: publisher? International Society for Photogrammetry and Remote Sensing.

Balk, D., Levy, M., Storeygard, A., Gaskell, J., Sharma, M. and Flor, R. (2005b). Child hunger in the developing world: an analysis of environmental and social correlates. *Food Policy* **30**, 584–611.

Balk, D., Gorokhovich, Y. and Levy, M. (2005c). *Estimation of Coastal Populations Exposed to the 26 December 2004 Tsunami*. Palisades, New York: Center for International Earth Science Information Network (CIESIN), Columbia University. Available at: http://www.ciesin.columbia.edu/pdf/tsunami_pop_exposure1.pdf.

Balk, D. and Yetman, G. (2005). *The global distribution of population: evaluating the gains in resolution refinement* (February 2005, draft). Palisades, New York: Center for International Earth Science Information Network (CIESIN), Columbia University. Available at: http://beta.sedac.ciesin.columbia.edu/gpw/docs/gpw3_documentation_final.pdf.

Beasley, M., Brooker, S., Ndinaromtan, M., Madjiouroum, E.M., Baboguel, M., Djenguinabe, E. and Bundy, D.A.P. (2002). First nationwide survey of the health of schoolchildren in Chad. *Tropical Medicine and International Health* **7**, 625–630.

Brooker, S., Donnelly, C.A. and Guyatt, H.L. (2000). Estimating the number of helminthic infections in the Republic of Cameroon from data on infection prevalence in schoolchildren. *Bulletin of the World Health Organization* **78**, 1456–1465.

Brooker, S., Hay, S.I., Issae, W., Hall, A., Kihamia, C.M., Lwambo, N.J.S., Wint, W., Rogers, D.J. and Bundy, D.A.P. (2001a). Predicting the distribution of urinary *schistosomiasis* in Tanzania using satellite sensor data. *Tropical Medicine and International Health* **6**, 998–1007.

Brooker, S., Miguel, E.A., Waswa, P., Namunyu, R., Moulin, S., Guyatt, H. and Bundy, D.A.P. (2001b). The potential of rapid screening methods for *Schistosoma mansoni* in western Kenya. *Annals of Tropical Medicine and Parasitology* **95**, 343–351.

Brooker, S., Beasley, M., Ndinarotan, M., Madjiouroum, E.M., Baboguel, M., Djenguinabe, E., Hay, S.I. and Bundy, D.A.P. (2002). Use of remote sensing and a geographical information system in a national helminth control programme in Chad. *Bulletin of the World Health Organization* **80**, 783–789.

Brooker, S., Pratap, S., Waikagul, J., Suvanee, S., Kojima, S., Takeuchi, T., Luong, T.V. and Looareesuwan, S. (2003). Mapping soil-transmitted helminth infections in Southeast Asia and implications for parasite control. *Southeast Asian Journal of Tropical Medicine and Public Health* **34**, 24–35.

Center for International Earth Science Information Network (CIESIN), Columbia University; International Food Policy Research Institute (IFPRI) and World Resources Institute (WRI). (2000). *Gridded Population of the World (GPW), Version 2*. Palisades, New York: CIESIN, Columbia University. Available at: http://sedac.ciesin.columbia.edu/plue/gpw.

Center for International Earth Science Information Network (CIESIN), Columbia University and Centro Internacional de Agricultura Tropical (CIAT). (2004). *Gridded Population of the World (GPW), Version 3 beta*. Palisades, New York: CIESIN, Columbia University. Available at: http://sedac.ciesin.columbia.edu/gpw.

Clarke, J.I. and Rhind, D.W. (1992). *Population Data and Global Environmental Change*. Human Dimensions of Global Environmental Change Programme Report 3, New York: International Social Science Council.

Craig, M.H., Snow, R.W. and Le Sueur, D. (1999). A climate-based distribution model of malaria transmission in sub-Saharan Africa. *Parasitology Today* **15**, 105–111.

Deichmann, U. and Eklundh, L. (1991). *Global Digital Datasets for Land Degradation Studies: A GIS Approach*. Nairobi, Kenya: United Nations Environment Programme, Global Resource Information Database, Case study no. 4.

Deichmann, U. (1994). *A Medium Resolution Population Database for Africa. Database Documentation and Digital Database*. Santa Barbara, California: National Center for Geographic Information and Analysis, University of California.

Deichmann, U. (1996a). *A Review of Spatial Population Database Design and Modeling*. NCGIA, Technical report 93-3. Santa Barbara, California: National Centre for Geographic Information and Analysis, University of California. Available at: http://www.ncgia.ucsb.edu/Publications/Tech_Reports/96/96-3.PDF.

Deichmann, U. (1996b). *African Population Database. Digital Database and Documentation*. Santa Barbara, California: National Center for Geographic Information and Analysis, University of California.

Deichmann, U. (1997). *Accessibility Indicators in GIS*. New York: United Nations Statistics Division, Department for Economic and Policy Analysis.

Deichmann, U., Balk, D. and Yetman, G. (2001). *Transforming Population Data for Interdisciplinary Usages: From Census to Grid*. Palisades, NY: CIESIN, Columbia University. Working paper available on-line at: http://sedac.ciesin.columbia.edu/plue/gpw/GPWdocumentation.pdf.

Dilley, M.R.S., Chen, B., Deichmann, U., Lerner-Lam, A. and Arnold, M. (2005). *Natural Disaster Hotspots: A Global Risk Analysis*. Washington, DC: The World Bank, Hazard Management Unit.

Dobson, J.E., Bright, E.A., Coleman, P.R., Durfee, R.C. and Worley, B.A. (2000). LandScan: a global population database for estimating populations at risk. *Photogrammetric Engineering and Remote Sensing* **66**, 849–857.

Elvidge, C.D., Baugh, K.E., Hobson, V.R., Kihn, E.A., Kroehl, H.W., Davis, E.R. and Cocero, D. (1997). Satellite inventory of human settlements using nocturnal radiation emissions: a contribution for the global tool chest. *Global Change Biology* **3**, 87–395.

Elvidge, C.D., Baugh, K.E., Dietz, J.B., Bland, T., Sutton, P.C. and Kroehl, H.W. (1999). Radiance calibration of DMSP-OLS low-light imaging data of human settlements. *Remote Sensing of Environment* **68**, 77–88.

Elvidge, C.D., Safran, J., Nelson, I.L., Tuttle, B.T., Hobson, V.R., Baugh, K.E., Dietz, J.B. and Erwin, E.H. (2004). Area and position accuracy of DMSP nighttime lights data. In: *Remote Sensing and GIS Accuracy Assessment* (R.S. Lunetta and J.G. Lyon, eds). Chapter 20, pp. 281–292. Boca Raton, Florida: CRC Press.

Fitrani, F., Hofman, B. and Kaiser, K. (2005). Unity in diversity? The creation of new local governments in a decentralising Indonesia. *Bulletin of Indonesian Economic Studies* **41**, 57–79.

Gallup, J.L. and Sachs, J.D. (2001). The economic burden of malaria. *American Journal of Tropical Medicine and Hygiene* **64**, 85–96.

Goodchild, M.F., Anselin, L. and Deichmann, U. (1993). A framework for the areal interpolation of socioeconomic data. *Environment and Planning A* **25**, 383–397.

Hay, S.I., Omumbo, J.A., Craig, M.H. and Snow, R.W. (2000). Earth observation, geographic information systems and *Plasmodium falciparum* malaria in sub-Saharan Africa. *Advances in Parasitology* **47**, 173–215.

Hay, S.I., Guerra, C.A., Tatem, A.J., Noor, A.M. and Snow, R.W. (2004). The global distribution and population at risk of malaria: past, present and future. *Lancet Infectious Diseases* **4**, 327–336.

Hay, S.I., Guerra, C.A., Tatem, A.J., Atkinson, P.M. and Snow, R.W. (2005a). Urbanization, malaria transmission and disease burden in Africa. *Nature Reviews Microbiology* **3**, 81–90.

Hay, S.I., Noor, A.M., Nelson, A. and Tatem, A.J. (2005b). The accuracy of human population maps for public health application. *Tropical Medicine and International Health* **10**, 1073–1086.

Haynes, K.E. and Fotheringham, A.S. (1984). *Gravity and Spatial Interaction Models*. London: Sage Publications.

Kabatereine, N.B., Brooker, S., Tukahebwa, E.M., Kazibwe, F. and Onapa, A. (2004). Epidemiology and geography of *Schistosoma mansoni* in Uganda: implications for planning control. *Tropical Medicine and International Health* **9**, 372–380.

Leddy, R. (1994). *Small Area Populations for the United States*. Paper presented at the Association of American Geographers Annual Meeting in San Francisco. Washington, DC: Geographic Studies Branch, International Programs Center, US Bureau of the Census.

Lindsay, S.W. and Thomas, C.J. (2000). Mapping and estimating the population at risk from lymphatic filariasis in Africa. *Transactions of the Royal Society of Tropical Medicine and Hygiene* **94**, 37–45.

Martin, D. and Bracken, I. (1991). Techniques for modelling population-related raster databases. *Environment and Planning A* **23**, 1069–1075.

Michelin Travel Publications. (2004). *Northwest Africa; Africa Northeast & Arabia; Central & Southern Africa, Madagascar; 1:4 million scale map sheets 741, 745 and 746*. France: Michelin Maps & Atlases.

Nelson, A. and Deichmann, U. (2004) *The African Population Database, Version 4*. New York: United Nations Environment Program (UNEP) and the Center for International Earth Science Information Network (CIESIN), Columbia University. Available at: http://www.na.unep.net/datasets/datalist.php3.

National Imagery and Mapping Agency (NIMA). (1997). *Vector Map Level 0 Digital Chart of the World 3rd Edition*. Fairfax, Virgina: NIMA. Available at: http://www.mapability.com/info/vmap0_index.html.

Noma, M., Nwoke, B.E.B., Nutall, I., Tambala, P.A., Enyong, P., Namsenmo, A., Remme, J., Amazigo, U.V., Kale, O.O. and Seketeli, A. (2002). Rapid epidemiological mapping of onchocerciasis (REMO): its application by the African programme for onchocerciasis control (APOC). *Annals of Tropical Medicine and Parasitology* **96**, 29–39.

Oak Ridge National Laboratory (ORNL). (2003). *LandScan Global Population Database*. Oak Ridge, Tenesse: Oak Ridge National Laboratory. Available at: http://www.ornl.gov/gist/.

Openshaw, S. (1991). A view on the GIS crisis in geography or using GIS to put Humpty-Dumpty back together again. *Environment and Planning A* **23**, 621–628.

Patel, N., Jones, K., Levy, M., Balk, D. and Daszak, P. (2006). Global trends in zoonotic disease emergence. In: *Wildlife Wealth, and Health: Tropical Forest Disturbance and Viral Disease Emergence* (R. Hardin, ed.). Cambridge, MA: Harvard University Press.

Pozzi, F., Small, C. and Yetman, G. (2003). Modeling the distribution of human population with nighttime satellite imagery and gridded population of the world. *Earth Observation Magazine* **12**, 1.

Reid, R.S., Kruska, R.L., Deichmann, U., Thornton, P.K. and Leak, S.G.A. (2000). Human population growth and the extinction of the tsetse fly. *Agriculture, Ecosystems and Environment* **77**, 227–236.

Rogers, D.J. and Randolph, S.E. (2000). The global spread of malaria in a future, warmer world. *Science* **289**, 1763–1766.

Sachs, J.D., Mellinger, A.D. and Gallup, J.L. (2001). The geography of poverty and wealth. *Scientific American* **284**, 70.

Schneider, A., Friedl, M.A., McIver, D.K. and Woodcock, C.E. (2003). Mapping urban areas by fusing multiple sources of coarse resolution remotely sensed data. *Photogrammetric Engineering and Remote Sensing* **69**, 1377–1386.

Small, C. and Cohen, J.E. (2004). Continental physiography, climate, and the global distribution of human population. *Current Anthropology* **45**, 269–277.

Small, C., Pozzi, F. and Elvidge, C.D. (2005). Spatial analysis of global urban extents from the DMSP-OLS night lights. *Remote Sensing of Environment* **96**, 277–291.

Snow, R.W., Craig, M.H., Deichmann, U. and Le Sueur, D. (1999a). A preliminary continental risk map for malaria mortality among African children. *Parasitology Today* **15**, 99–104.

Snow, R.W., Craig, M., Deichmann, U. and Marsh, K. (1999b). Estimating mortality, morbidity and disability due to malaria among Africa's non-pregnant population. *Bulletin of the World Health Organization* **77**, 624–640.

Snow, R.W., Craig, M.H., Newton, C.R.J.C. and Steketee, R.W. (2003). *The public health burden of* Plasmodium falciparum *malaria in Africa: deriving the numbers*. Working paper no. 11, Bethesda, Maryland: Disease Control Priorities Project, Fogarty International Center, National Institutes of Health.

Snow, R.W., Guerra, C.A., Noor, A.M., Myint, H.Y. and Hay, S.I. (2005). The global distribution of clinical episodes of *Plasmodium falciparum* malaria. *Nature* **434**, 214–217.

Sutton, P.D., Roberts, C., Elvidge and Baugh, K. (2001). Census from heaven: an estimate of the global human population using night-time satellite imagery. *International Journal of Remote Sensing* **22**, 3061–3076.

Tatem, A.J., Noor, A.M. and Hay, S.I. (2005). Assessing the accuracy of satellite derived global and national urban maps in Kenya. *Remote Sensing of Environment* **96**, 87–97.

Tobler, W.R. (1979). Smooth pycnophylactic interpolation of geographical regions. *Journal of the American Statistical Association* **74**, 519–530.

Tobler, W., Deichmann, U., Gottsgen, J. and Maloy, K. (1995). *The global demography project*. Technical Report TR-6-95. Santa Barbara, California: National Center for Geographic Information and Analysis (NCGIA), University of California.

Tobler, W., Deichmann, U., Gottsegen, J. and Maloy, K. (1997). World population in a grid of spherical quadrilaterals. *International Journal of Population Geography* **3**, 203–225.

Tufte, E.R. (1990). *Envisioning Information*, pp. 40–41. Cheshire, Connecticut: Graphics Press.

United Nations. (2001). *World Urbanization Prospects, 1999 Revision*. New York: United Nations Population Division, Department of Economic and Social Affairs.

van Leishout, M., Kovats, R.S., Livermore, M.T.J. and Martens, P. (2004). Climate change and malaria: analysis of the SRES climate and socio-economic scenarios. *Global Environmental Change* **14**, 87–99.

Defining the Global Spatial Limits of Malaria Transmission in 2005

C.A. Guerra[1], R.W. Snow[2,3] and S.I. Hay[1,2]

[1]*TALA Research Group, Tinbergen Building, Department of Zoology, University of Oxford, South Parks Road, Oxford OX1 3PS, UK*
[2]*Malaria Public Health & Epidemiology Group, Centre for Geographic Medicine, KEMRI/Wellcome Trust Research Laboratories, PO Box 43640, 00100 Nairobi, Kenya*
[3]*Centre for Tropical Medicine, John Radcliffe Hospital, University of Oxford, Oxford OX3 9DS, UK*

Abstract 157
1. Introduction 158
2. The Distribution of Malaria Risk from Travel Guidelines 159
3. The Biological Limits of Transmission 161
3.1. Altitudinal Mask 161
3.2. Climate Suitability Mask 167
3.3. Population Density Mask 167
4. Distinguishing *P. falciparum* and *P. vivax* risk 169
5. Regional Analysis 169
6. Discussion 173
7. Conclusions 174
Acknowledgements 175
References 175

ABSTRACT

There is no accurate contemporary global map of the distribution of malaria. We show how guidelines formulated to advise travellers on appropriate chemoprophylaxis for areas of reported *Plasmodium falciparum* and *Plasmodium vivax* malaria risk can be used to generate

ADVANCES IN PARASITOLOGY VOL 62
ISSN: 0065-308X $35.00
DOI: 10.1016/S0065-308X(05)62005-2

crude spatial limits. We first review and amalgamate information on these guidelines to define malaria risk at national and sub-national administrative boundary levels globally. We then adopt an iterative approach to reduce these extents by applying a series of biological limits imposed by altitude, climate and population density to malaria transmission, specific to the local dominant vector species. Global areas of, and population at risk from, *P. falciparum* and often-neglected *P. vivax* malaria are presented for 2005 for all malaria endemic countries. These results reveal that more than 3 billion people were at risk of malaria in 2005.

1. INTRODUCTION

During the halcyon days of global malaria eradication, mapping the precise spatial extent of the disease was central to the control efforts of the World Health Organization (WHO). Between the 1940s and 1970s, a huge investment was made in synthesising available information on the distribution of risk using various combinations of expert opinion, elevation, climate, presence/absence records of the disease and vectors, spleen rates, parasite rates, sporozoite rates, biting rates and haemoglobinopathy prevalence (Boyd, 1949; Pampana and Russell, 1955; WHO, 1966; Lysenko and Semashko, 1968; Dutta and Dutt, 1978). Since the 1970s, as the world's public health focus shifted from malaria eradication, an interest in mapping global malaria risk waned (Carter and Mendis, 2002; Hay *et al.*, 2004a).

Following a renewed commitment to financing comprehensive malaria control at a global scale, the significance of defining the disease burden has re-emerged as a priority (Hay *et al.*, 2004a; Snow *et al.*, 2005). This will allow regional and national requirements for drugs, insecticides, bed nets and other commodities to be assessed more accurately, and so malaria to be rolled back more effectively (Snow, 2004; Sachs, 2005). Despite an obvious need to map risk (Snow *et al.*, 1996), there remains no comprehensive definition of the spatial limits of malaria. Previous studies (Rogers and Randolph, 2000; Hay *et al.*, 2004b) have used information provided by the WHO on the extent of risks due to *P. falciparum* and *P. vivax* from advice to travellers

(WHO, 2003a, b). Precise details of how the WHO constructed these limits are difficult to obtain and there are several other public-domain sources of travel advice (CDC, 2003; IAMAT, 2004), which have not been harmonised with the WHO data (WHO, 2003b). Here we use geographic information systems (GIS) to triangulate and standardise international travel health guideline (ITHG) information and refine these limits with country-specific altitudinal exclusions, climate suitability criteria and population density, to make a new map of global malaria risk. The result is a species-specific estimate of the limits of malaria transmission in 2005.

2. THE DISTRIBUTION OF MALARIA RISK FROM TRAVEL GUIDELINES

ITHGs have been developed to advise travellers on appropriate malaria chemoprophylaxis. These guidelines are the only contemporary, global source of information on national and sub-national malaria risk. Three primary sources are available in the public-domain: the WHO's International Travel and Health guidelines (WHO-ITH) (WHO, 2005), the World Malaria Risk Chart of the International Association for Medical Assistance to Travellers (IAMAT-WMRC) (IAMAT, 2004) and the Health Information for International Travel ("Yellow Book") of the Centres for Disease Control and Prevention (CDC-YB) (CDC, 2003). These guidelines provide country-specific information that variously include (i) sub-national risk distribution; (ii) altitude-based transmission limits; (iii) risk definitions in specific urban and rural areas; (iv) *P. falciparum* to *P. vivax* ratios; (v) dominant vector species; (vi) anti-malarial drug resistance status; and (vii) prophylaxis regimens recommended. An example entry for Ecuador is as follows: "Malaria risk—*P. falciparum* (34%), *P. vivax* (66%)—exists throughout the year below 1500 m, with some risk in Cotopaxi, Loja and Los Ríos. Higher transmission risk is found in El Oro, Esmeraldas and Manabí. There is no risk in Guayaquil or Quito." (WHO, 2005). Despite sometimes being incomplete, as in this example, it should be acknowledged that ITHGs are inclusive, rather than exclusive of geographic areas of malaria risk, so that information given to potentially non-immune travellers is risk-averse. The empirical data used as input to these national entries are rarely detailed.

To map these data we focussed on the three main criteria: administrative boundaries, altitudinal limits and urban centres (Table 1). We combined the sub-national description of malaria with databases of administrative areas within countries to define crude spatial limits. We obtained first- and, occasionally, second-level sub-national administrative boundaries for all malarious countries from the Food and Agriculture Organization's GeoNetwork portal (http://www.fao.org/geonetwork/) ($n = 94$), the International Centre for Tropical Agriculture (CIAT) (http://www.ciat.cgiar.org/) ($n = 6$), and the Environmental Systems Research Institute (ArcView Data & Maps CD, ESRI, Redlands, California, USA) ($n = 4$).

We defined all classifications of malaria risk in the ITHG entries as malaria presence, except those of "no risk", "negligible risk" and "sporadic cases", which we classified as absent. Descriptions of sub-national malaria risk that were not geographically specific were impossible to map and were ignored. An exception was made if malaria risk was described as present in $\geqslant 50\%$ of the administrative area, in which case transmission was considered possible throughout that administrative unit. Where data were available from more than one source (Table 1), we used the finest spatial resolution and most comprehensive information. A digital elevation model (DEM) at approximately 1×1 km spatial resolution (Hastings and Dunbar, 1998) was

Table 1 Comprehensiveness of the information provided by international travel and health guidelines. Figures indicate the number of entries per category that prove useful for mapping.

Source	Administrative	Altitude	Urban	All
WHO-ITH[a]	27	20	14	61
IAMAT-WMRC[b]	30	41	31	102
CDC-YB[c]	30	14	26	70
Unique[d]	42	42	37	121

[a]International Travel and Health Guidelines of the WHO (WHO, 2005).
[b]International Association for Medical Assistance to Travellers' World Malaria Risk Chart (IAMAT, 2004).
[c]Centres for Disease Control and Prevention's Yellow Book (CDC, 2003).
[d]The maximum number of entries per criteria regardless of source.

used to implement national altitudinal transmission limits, reported in the ITHGs in metres above sea level. We could not map non-specific "highland" or "lowland" descriptions. If the ITHG sources provided conflicting limits, we used the higher altitude threshold. Finally, the ITHGs reported 70 cities as malaria free. These were geo-referenced using electronic geographic databases (Microsoft Corporation, 2005; The Getty Research Institute, 2005; University of California, 2005), co-located to their urban extents as defined by the Global Rural-Urban Mapping Project (GRUMP) (CIESIN/IFPRI/WB/CIAT, 2004) and then excluded.

Of the 107 countries reporting some degree of malaria risk, we mapped 104 according to our ITHG exclusion criteria (Table 2). Uzbekistan reported only "sporadic cases" and was not mapped as a malaria endemic country (MEC) in this paper. For Algeria, no corresponding administrative data could be obtained, and for North Korea there was insufficient detail in the sub-national description of risk. Despite the ITHGs being independent documents, there was relatively little complementary information: of a potential 318 entries (106 MECs × 3, i.e. risk information defined by administrative boundaries, altitude or urbanisation), there were only 121 unique reports, with IAMAT-WMRC the most comprehensive and WHO-ITH the least (Table 1).

3. THE BIOLOGICAL LIMITS OF TRANSMISSION

3.1. Altitudinal Mask

Temperature is inversely related to altitude, dropping by approximately 0.98°C for every 100-metre increase above absolute sea level (Henderson-Sellers and Robinson, 1991). Mosquitoes and malaria transmission are thus sensitive to altitude (Cox *et al.,* 1999). Altitudinal limits from the ITHGs were available for 42 countries. The majority of the countries (44/62) for which no information was available were in Africa and we assumed no altitudinal limits for most of these (see Section 3.2). For the 18 remaining non-African MECs, we defined limits by those of neighbouring countries with similar dominant vector species. To identify the latter, we used a global map

Table 2 Country summary data of area and population at risk (PAR) extractions

Country[a]	ITHGs criteria[b]			Pfr[c]	Ad1[d]	Area[e]					Population[f]		
	Ad	Alt	Urb			Total	WHO 2002	ITHG	ALT-MASK	POP-MASK	Pf	Pv	Pf+Pv
AFRO													
Algeria	Yes	No	No	0.610	48	2.32	0.02	n/a	0.01	0.00	0.00	0.00	0.00
Angola	No	No	No	1.000	18	1.25	1.25	1.25	1.24	0.93	12.92	0.00	0.00
Benin	No	No	No	1.000	12	0.12	0.12	0.12	0.12	0.11	6.29	0.00	0.00
Botswana	Yes	No	No	1.000	10	0.58	0.17	0.18	0.18	0.04	0.16	0.00	0.00
Burkina Faso	No	No	No	1.000	45	0.27	0.27	0.27	0.27	0.27	13.49	0.00	0.00
Burundi	No	No	No	1.000	17	0.02	0.02	0.02	0.02	0.02	6.03	0.00	0.00
Cameroon	No	No	No	1.000	10	0.47	0.47	0.47	0.46	0.44	12.56	0.00	0.00
Cape Verde	Yes	No	No	1.000	17	0.00	0.00	0.00	0.00	0.00	0.25	0.00	0.00
Central Af. Republic	No	No	No	1.000	17	0.62	0.62	0.62	0.62	0.27	3.22	0.00	0.00
Chad	No	No	No	1.000	14	1.26	0.81	1.26	0.89	0.52	9.05	0.00	0.00
Comoros	No	No	No	0.950	3	0.00	0.00	0.00	0.00	0.00	0.61	0.00	0.00
Congo	No	No	No	1.000	10	0.34	0.34	0.34	0.34	0.24	3.43	0.00	0.00
Côte d'Ivoire	No	No	No	1.000	16	0.32	0.32	0.32	0.32	0.32	14.10	0.00	0.00
Dem. Rep. Congo	No	No	No	1.000	11	2.34	2.33	2.34	2.31	2.06	46.88	0.00	0.00
Equatorial Guinea	No	No	No	1.000	7	0.03	0.03	0.03	0.03	0.02	0.50	0.00	0.00
Eritrea	No	Yes	Yes	0.644	9	0.12	0.12	0.12	0.12	0.12	0.00	0.00	4.21
Ethiopia	No	Yes	Yes	0.622	11	1.13	0.94	0.94	0.89	0.79	0.00	0.62	44.09
Gabon	No	No	No	1.000	9	0.27	0.27	0.27	0.27	0.06	1.30	0.00	0.00
Gambia	No	No	No	1.000	7	0.01	0.01	0.01	0.01	0.01	1.08	0.00	0.00
Ghana	No	No	No	1.000	10	0.24	0.24	0.24	0.24	0.24	18.38	0.00	0.00
Guinea	No	No	No	1.000	8	0.25	0.25	0.25	0.25	0.23	8.02	0.00	0.00
Guinea-Bissau	No	No	No	1.000	9	0.03	0.03	0.03	0.03	0.03	1.39	0.00	0.00
Kenya	No	Yes	Yes	1.000	8	0.57	0.53	0.52	0.51	0.42	23.67	0.00	0.00
Liberia	No	No	No	1.000	9	0.10	0.10	0.10	0.10	0.09	2.39	0.00	0.00

Madagascar	No	No	No	1.000	6	0.59	0.59	0.59	0.59	0.59	15.75	0.00	0.00
Malawi	No	No	No	1.000	3	0.11	0.11	0.11	0.11	0.10	11.78	0.00	0.00
Mali	No	No	No	1.000	9	1.26	0.66	1.26	0.99	0.60	11.82	0.00	0.00
Mauritania	Yes	No	No	1.000	13	1.04	0.28	0.76	0.59	0.23	1.08	0.00	0.00
Mauritius	Yes	No	No	0.000	9	0.00	0.00	0.00	0.00	0.00	0.00	0.57	0.00
Mayotte	No	No	No	1.000	2	0.00	0.00	0.00	0.00	0.00	0.31	0.00	0.00
Mozambique	No	No	No	1.000	11	0.78	0.78	0.78	0.78	0.71	17.97	0.00	0.00
Namibia	Yes	No	No	1.000	13	0.83	0.15	0.45	0.34	0.17	1.14	0.00	0.00
Niger	No	No	No	1.000	8	1.19	0.57	1.19	0.85	0.36	12.20	0.00	0.00
Nigeria	No	No	No	1.000	37	0.91	0.91	0.91	0.91	0.89	108.52	0.00	0.00
Rwanda	No	No	No	1.000	12	0.03	0.03	0.03	0.02	0.02	5.50	0.00	0.00
S. Tome & Principe	No	No	No	0.950	2	0.00	0.00	0.00	0.00	0.00	0.15	0.00	0.00
Senegal	No	No	No	1.000	10	0.20	0.20	0.20	0.20	0.17	8.24	0.00	0.00
Sierra Leone	No	No	No	1.000	4	0.07	0.07	0.07	0.07	0.07	4.41	0.00	0.00
South Africa	Yes	No	No	1.000	9	1.22	0.08	0.30	0.17	0.10	11.12	0.00	0.00
Swaziland	No	No	No	1.000	4	0.02	0.00	0.02	0.02	0.02	0.92	0.00	0.00
Togo	No	No	No	1.000	5	0.06	0.06	0.06	0.06	0.06	5.17	0.00	0.00
Uganda	No	No	Yes	1.000	56	0.21	0.21	0.21	0.20	0.19	24.44	0.00	0.00
U. Rep. of Tanzania	No	Yes	No	1.000	25	0.88	0.88	0.86	0.86	0.56	28.97	0.00	0.00
Zambia	No	No	No	1.000	9	0.75	0.75	0.75	0.75	0.75	11.25	0.00	0.00
Zimbabwe	No	Yes	Yes	1.000	10	0.39	0.39	0.30	0.39	0.35	10.68	0.00	0.00
						23.19	**16.01**	**18.54**	**17.13**	**13.17**	**477.15**	**1.19**	**48.31**
AMRO													
Argentina	Yes	Yes	No	0.000	23	2.78	0.01	0.14	0.14	0.10	0.00	2.19	0.00
Belize	No	Yes	Yes	0.000	6	0.02	0.02	0.02	0.02	0.02	0.00	0.23	0.00
Bolivia	Yes	Yes	Yes	0.046	9	1.08	0.27	0.76	0.76	0.38	0.00	1.71	0.53
Brazil	Yes	Yes	Yes	0.214	27	8.52	4.82	5.08	5.08	1.61	0.00	2.40	14.94
Colombia	Yes	Yes	Yes	0.384	32	1.14	0.95	0.90	0.90	0.48	0.00	0.87	12.37
Costa Rica	Yes	Yes	Yes	0.002	7	0.05	0.05	0.03	0.03	0.03	0.00	0.70	0.00
Dominican Republic	Yes	Yes	No	0.997	31	0.05	0.05	0.01	0.01	0.01	0.68	0.00	0.00
Ecuador	Yes	Yes	Yes	0.212	22	0.26	0.11	0.12	0.12	0.09	0.00	0.00	3.35
El Salvador	Yes	Yes	No	0.000	14	0.02	0.02	0.00	0.00	0.00	0.00	1.09	0.00

(continued)

Table 2 (*continued*)

Country[a]	ITHGs criteria[b]			Pfr[c]	Ad1[d]	Area[e]					Population[f]		
	Ad	Alt	Urb			Total	WHO 2002	ITHG	ALT-MASK	POP-MASK	Pf	Pv	Pf + Pv
French Guiana	No	No	No	0.696	2	0.08	0.02	0.04	0.04	0.00	0.00	0.00	0.05
Guatemala	Yes	Yes	Yes	0.051	22	0.11	0.04	0.07	0.07	0.06	0.00	0.00	3.80
Guyana	No	No	Yes	0.470	10	0.21	0.20	0.21	0.20	0.08	0.00	0.00	0.64
Haiti	Yes	Yes	No	1.000	9	0.03	0.03	0.01	0.01	0.01	3.15	0.00	0.00
Honduras	No	Yes	Yes	0.035	18	0.11	0.11	0.09	0.09	0.08	0.00	4.10	0.00
Mexico	Yes	Yes	No	0.004	32	1.96	0.82	0.41	0.41	0.40	0.00	15.79	0.00
Nicaragua	No	Yes	Yes	0.133	16	0.12	0.13	0.09	0.09	0.06	0.00	0.00	1.83
Panama	Yes	Yes	Yes	0.150	10	0.08	0.03	0.02	0.02	0.02	0.00	0.00	0.19
Paraguay	Yes	No	No	0.000	18	0.40	0.02	0.05	0.05	0.05	0.00	1.26	0.00
Peru	Yes	Yes	Yes	0.171	25	1.29	0.54	0.76	0.76	0.38	0.00	0.40	5.23
Suriname	Yes	Yes	Yes	0.787	10	0.15	0.11	0.13	0.13	0.02	0.01	0.00	0.02
Venezuela	Yes	Yes	Yes	0.088	23	0.92	0.20	0.38	0.38	0.15	0.00	1.04	3.02
						19.39	**8.56**	**9.33**	**9.32**	**4.03**	**3.83**	**31.78**	**45.98**
EMRO													
Afghanistan	No	Yes	No	0.385	32	0.64	0.51	0.38	0.38	0.37	1.77	0.14	12.44
Djibouti	No	No	No	0.980	5	0.02	0.02	0.02	0.02	0.02	0.22	0.00	0.00
Egypt	Yes	No	No		4	0.98	0.01	0.01	n/a	n/a	n/a	n/a	n/a
Iran	No	Yes	No	0.200	24	1.61	0.56	1.02	1.02	0.98	0.00	0.00	38.83
Iraq	Yes	Yes	Yes	0.000	19	0.44	0.29	0.10	0.10	0.10	0.00	7.95	0.00
Morocco	Yes	No	Yes	0.000	15	0.41	0.01	0.01	0.00	0.00	0.00	0.06	0.00
Oman	Yes	Yes	No	0.000	8	0.31	0.02	0.00	0.00	0.00	0.00	0.03	0.00
Pakistan	No	Yes	No	0.365	5	0.88	0.85	0.74	0.74	0.68	0.00	0.00	122.99
Saudi Arabia	Yes	No	Yes	0.471	14	1.93	0.18	0.86	0.85	0.31	1.20	0.13	12.21
Somalia	No	No	No	0.722	18	0.64	0.64	0.64	0.58	0.54	0.00	0.00	7.52
Sudan	No	No	No	0.851	18	2.49	1.86	2.49	2.13	1.57	26.31	0.00	2.83

Syrian Arab Rep.	No	Yes	No	0.540	12	0.19	0.03	0.04	0.04	0.04	0.00	0.00	4.02
Yemen	No	Yes	Yes	0.956	21	0.46	0.34	0.43	0.43	0.25	13.20	0.00	1.78
						10.99	**5.30**	**6.75**	**6.29**	**4.86**	**42.71**	**8.32**	**202.62**
EURO													
Armenia	Yes	No	No	0.000	11	0.03	0.01	0.00	0.00	0.00	0.00	0.27	0.00
Azerbaijan	No	No	No	0.000	63	0.08	0.02	0.00	0.00	0.00	0.00	0.17	0.00
Georgia	Yes	No	Yes	0.000	14	0.07	0.01	0.01	0.01	0.01	0.00	0.53	0.00
Kyrgyzstan	Yes	No	No	0.000	7	0.19	n/a	0.08	0.03	0.03	0.00	1.25	0.00
Tajikistan	Yes	No	No	0.003	6	0.14	0.01	0.11	0.04	0.04	0.00	3.23	0.00
Turkey	Yes	No	No	0.000	81	0.78	0.13	0.19	0.13	0.13	0.00	13.64	0.00
Turkmenistan	Yes	No	No	0.000	6	0.46	0.02	0.09	0.09	0.09	0.00	1.16	0.00
Uzbekistan	No	No	No	n/a	n/a	n/a	n/a	n/a	n/a	n/a	n/a	n/a	n/a
						1.76	**0.19**	**0.48**	**0.29**	**0.30**	**0.00**	**20.26**	**0.00**
SEARO													
Bangladesh	No	No	Yes	0.407	6	0.14	0.14	0.14	0.14	0.13	0.00	0.00	124.61
Bhutan	Yes	Yes	No	0.387	18	0.04	0.02	0.01	0.01	0.01	0.00	0.00	0.91
India	Yes	Yes	Yes	0.278	34	3.09	2.93	2.94	2.94	2.86	0.00	88.26	857.93
Indonesia	No	Yes	Yes	0.385	27	1.90	1.71	1.75	1.75	1.44	0.00	0.00	151.08
Korea, D. P. R.	No	No	No		12	0.12	0.00	n/a	n/a	n/a	n/a	n/a	n/a
Myanmar	Yes	Yes	Yes	0.788	16	0.67	0.66	0.53	0.53	0.47	0.00	0.00	38.36
Nepal	Yes	Yes	Yes	0.090	5	0.15	0.08	0.06	0.06	0.06	0.00	7.66	9.64
Sri Lanka	Yes	Yes	Yes	0.222	9	0.07	0.06	0.06	0.06	0.06	0.00	0.00	10.32
Thailand	No	No	Yes	0.469	76	0.52	0.35	0.51	0.49	0.50	0.00	0.00	58.08
Timor-Leste	No	No	No	0.534	14	0.01	0.01	0.01	0.01	0.01	0.00	0.00	0.62
						6.70	**5.97**	**6.01**	**5.99**	**5.54**	**0.00**	**95.93**	**1 251.55**
WPRO													
Cambodia	No	No	Yes	0.870	24	0.18	0.18	0.18	0.18	0.16	2.03	0.00	10.79
China	Yes	Yes	Yes	0.100	32	9.44	1.24	2.36	2.36	2.32	0.00	453.90	287.79
Lao P. D. R.	No	No	Yes	0.960	18	0.23	0.23	0.23	0.23	0.22	3.56	0.00	2.30
Malaysia	No	Yes	No	0.565	14	0.33	0.33	0.27	0.27	0.26	0.00	0.00	9.69
Papua New Guinea	No	Yes	No	0.727	20	0.46	0.38	0.41	0.41	0.41	0.47	0.00	3.69
Philippines	Yes	Yes	Yes	0.605	16	0.30	0.30	0.21	0.21	0.21	0.00	0.00	46.95

(continued)

Table 2 (*continued*)

Country[a]	ITHGs criteria[b]			Pfr[c]	Ad1[d]	Area[e]					Population[f]		
	Ad	Alt	Urb			Total	WHO 2002	ITHG	ALT-MASK	POP-MASK	Pf	Pv	Pf + Pv
Republic of Korea	No	No	No	0.000	15	0.10	0.00	0.03	0.03	0.02	0.00	4.86	0.00
Solomon Islands	Yes	Yes	No	0.646	9	0.03	0.03	0.02	0.02	0.02	0.00	0.00	0.31
Vanuatu	No	No	Yes	0.525	6	0.01	0.01	0.01	0.01	0.01	0.00	0.00	0.18
Viet Nam	No	Yes	Yes	0.590	61	0.33	0.33	0.32	0.32	0.32	0.00	0.00	69.76
						11.42	**3.03**	**4.05**	**4.04**	**3.95**	**6.06**	**458.76**	**431.47**
Global						**73.44**	**39.06**	**45.15**	**43.06**	**31.85**	**529.75**	**616.25**	**1 979.93**

[a]The data are presented alphabetically by WHO regional office and country name with totals shown in bold at the end of each section and at the end of the table for the World.

[b]Refers to presence or absence of any of the main three mapping criteria used (Ad, administrative, Alt, altitude and Urb, urban).

[c]The mean *P. falciparum* ratio as used in Figures 1 and 2A–R (see colour plate section).

[d]The number of administrative one level divisions per country.

[e]Area totals are presented for each country as per the WHO 2002 boundaries, the ITHGs and the ITHGs with the altitudinal mask (ALT-MASK) and the population mask (POP-MASK) exclusions in millions of km^2.

[f]Populations in 2005 living predominantly under *P. falciparum* (Pf) and *P. vivax* (Pv) and mixed (Pf + Pv) risk are also presented in millions. Populations were projected to 2005 from GRUMP at $\sim 1 \times 1\ km^2$ spatial resolution (CIESIN/IFPRI/WB/CIAT, 2004).

developed by Kiszewski *et al.* (2004) and mapped the altitudinal limits using the same 1×1 km spatial resolution DEM (Hastings and Dunbar, 1998).

3.2. Climate Suitability Mask

Since the information in ITHG reports for African countries was so sparse (Table 2), we used the MARA/ARMA climate suitability model to further adjust the malaria transmission limits on this continent (Craig *et al.,* 1999). The MARA model describes climatic conditions (or fuzzy climate suitability, FCS) that range from unsuitable (0) to completely suitable (1) for stable *P. falciparum* transmission. Since the MARA model includes climatology-derived temperature limits, this can also be used as a surrogate altitudinal mask. To do this we have assumed that FCS values of zero are incompatible with malaria risk, as supported by a recent analysis of parasite prevalence and FCS values in East Africa (Omumbo *et al.,* 2004). The spatial resolution of the MARA model was too coarse to apply exclusions to the territories of Cape Verde, Comoros, Mauritius, Mayotte and Sao Tome and Principe. For Comoros, we assumed the same altitudinal limit as that of Ethiopia (2000 m), based on their similar dominant vectors. Altitude masks were unnecessary for the remaining low-lying island states as they have no areas above 1800 m, which is at or below the lowest altitude threshold reported elsewhere in Africa for the same dominant vector species compositions (Kiszewski *et al.,* 2004).

3.3. Population Density Mask

Two population density extremes were applied to refine the spatial limits of transmission further. First masked, are those areas where environmental conditions may support malaria, but where there are too few people, so the human malaria parasites cannot complete their life-cycle and do not pose any public health concern; these include, for example, dense forests and true deserts. We therefore used the 1×1 km resolution GRUMP population density surface, which allows for equal area corrections (CIESIN/IFPRI/WB/CIAT, 2004) (see below), to exclude all areas in the remaining distribution with a population density of <1 person per km^2.

Second are those areas where population density is so high that conditions become unsuitable for transmission through the process of urbanisation (Hay *et al.,* 2000; Robert *et al.,* 2003; Omumbo *et al.,* 2005). Urbanisation has been shown to reduce malaria transmission on average by an order of magnitude across Africa (Hay *et al.,* 2005). There is no reason to think that the same fundamental processes of reduced (i) Anopheline diversity; (ii) biting rates; (iii) sporozoite rates; (iv) transmission; and thus (v) human malaria infections in urban versus rural areas do not apply globally. This is certainly true for the 70 cities cited as malaria free in ITHGs. Moreover, the 24 cities that report urban malaria often refer to infection risk on their peripheries. A potential confounder to this global trend could be the presence of the urban malaria vector *Anopheles stephensi* in the Indian subcontinent (Rowland *et al.,* 2002). A detailed look at the evidence indicates that vector densities and sporozoite rates show similar declines from rural, through peri-urban, to urban localities in Delhi (Sharma *et al.,* 1993), Gurgaon (Sharma, 1995) and Karachi (Nalin *et al.,* 1985) to those in Africa.

We projected population counts for the year 2000 (CIESIN/IFPRI/WB/CIAT, 2004) to 2005 by applying national, medium variant, intercensal growth rates by country (UNPD, 2004) before deriving contemporary population densities using an area-by-pixel surface (CIESIN/IFPRI/WB/CIAT, 2004). We geo-referenced (Microsoft Corporation, 2005; The Getty Research Institute, 2005; University of California, 2005) cities with populations equal to or greater than one million people (UNSD, 2001) in MECs and identified their urban extents in GRUMP ($n = 204$). We then investigated population density frequency statistics within each of these urban extents. Significant regional differences in population density were apparent, so a conservative threshold of intensity of urbanisation was used, corresponding to the median of population density means associated with the urban extents by region. The medians were 4218, 1533 and 2513 persons per km^2 for Africa, the Americas and Asia-Europe, respectively. The same median as Asia-Europe was used in Oceania MECs, where no cities of $\geqslant 1$ million people currently exist. We thus masked as no malaria risk all those areas that could be unambiguously classified as intensely urban relative to their corresponding region.

The resulting map, after applying the three masks described here, is shown in Figure 1 (Figure 1 is Plate 5.1 in the Separate Color Plate Section).

4. DISTINGUISHING *P. FALCIPARUM* AND *P. VIVAX* RISK

Global information of the distribution of *P. falciparum* and *P. vivax* is not comprehensively detailed in the ITHGs. Sub-national statistics are available from other sources (FDRE, 2002; PAHO, 2003; Hay *et al.,* 2004a; Sintasath, 2004; Kolaczinski *et al.,* 2005). For countries or areas within countries where no sub-national data were available, national average data were used from ITHGs or other sources (PAHO, 2003; Korenromp, 2005) (Table 2). Specific data for Mayotte were not available, so we assumed 100% *P. falciparum* risk based on ITHG descriptions. We defined areas reporting $\geqslant$95% *P. falciparum* cases as predominantly *P. falciparum* and those $\leqslant$5% *P. falciparum* as predominantly *P. vivax* endemic. The remainder of the distribution is of mixed (*P. falciparum* and *P. vivax*) endemicity. *P. ovale* and *P. malariae* were not considered here, as these are relatively rare malaria parasites and infrequently reported in national statistics. These divisions are shown globally (Figure 1) and in greater detail by WHO geographic region (Figures 2A–R) (Figures 2A–R are Plate 5.2A–R in the Separate Color Plate section). We now discuss the implications of the biological limits to transmission and the parasite species distributions by region and highlight some known anomalies.

5. REGIONAL ANALYSIS

Before the biological exclusions were applied, the global malaria area at risk of malaria was 45.15 million km^2, considerably higher than the 39.06 million km^2 derived from the WHO 2002 boundaries (Figure 3A, Table 2). The altitude and climate mask reduced this slightly to 43.06 million km^2. The largest percentage reductions were in countries of the

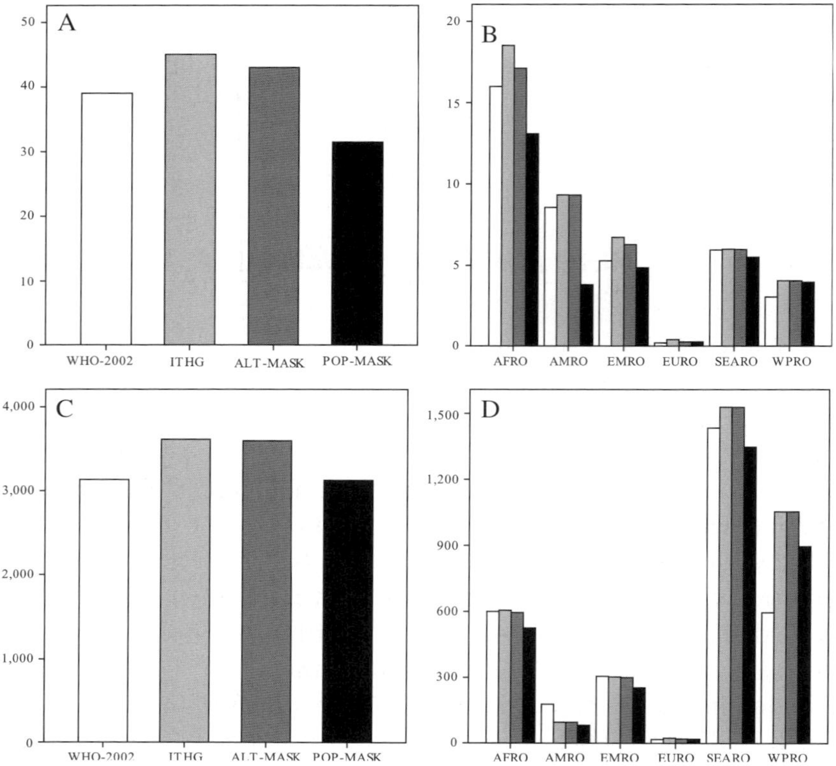

Figure 3 A–D Bar charts showing area (A and B) and population at risk (PAR) (C and D) of malaria according to the WHO 2002 limits (white), ITHGs (grey), and the progressive implementation of our altitudinal (dark grey) population (black) masks, globally and stratified by WHO regional office. The scales of area (km^2) and population are in millions. Figures were calculated using ArcView 3.2 (ESRI, Redlands, California, USA). Area extractions were undertaken on an equal area projection.

European Regional Office (EURO) (39.9%) and African Regional Office (AFRO) (7.5%) due to the mountainous areas and large expanses of zero climate suitability in the northern Sahel, respectively (Figure 3B; Table 2). The population mask had a very considerable effect in reducing the global area at risk further, to 31.85 million km^2 (43.4% of the MECs total land area) (Figures 1 and 3A, Table 2). This incremental reduction was most noticeable in countries of the American Regional Office (AMRO) (56.8%), AFRO (23.1%) and the Eastern

Mediterranean Regional Office (EMRO) (22.8%), mainly due to the exclusion of large, relatively unpopulated areas of forest and desert land. In the final coverage, the rank of regions by area at risk was AFRO (13.17 million km^2, 56.8% of the regional office MEC land area), the South East Asian Regional Office (SEARO) (5.54 million km^2, 82.7% of the regional office MEC land area), EMRO (4.86 million km^2, 44.2% of the regional office MEC land area), AMRO (4.03 million km^2, 20.8% of the regional office MEC land area), the Western Pacific Regional Office (WPRO) (3.95 km^2, 34.6% of the region's land area) and EURO (0.30 million km^2, 17.0% of the regional office MEC land area) (Figure 3B, Table 2). The global malarious area was thus progressively reduced by iterations of biological exclusions with our final distribution model 18.5% smaller (7.21 million km^2) in area than that which would be derived from using the WHO 2002 limits (WHO, 2003b) (Figure 3A).

The global heterogeneity in the distribution of human population (Cohen and Small, 1998) generated some rather striking contrasts in population at risk (PAR) extractions. Global PAR derived from the WHO 2002 boundaries is 3.133 billion persons, much less than the 3.616 billion suggested by the crude ITHG limits (Figure 3C). Our altitude and climate mask marginally reduced this figure further to 3.596 billion people. After implementing the population mask, the final global PAR is estimated at 3.126 billion. This was almost entirely due to the removal of the highly populous urban areas of low to zero malarial risk. This final PAR estimate is very similar to that extracted from the WHO 2002 boundaries (only 0.22% lower) but critically different in regional distribution of PAR estimates (Figure 3D, Table 2). Extractions from the final coverage show the malarious regions by rank PAR were SEARO (1.347 billion persons), WPRO (0.896 billion persons), AFRO (0.527 billion persons), EMRO (0.254 billion persons), AMRO (0.082 billion persons) and EURO (0.020 billion persons). These changes in rank show the important contribution that the very large population concentrations in SEARO and WPRO make to the PAR estimates. It is essential to note at this stage that all risk is not equal, and a more detailed discussion of the *P. falciparum* and *P. vivax* partitioning by region follows.

In AFRO (Figures 2A–C), we can see how the various iterations have refined the transmission limits for malaria infection particularly to the arid north and south of the continent, although significant areas have been excised from the mid-latitude tropical forest range. A small focus of malaria risk emerged in Illizi department, Algeria, after our climate masking, which we assumed to correspond to the focus reported by ITHGs in Ihrir (CDC, 2003). Hence, our final map allows for a small area of malaria risk in Algeria. The distribution of almost exclusively *P. falciparum* on the African continent is remarkable, and fits with received wisdom about the evolution to fixation of the Duffy negativity blood group allele, making these populations refractory to *P. vivax* infection (Livingstone, 1984; Zimmerman, 2004). Information on parasite species ratio is poor in AFRO, however, and the stark transitions between southern Ethiopia and neighbouring countries (as well as Sudan and Somalia from EMRO (Figures 2G–I)), suggest that a more detailed investigation of the distribution of *P. vivax* is warranted in these areas on the latitidudinal margins of risk.

In AMRO (Figures 2D–F), by contrast, *P. vivax* is by far the dominant parasite (Roberts *et al.,* 2002). The most obvious exclusions are those of the low-populated tropical forests. The regional distribution of risk compares favourably with other sources (Roberts *et al.,* 2002; PAHO, 2003), but the biological exclusions failed to capture completely the situation in Argentina and Paraguay, where incidence of malaria is reportedly low (PAHO, 2003), yet our map indicates non-negligible areas and PARs for these countries (Table 2).

EMRO (Figures 2G–I) is a truly heterogeneous region encompassing Morocco, Somalia and Sudan, as well as the Middle Eastern block through to Pakistan (Beljaev, 2002). Summary is therefore difficult but *P. falciparum* is highest in the south-west with risk declining as one travels north and east. Areas of "very limited risk" reported by ITHGs in Egypt (CDC, 2003; WHO, 2005) are not captured by our climate mask and were excluded. Our final map, therefore, does not allow for malaria risk in Egypt, where no cases have been reported since 1998 (WHO, 2005) despite high malaria risk being documented in Fayoum governorate (Hassan *et al.,* 2003).

The EURO MECs (Figures 2J–L) are essentially Turkey and the southern states newly independent from the former Soviet Union (Sabitinelli, 2002). Their small global fraction of malaria PAR is largely due to *P. vivax* infections.

SEARO (Figures 2M–O) is best characterised as endemic for predominantly *P. vivax*, but with significant foci of *P. falciparum* transmission (Sharma, 2002). A known anomaly in our coverage is evident from the fact that malaria has long been endemic in the Korean peninsula (Feighner *et al.,* 1998). After a period of decline following the Korean War, malaria re-emerged in the Demilitarised Zone, probably due to an epidemic in North Korea since 1993 (Lee *et al.,* 2002). ITHGs, however, are ambiguous about the areas of risk in this country, making them impossible to map.

In WPRO (Figures 2P–R), *P. vivax* dominance gives way to *P. falciparum* as one moves south and east (Schapira, 2002). China dominates the PAR extractions for this region and introduces a warming. An overestimation in area, and hence PAR, by our map is possible in this country, where the average spatial resolution of heavily populated first administrative units is poor and ITHG subnational descriptions fail to capture lower-level administrative detail.

6. DISCUSSION

The most widely cited map of the current global malaria distribution is WHO 2002 (WHO, 2003b). The information source is cited as (WHO, 2003a) but there are important discrepancies between the map (WHO, 2003b) and suggested source data. For example, the spatial limits according to ITHGs are 6.09 million km^2 larger than the WHO 2002 boundary (Table 2). These geographical inconsistencies and a lack of detailed information of their origins make them of unknown fidelity in risk mapping. In contrast, the methods presented here are implemented with public-domain data and are hence easily reproducible, the maps used to generate individual country data are presented in detail, and the PAR numbers are made available for scrutiny. In addition, and for the first time, parasite species' distributions are defined globally (including the often neglected *P. vivax*) and are suggested as a more comprehensive map against which to measure PAR of malaria in 2005.

Our map comes with obvious caveats, since we have implemented crude rules at the global scale. Noticeable anomalies (e.g. Argentina,

Egypt, Morocco, North Korea, Paraguay, and possibly China) have been highlighted and it is clear that, at a sub-national level, discrepancies will be found as, because risk is not a static phenomenon, and global information is incomplete. We therefore propose these maps as a working template for future refinement.

Reconnaissance of global malaria data is required to refine distributions at the margins and this can only be done with malaria-risk data at higher spatial resolution. These data probably exist at the country level and need to be collated at the global scale within a GIS framework. In terms of PAR, the top ten countries globally are India, China, Indonesia, Bangladesh, Pakistan, Nigeria, Vietnam, Thailand, Democratic Republic of the Congo and the Philippines (Table 2). Pragmatically, Error in PAR estimates globally would be reduced most substantially by focussing on improving distribution limits and *P. falciparum* ratios for these territories.

Modelling risks within these margins is also critical and is the subject of planned future work. Local endemicity within these transmission limits will be substantially mediated by the influences of land-use changes such as deforestation (Walsh *et al.,* 1993) and urbanisation (Hay *et al.,* 2005), prevalence of other conditions such as HIV/AIDS, tuberculosis and malnutrition (Bates *et al.,* 2004), as well as local control and intervention efforts (Korenromp, 2005). Our immediate goal is to validate historical malaria endemicity maps (Lysenko and Semashko, 1968) using empirical data and to generate plausible scenarios of some of the above-mentioned confounding influences to satisfactorily adjust endemicity to modern day risk. This will fill pressing needs to estimate global incidence of malaria and commodity burdens. Our longer-term goal is to construct an independent global map of modern day endemicity within the boundaries we have defined and will continue to refine.

7. CONCLUSIONS

While there will remain uncertainties about the precise global extent of malaria, we have reduced these as far as possible without systematic country-specific surveys and information. Urgent attention is required to reduce the uncertainty surrounding factors that affect the spatial extent of malaria on a global scale and these would sensibly target the most populous malarious nations. This will allow the international community to better define the needs for therapeutic and disease prevention commodities so that well-intentioned governments and UN agencies can make requests for sufficient financial resources.

ACKNOWLEDGEMENTS

We thank Simon Brooker, Alastair Graham, Sarah Hay, Eline Korenromp, Abdisalan Noor, Sarah Randolph and Andy Tatem for comments on earlier drafts of the manuscript. The WHO/Roll Back Malaria Department Monitoring and Evaluation team and staff from the WHO regional office provided passive case detection data from which some *P. falciparum* and *P. vivax* ratios were derived. Lorena Lucioparedes is thanked for geo-referencing UN cities. CAG is partially funded by the Fundación para la Ciencia y Tecnología (FUNDACYT), Quito, Ecuador. RWS is a Wellcome Trust Senior Research Fellow (#058992) and acknowledges the support of the Kenyan Medical Research Institute (KEMRI). SIH is funded by a Research Career Development Fellowship from the Wellcome Trust (#069045). This paper is published with the permission of the director of KEMRI.

REFERENCES

Bates, I., Fenton, C., Gruber, J., Lalloo, D., Medina Lara, A., Squire, S.B., Theobald, S., Thomson, R. and Tolhurst, R. (2004). Vulnerability to malaria, tuberculosis, and HIV/AIDS infection and disease. Part 1: determinants operating at individual and household level. *Lancet Infectious Diseases* **4**, 267–277.

Beljaev, A.E. (2002). Determinants of malaria in the Middle East and North Africa. In: *The Contextual Determinants of Malaria* (E.A. Casman and H. Dowlatabadi, eds), pp. 137–166. Washington, DC: Resources for the Future Press.

Boyd, M.F. (1949). *Malariology*. Philadelphia: W.B. Saunders.

Carter, R. and Mendis, K.N. (2002). Evolutionary and historical aspects of the burden of malaria. *Clinical Microbiology Reviews* **15**, 564–594.

CDC. (2003). *Health information for international travel 2003–2004*. Atlanta: Centers for Disease Control and Prevention.

CIESIN/IFPRI/WB/CIAT. (2004). *Global Rural Urban Mapping Project (GRUMP): Gridded Population of the World, version 3, with urban reallocation (GPW-UR)*. Center for International Earth Science Information Network, Columbia University; International Food Policy Research Institute; The World Bank; and Centro Internacional de Agricultura Tropical. Palisades, NY. Socioeconomic Data and Applications center (SEDAC), Columbia University. Available at http://sedac.ciesin.columbia.edu/gpw. (downloaded October 2004).

Cohen, J.E. and Small, C. (1998). Hypsographic demography: the distribution of human population by altitude. *Proceedings of the National Academy of Sciences of the United States of America* **95**, 14009–14014.

Cox, J., Craig, M., Le Sueur, D. and Sharp, B. (1999). *Mapping malaria risk in the highlands of Africa. Technical report*. Durban: MARA/HIMAL.

Craig, M.H., Snow, R.W. and le Sueur, D. (1999). A climate-based distribution model of malaria transmission in sub-Saharan Africa. *Parasitology Today* **15**, 105–111.

Dutta, H.M. and Dutt, A.K. (1978). Malarial ecology: a global perspective. *Social Science & Medicine* **12**, 69–84.

FDRE. (2002). *Health and Health Related Indicators*. Addis Ababa, Ethiopia: Planning and Programming Department, Ministry of Health, Federal Democratic Republic of Ethiopia.

Feighner, B.H., Pak, S.I., Novakoski, W.L., Kelsey, L.L. and Strickman, D. (1998). Reemergence of *Plasmodium vivax* malaria in the republic of Korea. *Emerging Infectious Diseases* **4**, 295–297.

Hassan, A.N., Kenawy, M.A., Kamal, H., Abdel Sattar, A.A. and Sowilem, M.M. (2003). GIS-based prediction of malaria risk in Egypt. *Eastern Mediterranean Health Journal* **9**, 548–558.

Hastings, D.A. and Dunbar, P.K. (1998). Development and assessment of the global land 1 km base elevation digital elevation model (GLOBE). *International Archives of Photogrammetry and Remote Sensing* **32**, 218–221.

Hay, S.I., Guerra, C.A. and Snow, R.W. (2004b). *Determination of populations at malaria risk. Report on Agreement to Perform Work, M50/370/19*. Geneva: World Health Organization.

Hay, S.I., Guerra, C.A., Tatem, A.J., Noor, A.M. and Snow, R.W. (2004a). The global distribution and population at risk of malaria: past, present, and future. *Lancet Infectious Diseases* **4**, 327–336.

Hay, S.I., Guerra, C.A., Tatem, A.J., Atkinson, P.M. and Snow, R.W. (2005). Urbanization, malaria transmission and disease burden in Africa. *Nature Reviews Microbiology* **3**, 81–90.

Hay, S.I., Rogers, D.J., Toomer, J.F. and Snow, R.W. (2000). Annual *Plasmodium falciparum* entomological inoculation rate (EIR) across Africa: literature survey, internet access and review. *Transactions of the Royal Society of Tropical Medicine and Hygiene* **94**, 113–127.

Henderson-Sellers, A. and Robinson, P.J. (1991). *Contemporary Climatology*. London, UK: Longman Scientific and Technical.

IAMAT. (2004). *World Malaria Risk Chart*. Guelph, Canada: International Association for Medical Assistance to Travellers.

Kiszewski, A., Mellinger, A., Spielman, A., Malaney, P., Sachs, S.E. and Sachs, J. (2004). A global index representing the stability of malaria transmission. *American Journal of Tropical Medicine and Hygiene* **70**, 486–498.

Kolaczinski, J., Graham, K., Fahim, A., Brooker, S. and Rowland, M. (2005). Malaria control in Afghanistan: progress and challenges. *Lancet* **365**, 1506–1512.

Korenromp, E. (2005). Malaria incidence estimates at country level for the year 2004 – proposed estimates and draft report. Geneva, Switzerland: World Health Organization (WHO).

Lee, J.S., Lee, W.J., Cho, S.H. and Ree, H.I. (2002). Outbreak of vivax malaria in areas adjacent to the demilitarized zone, South Korea, 1998. *American Journal of Tropical Medicine and Hygiene* **66**, 13–17.

Livingstone, F.B. (1984). The Duffy blood-groups, vivax malaria, and malaria selection in human-populations—a review. *Human Biology* **56**, 413–425.

Lysenko, A.Y. and Semashko, I.N. (1968). Geography of malaria: a medico-geographic profile of an ancient disease [in Russian]. In: *Medicinskaja Geografija* (A.W. Lebedew, ed.), pp. 25–146. Moscow: Academy of Sciences.

Microsoft Corporation. (2005). *Encarta 2005 Premium Suite*. Seattle: Microsoft Corporation.

Nalin, D.R., Mahood, F., Rathor, H., Muttalib, A., Sakai, R., Chowdhry, M.A., Safdar, G., ul, Haq, I., Munir, M., Suleiman, M., Bahir, M. and Mujtaba, S.M. (1985). A point survey of periurban and urban malaria in Karachi. *Journal of Tropical Medicine and Hygiene* **88**, 7–15.

Omumbo, J.A., Guerra, C.A., Hay, S.I. and Snow, R.W. (2005). The influence of urbanisation on measures of *Plasmodium falciparum* infection prevalence in East Africa. *Acta Tropica* **93**, 11–21.

Omumbo, J.A., Hay, S.I., Guerra, C.A. and Snow, R.W. (2004). The relationship between the *Plasmodium falciparum* parasite ratio in childhood and climate estimates of malaria transmission in Kenya. *Malaria Journal* **3**, 17.

PAHO. (2003). *Report on the status of malaria programs in the Americas (based on 2002 data) CD44/INF/3—44th Directing Council/55th Session of the Regional Committee*. Washinghton, DC: Pan American Health Organization.

Pampana, E.J. and Russell, P.F. (1955). *Le Paludisme: problème mondial*, Vol. 69. Geneva: World Health Organization pp. 317–321.

Robert, V., Macintyre, K., Keating, J., Trape, J.F., Duchemin, J.B., Warren, M. and Beier, J.C. (2003). Malaria transmission in urban sub-Saharan Africa. *American Journal of Tropical Medicine and Hygiene* **68**, 169–176.

Roberts, D.R., Masuoka, P. and Au, A.Y. (2002). Determinants of malaria in the Americas. In: *The Contextual Determinants of Malaria* (E.A. Casman and H. Dowlatabadi, eds), pp. 35–58. Washington, DC: Resources for the Future Press.

Rogers, D.J. and Randolph, S.E. (2000). The global spread of malaria in a future, warmer world. *Science* **289**, 1763–1766.

Rowland, M., Mohammed, N., Rehman, H., Hewitt, S., Mendis, C., Ahmad, M., Kamal, M. and Wirtz, R. (2002). Anopheline vectors and malaria transmission in eastern Afghanistan. *Transactions of the Royal Society of Tropical Medicine and Hygiene* **96**, 620–626.

Sabitinelli, G. (2002). Determinants of malaria in WHO European region. In: *The Contextual Determinants of Malaria* (E.A. Casman and H. Dowlatabadi, eds), pp. 66–92. Washington, DC: Resources for the Future Press.

Sachs, J.D. (2005). Achieving the millennium development goals—the case of malaria. *The New England Journal of Medicine* **352**, 115–117.

Schapira, A. (2002). Determinants of malaria in Oceania and East Asia. In: *The Contextual Determinants of Malaria* (E.A. Casman and H. Dowlatabadi, eds), pp. 93–109. Washington, DC: Resources for the Future Press.

Sharma, R.S. (1995). Urban malaria and its vectors *Anopheles stephensi* and *Anopheles culicifacies* (Diptera: Culicidae) in Gurgaon, India. *Southeast Asian Journal of Tropical Medicine and Public Health* **26**, 172–176.

Sharma, S.N., Subbarao, S.K., Choudhury, D.S. and Pandey, K.C. (1993). Role of *An. culicifacies* and *An. stephensi* in malaria transmission in urban Delhi. *Indian Journal of Malariology* **30**, 155–168.

Sharma, V.P. (2002). Determinants of malaria in South Asia. In: *The Contextual Determinants of Malaria* (E.A. Casman and H. Dowlatabadi, eds), pp. 110–132. Washington, DC: Resources for the Future Press.

Sintasath, D. (2004). National malaria prevalence survey (2000–2001). Washington DC: Environmental Health Project, Ministry of Health of Eritrea and the Office of Health, Infectious Diseases and Nutrition, Bureau of Global Health, U.S. Agency for International Development.

Snow, R.W. (2004). The invisible victims. *Nature* **430**, 934–935.

Snow, R.W., Guerra, C.A., Noor, A.M., Myint, H.Y. and Hay, S.I. (2005). The global distribution of clinical episodes of *Plasmodium falciparum* malaria. *Nature* **434**, 214–217.

Snow, R.W., Marsh, K. and le Sueur, D. (1996). The need for maps of transmission intensity to guide malaria control in Africa. *Parasitology Today* **12**, 455–457.

The Getty Research Institute. (2005). *Getty Thesaurus of Geographic Names Online*. Los Angeles, CA: The Getty Research Institute.

University of California. (2005). *Alexandria Digital Library Project*. Santa Barbara: University of California.

UNPD. (2004). *World Population Prospects: population database*. New York: United Nations Population Division, http://esa.un.org/unpp/.

UNSD. (2001). *Demographic Yearbook*. New York: United Nations Statistics Division.

Walsh, J.F., Molyneux, D.H. and Birley, M.H. (1993). Deforestation: effects on vector-borne disease. *Parasitology* **106**, S55–S75.

WHO. (1966). Malaria eradication in 1965. *World Health Organization Chronicle* **20**, 286–300.

WHO. (2003a). *International Travel and Health: Situation as on 1 January 2003*. Geneva: World Health Organization.

WHO. (2003b). *Worldwide Malaria Distribution in 2002*. Geneva: Public Health Mapping Group, World Health Organizaion.

WHO. (2005). *International Travel and Health: Situation as on 1 January 2005*. Geneva: World Health Organization.

Zimmerman, P.A. (2004). The enigma of *Plasmodium vivax* malaria and erythrocyte Duffy negativity. In: *Infectious Disease and Host-Pathogen Evolution* (K.R. Dronamraju, ed.), pp. 141–172. Cambridge: Cambridge University Press.

The Global Distribution of Yellow Fever and Dengue

D.J. Rogers[1], A.J. Wilson[1], S.I. Hay[1,2] and A.J. Graham[1]

[1]*TALA Research Group, Tinbergen Building, Department of Zoology, University of Oxford, South Parks Road, Oxford OX1 3PS, UK*
[2]*Malaria Public Health & Epidemiology Group, Centre for Geographic Medicine, KEMRI, PO Box 43640, 00100 Nairobi GPO, Kenya*

Abstract 182
1. Introduction 182
2. The Pathogens 183
   2.1. Yellow Fever 183
   2.2. Dengue 189
3. Materials and Methods 193
   3.1. Existing Maps 193
   3.2. Archive and Literature Searches 194
   3.3. Yellow Fever Data 196
   3.4. Dengue Fever Data 197
   3.5. Environmental Data from Satellites 198
   3.6. The Modelling Approach 199
4. Results 200
   4.1. Risk Maps for Yellow Fever and Dengue 200
   4.2. Overall Model Accuracy 201
   4.3. Importance of Individual Variables 202
   4.4. Variability of Bootstrap Results 203
   4.5. Populations at Risk 205
5. Discussion 208
6. Conclusion 209
Acknowledgements 210
References 211

ADVANCES IN PARASITOLOGY VOL 62
ISSN: 0065-308X $35.00
DOI: 10.1016/S0065-308X(05)62006-4

ABSTRACT

Yellow fever has been subjected to partial control for decades, but there are signs that case numbers are now increasing globally, with the risk of local epidemic outbreaks. Dengue case numbers have also increased dramatically during the past 40 years and different serotypes have invaded new geographical areas. Despite the temporal changes in these closely related diseases, and their enormous public health impact, few attempts have been made to collect a comprehensive dataset of their spatial and temporal distributions. For this review, records of the occurrence of both diseases during the 20th century have been collected together and are used to define their climatic limits using remotely sensed satellite data within a discriminant analytical model framework. The resulting risk maps for these two diseases identify their different environmental requirements, and throw some light on their potential for co-occurrence in Africa and South East Asia.

1. INTRODUCTION

Yellow fever virus is the type virus of the family *Flaviviridae* (from the Latin *flavus*, meaning yellow), and is thought to have originated in West Africa (Cliff *et al.*, 2004). It was one of the earliest viruses to be identified and linked to human disease. Although substantial variation exists among strains, they can be grouped into monophyletic geographical variants, called topotypes. African isolates are usually grouped into two topotypes, associated with East and West Africa (Deubel *et al.*, 1986; WHO, 2001), although some studies have argued for up to five (Mutebi *et al.*, 2001). Two more have been identified from South America, although one has not been recovered since 1974, suggesting that it may be extinct in the wild. There is no evidence for a difference in virulence between the topotypes (WHO, 2001).

Dengue virus is also a member of the family Flaviviridae, and is closely related to yellow fever virus. There are four serotypically distinct types of dengue virus (DEN-1, DEN-2, DEN-3 and DEN-4). Although recombination occurs between strains of the same serotype, no interserotype recombination has been observed (Gubler and Kuno, 1997).

2. THE PATHOGENS

2.1. Yellow Fever

2.1.1. History

Yellow fever virus was probably introduced into the New World via ships carrying slaves from West Africa. The first recorded epidemics of yellow fever occurred in Mexico and Guadeloupe in 1648. Throughout the 18th and 19th centuries, regular and devastating epidemics of yellow fever occurred across the Caribbean, Central and South America, the southern United States and Europe. The impact of yellow fever prompted some American colonies to refuse entry to ships from infected areas (Pearson and Miles, 1980), and later led to the establishment of formal quarantine arrangements (of the sort already in place in Europe as a consequence of the plague). Despite these measures, urban epidemics continued, and in 1793, approximately one in ten of the inhabitants of Philadelphia, USA (then home of the federal government) died during an epidemic of yellow fever. Mortality from yellow fever and malaria caused the failure of the French Panama Canal project in the 1880s and 1890s (McCullough, 1977; Gallup and Sachs, 2000). The Yellow Fever Commission, founded as a consequence of excessive disease mortality during the Spanish–American War (1898), concluded that the best way to control the disease was to control the mosquito. William Gorgas successfully eradicated yellow fever from Havana by destroying larval breeding sites and this strategy of source reduction was then successfully used to reduce disease problems and thus finally permit the construction of the Panama Canal in 1904. Success was due largely to a top-down, military approach involving strict supervision and discipline (Gorgas, 1915). In 1946, an intensive *Aedes aegypti* eradication campaign was initiated in the Americas, which succeeded in reducing vector populations to undetectable levels throughout most of its range.

The production of an effective vaccine in the 1930s led to a change of emphasis from vector control to vaccination for the control of yellow fever. Vaccination campaigns almost eliminated urban yellow fever but incomplete coverage, as with incomplete anti-vectorial measures previously, meant the disease persisted, and outbreaks in remote forest areas continued (Barros and Boecken, 1996; Vasconcelos *et al.*, 2001).

2.1.2. Symptoms

The symptoms of yellow fever are very variable and depend on the severity of infection. Although a small proportion of infections are asymptomatic, victims typically develop a number of influenza-like symptoms including fever, joint pains and headache between three and six days after infection. Three or four days after the appearance of these symptoms they may disappear, and in most cases convalescence begins (Monath, 2001). In other cases, after a remission of 6–12 hours, febrile symptoms return accompanied by nausea, vomiting, epigastric pain, renal failure, jaundice (hence the common name for the disease) and haemorrhaging (Monath, 2001; WHO, 2001). Half of patients at this stage die within 10–14 days, but the remainder recover without significant organ damage. Immunity is lifelong following infection.

2.1.3. Epidemiology

Yellow fever virus circulates in both urban and sylvatic settings (Figure 1), involving several mosquito and vertebrate species. In the sylvatic cycle, mosquitoes such as *Aedes africanus* (in Africa) or *Haemagogus* species (in the Americas) act as the main vectors and monkeys as the primary host. Vertical transmission also occurs within the mosquito population, and may have an important role in maintaining the sylvatic cycle (Aitken *et al.*, 1979; Fontenille *et al.*, 1997). Infected mosquitoes occasionally bite unvaccinated forest workers such as hunters and loggers. This route of transmission of yellow fever is viewed as an occupational disease in parts of South America, and seldom causes epidemics.

Peridomestic mosquitoes biting both humans and monkeys are capable of sustaining small-scale epidemics of yellow fever in rural human populations. This role is played by a number of species in Africa (e.g. *Aedes simpsoni*), resulting in the epidemic form of yellow

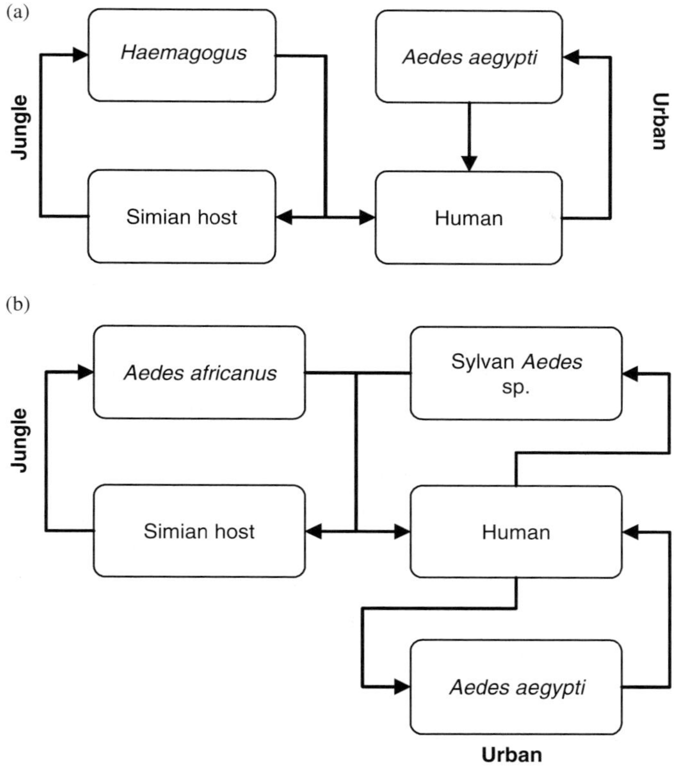

Figure 1 Yellow fever transmission cycles for (a) South America and (b) Africa. Simian host species in South America include *Alouatta* sp., *Ateles* sp., *Callithrix* sp., *Cebus* sp. and *Saimiri* sp. while those in Africa include *Colobus abyssinicus*, *C. polycomos*, *C. badius*, *Cercopithicus* sp., *Cercocebus* sp., *Erythrocebus* sp., *Papio papio*, *P. anubis* and *Pan troglodytes*.

fever most commonly seen in recent decades. It is thought that *Aedes albopictus*, the Asian tiger mosquito, is capable of adopting a similar role (Gratz, 2004). From its natural distribution in south-east Asia, this species has been introduced and has become established in much of Central and South America, the Pacific, Australasia, Africa and areas of Europe during the 1980s and 1990s (CDC, 2001). When anthropophilic mosquitoes such as *Aedes aegypti* become infected, rapid human–human transmission may begin. In these circumstances, an epidemic of yellow fever may spread quickly in dense urban areas. Urban epidemics are the form most feared by public health authorities, and may incapacitate significant proportions of the population of a city.

2.1.4. Distribution and Impacts

It was acknowledged by the Health Organization of the League of Nations (the forerunner to the World Health Organization (WHO)) that yellow fever was a severe burden on endemic countries. The work of Soper and the Brazilian Cooperative Yellow Fever Service (Soper, 1934, 1935a, b) began to determine the geographical extent of the disease, specifically in Brazil. Regional maps of disease outbreaks were published by Sawyer (1934), but it was not until after the formation of the WHO that a global map of yellow fever endemicity was first constructed (van Rooyen and Rhodes, 1948). This map was based on expert opinion (United Nations Relief and Rehabilitation Administration/Expert Commission on Quarantine) and serological surveys. The present-day distribution map for yellow fever is still essentially a modified version of this map.

At the beginning of the 21st century yellow fever has been estimated to affect as many as 200 000 people annually in the tropics of Africa and South America (Vainio and Cutts, 1998), and causes an estimated 30 000 deaths each year (Division of Epidemiological Surveillance and Health Situation and Trend Assessment, 1992). Approximately 2.5 billion people live within the current range of *Aedes aegypti* (Gubler, 1998), and must be considered at risk of either or both yellow fever and dengue. The rapid spread of *Aedes albopictus* in recent years has also increased the risk of epidemics (Knudsen, 1995; Tatem *et al.*, this volume, pp. 293–343). Figure 2 presents the known global distribution of these two important vector species. In the past, the mortality rate arising from infection with yellow fever was much higher than it is today, due to inappropriate health care, or a lack of it altogether, but such death rates are now avoidable, although they are not always avoided. Urban health services may be overwhelmed by large numbers of patients, and this is likely to occur when levels of human immunity are low.

Yellow fever is conspicuously absent from Asia, despite multiple opportunities for introduction. This absence does not seem to be

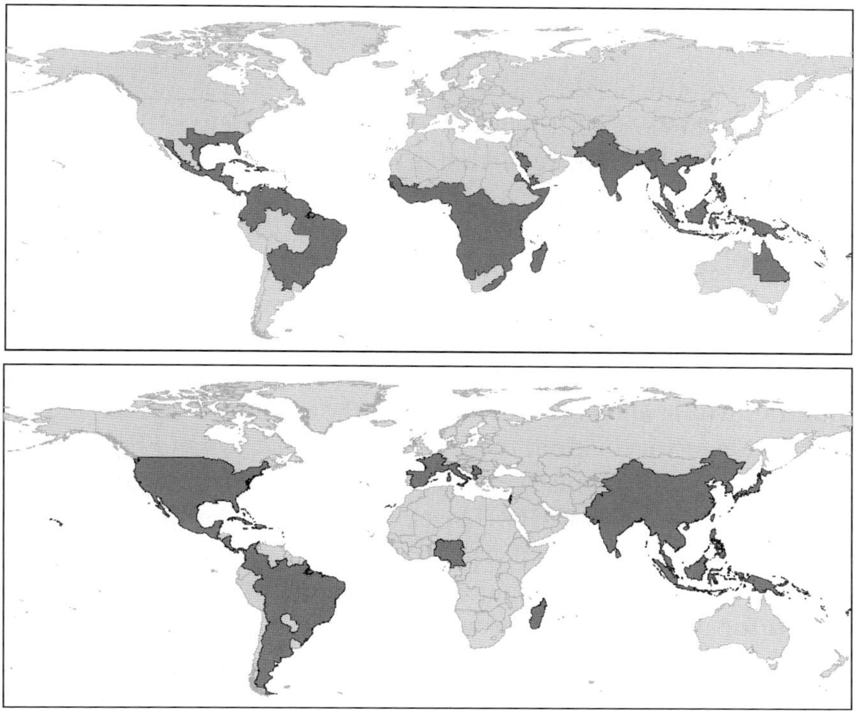

Figure 2 Global distribution of *Aedes aegypti* (top) and *Aedes albopictus* (bottom), two important vector species of yellow fever and dengue. *Aedes albopictus* distributions are provided at national scale. Australia, New Zealand and South Africa have all reported mosquito interception at ports (see Tatem *et al.*, this volume, pp. 293–343). *Source*: Center for International Earth Science Information Network (CIESIN) [http://www.ciesin.org/docs/001-613/map15.gif], supplemented with information from Gratz (2004); Gubler (2003); Lounibos (2002); Medlock *et al.* (2005); Moore (1999); and Moore and Mitchell (1997).

attributable to differences in the susceptibility of vectors, and all components of a suitable transmission cycle appear to be present (Vainio and Cutts, 1998). Race appears to affect susceptibility to dengue (Guzman *et al.*, 1990), but there are no adequate comparable epidemiological studies of yellow fever susceptibility (Vainio and Cutts, 1998). Although there is some evidence that other flaviviruses may offer cross-protection against yellow fever (Gordon-Smith *et al.*, 1962), why yellow fever does not occur in Asia is still unexplained.

2.1.5. Control

The first live-attenuated vaccine for yellow fever was developed between 1934 and 1935 in French West Africa (Durieux, 1956), and its use achieved a dramatic reduction in incidence within about five years of its introduction (Figure 3). Unfortunately, it was associated with a high risk of encephalitic reaction in children (3–4/1000, with a fatality rate of 38%), and its production was discontinued in 1980. The 17D live-attenuated vaccine still in use today was developed in 1936, and a single dose confers immunity for at least ten years in 95% of the cases. In a bid to contain the spread of the disease, travellers to countries within endemic areas or those thought to be 'at risk' require a certificate of vaccination; these countries are shown in Figure 4. The yellow fever certificate is the only internationally regulated certification supported by the WHO. The effectiveness of the vaccine reduces the need for anti-vectorial campaigns directed specifically against yellow fever. As the same major vector is involved, control of *Aedes aegypti* for dengue reduction will also reduce yellow fever transmission where both diseases co-occur, especially within urban settings.

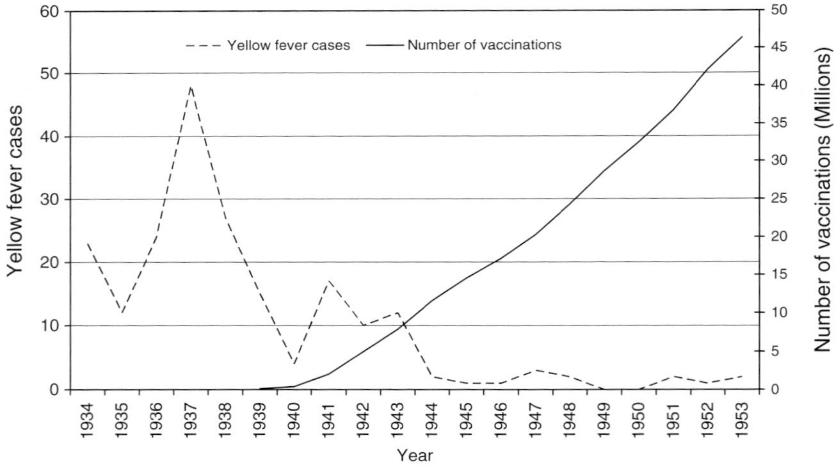

Figure 3 Plot of the number of yellow fever cases and the number of vaccinations in French West Africa for the period of 1934–1953, demonstrating the effect of the French vaccine on case numbers. *Source*: Vainio and Cutts (1998).

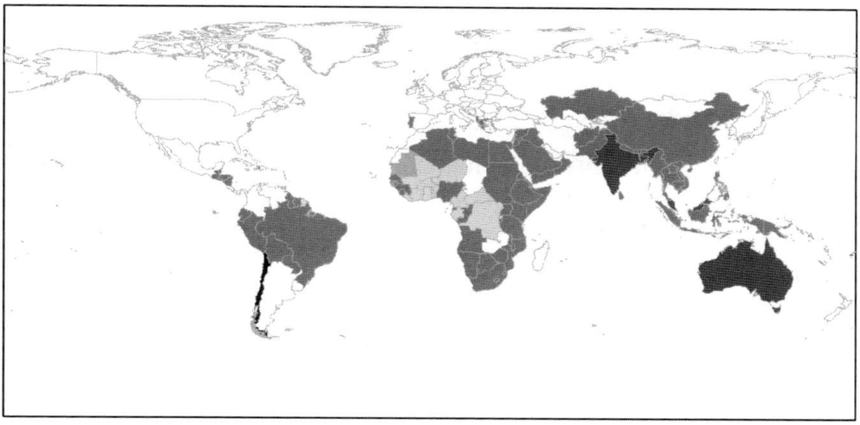

Figure 4 Yellow fever vaccination certificate requirements by country. □ E1 requirements, ▫E2 requirements ▫E3 requirements ▪E4 requirements ■ E5 requirements. There are five levels of certification: E1—immunisation is an essential requirement for entry to the country concerned and a certificate is required, except for infants under one year, E2—immunisation is an essential requirement for entry to the country concerned and a certificate is required (except for infants under one year) unless arriving from non-infected areas and staying for less than two weeks, E3—immunisation is an essential requirement for entry to the country concerned and a certificate is required if the traveller arrives from an infected country or area, E4—immunisation is an essential requirement for entry to the country concerned and a certificate is required if arriving within six days of having visited an infected country, E5—immunisation is an essential requirement for entry to the country concerned and a certificate is required for entry to the country from endemic areas, travelling to Easter island. Source: From data obtained at [http://www.dh.gov.uk/PolicyAndGuidance/HealthAdviceForTravellers/GeneralHealthAdvice/Diseases/DiseasesArticle/fs/en?].

2.2. Dengue

2.2.1. History

Probable epidemics of dengue fever have been recognised since at least 1779 (Rush, 1789), and have been recorded from Africa, Asia, Europe and the Americas since the early 19th century (Armstrong, 1923). Although it is rarely fatal, up to 90% of the population of an infected area can be incapacitated during the course of an epidemic (Armstrong, 1923;

Siler *et al.*, 1926). Widespread movements of troops and refugees during and after World War II introduced vectors and viruses into many new areas, and this trend has continued (Calisher, 2005) with the growth of global transport networks (Tatem *et al.*, this volume, pp. 293–343). By the end of the 20th century, annual epidemics of dengue were occurring in many parts of Central and South America (Pinheiro, 1989; Rodriguez-Roche *et al.*, 2005), throughout the Pacific Islands (Effler *et al.*, 2005) and South East Asia and with occasional outbreaks in North Australia (Doherty *et al.*, 1967) and Africa.

2.2.2. Symptoms

Infection with any of the four dengue serotypes may result in a spectrum of clinical manifestations. After an incubation period of around five to six days (Siler *et al.*, 1926; Innis *et al.*, 1988), patients develop symptoms including joint pain, fever and headaches (Halstead, 1997; Nisalak *et al.*, 2002). Dengue fever has unsurprisingly been mistaken for yellow fever as well as other diseases including influenza, measles, typhoid and malaria (Hare, 1898; Siler *et al.*, 1926; Lopez-Correa *et al.*, 1979; Holmes *et al.*, 1998). Although highly uncomfortable, dengue fever is rarely fatal and survivors appear to have lifelong immunity to the homologous serotype.

Far more serious is dengue haemorrhagic fever (DHF), where additional symptoms develop, including haemorrhaging and shock. The mortality from DHF can exceed 30% if appropriate care is unavailable. The most significant risk factor for DHF is when secondary infection with a different serotype occurs in people who have already had, and recovered from, a primary dengue infection. It is suggested that virus infection is enhanced by the presence of pre-existing heterotypic antibodies (Halstead, 1970, 1988; Halstead *et al.*, 1980). Under this 'antibody-dependent enhancement' hypothesis, pre-existing antibodies bind to the heterotypic dengue virus particles but fail to neutralise them once cross-reactive antibodies have decayed below a certain level. These infectious antibody–virus complexes bind to receptors on macrophages more easily, resulting in a higher level of viral uptake. The presence of antibodies to a heterotypic serotype has been demonstrated to increase the ability of dengue virus particles to infect monocytes in both *in vitro*

(Halstead and O'Rourke, 1977) and *in vivo* (Halstead, 1979) primate models, and enhancement appears in the sera of recently infected patients—i.e. those with neutralising levels of antibodies—if the sera are diluted (Kliks *et al.*, 1989). As yet, a link has not been firmly established between increased viraemia and the pathophysiological changes seen in DHF (Halstead, 1979) and others have suggested that these may be related to T-cell activation (Rothman and Ennis, 1999; Zivna *et al.*, 2002; Mongkolsapaya *et al.*, 2003).

2.2.3. Epidemiology

As with yellow fever, dengue is thought to have originally been a sylvatic virus, and a complex sylvatic cycle involving multiple mosquito and vertebrate species still exists in the forests of South-East Asia (Knudsen, 1977) and West Africa (Diallo *et al.*, 2003). Dengue has adapted to changes in human demography very effectively. The main vector of dengue is the anthropophilic *Aedes aegypti*, which is found in close association with human settlements throughout the tropics, breeding mainly in containers in and around buildings (Christophers, 1960; Sheppard *et al.*, 1969; Southwood *et al.*, 1972; Trpis and Hausermann, 1986), and feeding almost exclusively on humans (Christophers, 1960; Scott *et al.*, 1993). As a result, dengue is essentially a disease of tropical urban areas, although occasional sylvatic outbreaks do still occur in areas of West Africa (Diallo *et al.*, 2003). Both vertical (Rosen *et al.*, 1985) and sexual (Rosen, 1987) transmissions of dengue are possible in *Aedes aegypti*, but they occur at extremely low rates and are not thought to be epidemiologically significant. *Aedes albopictus* also transmits dengue in parts of Asia but is not as important, at present, as is *Aedes aegypti*.

DHF first came to attention in the 1950s and has since spread rapidly throughout the tropical world (Halstead, 2005). The first DHF epidemic was reported in Manila in 1953–1954, but cases of probable DHF can be found in the literature stretching back to 1779 (Rush, 1789; Halstead, 1997). Before 1970, only nine countries had experienced DHF epidemics, but by 1995 this number had increased fourfold (WHO, 2001). The appearance of DHF stimulated large amounts of dengue research, which established the existence of the four serotypes and the range of competent vectors, and led to the adoption of *Aedes aegypti* control programmes in some areas (particularly South-East Asia) (Kilpatrick *et al.*, 1970).

There is currently some debate over the taxonomy and nomenclature of Aedine mosquitoes (Diptera: Culicidae). A recent phylogenetic analysis suggested elevating the subgenus *Stegomyia* to full generic status, and *Aedes* to a supergenus (Reinert *et al.*, 2004), in which case *Aedes aegypti* would become *Stegomyia aegypti.* This proposal has attracted criticisms of both a theoretical and a practical nature, and the Walter Reed Biosystematics Unit has established a Mosquito Taxonomy Review Committee to help resolve the issue [http://wrbu.si.edu/forums/]. In this review, and others in this volume, the more familiar name, *Aedes*, is retained to avoid confusion until the situation is resolved.

2.2.4. Distribution and Impacts

Dengue case numbers have increased considerably since the 1960s; by the end of the 20th century an estimated 50 million cases of dengue fever and 500 000 cases of DHF were occurring every year (WHO, 2001). There have been several attempts to estimate the economic impact of dengue: the 1977 epidemic in Puerto Rico was thought to have cost between \$6.1 and \$15.6 million (\$26–\$31 per clinical case) (Von Allmen *et al.*, 1979), while the 1981 Cuban epidemic (with a total of 344 203 reported cases) cost about \$103 million (around \$299 per case) (Kouri *et al.*, 1989). Costs of non-epidemic dengue, including dengue fever, are difficult to estimate (Clark *et al.*, 2005; Meltzer *et al.*, 1998). Dengue is essentially an urban problem in the tropics, and urban populations are projected to increase from three to five billion by 2030, with much of this increase occurring in less-developed countries (United Nations, 2004; Hay *et al.*, 2005).

Large numbers of asymptomatic dengue infections have been confirmed in South-East Asia (Sangkawibha *et al.*, 1984; Burke *et al.*, 1988; Chen *et al.*, 1996; Porter *et al.*, 2005), the South Pacific (Maguire *et al.*, 1974) and South America (Kochel *et al.*, 2002; Teixeira *et al.*, 2002), and are suspected in areas of Africa (Saluzzo *et al.*, 1986). These asymptomatic cases may conceal the true extent of infection incidence and virus diversity in a population.

2.2.5. Control

There is no cure for dengue fever or for DHF. Currently, the only treatment is symptomatic, but this can reduce mortality from DHF to less than 1% (WHO, 2002). Unfortunately, the extent of dengue epidemics means that local public health services are often overwhelmed by the demands for treatment.

Vaccine development has been complicated by the potential risk of vaccination resulting in antibody-dependent enhancement of future heterotypic infection (Vaughn, 2000; Halstead and Deen, 2002), although a tetravalent vaccine is presently undergoing clinical trials (Bhamarapravati *et al.*, 2003).

The most common historical approach to limiting dengue was source control of *Aedes aegypti.* Although such programmes are capable of achieving significant reductions in the house index (the percentage of buildings positive for immature vectors) (Mendes Luz *et al.*, 2003), they are labour-intensive, expensive to sustain and appear to be less effective than expected at preventing dengue transmission. An informative review of the pitfalls and progress experienced by *Aedes aegypti* control programmes is presented in Reiter and Gubler (1997).

Modern *Aedes aegypti* control programmes such as those in Singapore (Goh, 1995, 1997; Wilder-Smith *et al.*, 2004) and Cuba (Spiegel *et al.*, 2002) are also based on locating and eliminating domestic breeding sites, but growing prosperity in many areas has led communities to resent the invasion of their homes by control officials. Furthermore, short-term success generally lowers public perceptions of disease risk, and generates increased hostility to continued control attempts.

3. MATERIALS AND METHODS

3.1. Existing Maps

No serious attempts to update the yellow fever distribution map have been made since the first global map was produced in 1948, although part of the map for Africa was slightly amended in 1986 (WHO, 1986). The most recent global map, for the situation in 2003, was released in 2004 [http://www.who.int/ith/en]. Despite the implementation of vaccination programmes, attempts to eradicate the mosquito vectors and a substantially altered human population distribution and socio-economic status over the last 50 years, this map is still based on the earlier maps with relatively few significant changes. Endemicity is reported mostly at country level, and some countries are included that have not had reported outbreaks for at least two or three decades. While it is obviously wise to avoid underestimating the extent of a zoonosis such as yellow fever, overestimation of the current situation may lead to a dilution of control efforts through their application in places where they are not really required.

The latest distribution map for 'all cause dengue' (dengue risk irrespective of serotype) reports the global situation in 2003 [http://www.who.int/ith/en], but it is not clear how this map was derived. As with the yellow fever map, the dengue map provides an indication of the historical occurrence of this disease and again includes large areas (especially in Africa) that may indeed have the disease, but have never reported any cases of it. Thus, both the yellow fever and dengue maps could be improved upon, most efficiently through the use of modern mapping and modelling techniques. Attempts using climate data and logistic regression have been made for dengue with some success (Hales *et al.*, 2002), but the global climate datasets used for this modelling have relatively coarse spatial resolution, as do the resultant maps. This paper applies discriminant analytical methods within an information theoretic approach (Rogers, this volume, pp. 1–35) to point records gleaned from archived reports and literature surveys, and high spatial resolution environmental information derived from satellites to create risk maps of environmental suitability for both diseases.

3.2. Archive and Literature Searches

A detailed search of the WHO library archives identified point records for yellow fever outbreaks between 1900 and 1959 (Boyce, 1906; Great Britain Colonial Office Yellow Fever Commission, 1915a,b and c; Noguchi *et al.*, 1924; General Government of French West Africa, 1929;

Sawyer, 1934; Soper, 1934, 1935a, b, 1955; Jorge, 1938; Bustamante, 1958; WHO, 1971). Serological data were not included because they confuse populations infected naturally with those protected by vaccination coverage. The location data were either digitised using ESRI ArcView 3.2 or geo-located using an online gazetteer [http://www.traveljournals.net/explore/index.html]. Between 1900 and 1959, a total of 450 locations of yellow fever outbreaks in Africa were identified, with a further 707 locations in the Americas.

Point records for yellow fever and dengue infections between 1960 and 2005 were gathered by querying the PubMed archives [http://www.ncbi.nlm.nih.gov/entrez/query.fcgi] in conjunction with country names obtained from the GEOnet Names Server (GNS) [http://earth-info.nga.mil/gns/html/cntry_files.html]. This query was initially restricted to articles in English and with abstracts. The references were loaded into RefViz [http://www.adeptscience.co.uk/products/refman/refviz/info.html], which automatically groups them into clusters depending upon their key words. Clusters judged to be unlikely to contain spatial information were deleted. The abstracts of papers surviving this initial cull were then checked and references discarded that did not contain data that could be geo-referenced. A second reference search was later undertaken for both diseases, with no language restrictions and only in those countries that looked promising after the first search round. RefViz was again applied to filter the results. For yellow fever, a total of 663 references were initially downloaded from PubMed, reduced by 395 when unpromising keyword groups were deleted and a further 96 when the abstracts were read. The remaining 172 references provided 281 geo-referenced points of yellow fever occurrence. For dengue, a total of 1735 references were downloaded from PubMed, reduced by 781 when unpromising keyword groups were considered and by a further 77 after reading their abstracts. The remaining 877 references provided 897 geo-referenced dengue points.

The geo-referenced data points are shown in Figure 5 (yellow fever) and 6 (dengue). It is convenient to split the yellow fever map into two, for the periods 1900–1959 and 1960–2005, but the records for dengue are more sparse in the early years of the last century and so only a single map is shown, covering the period 1960–2005.

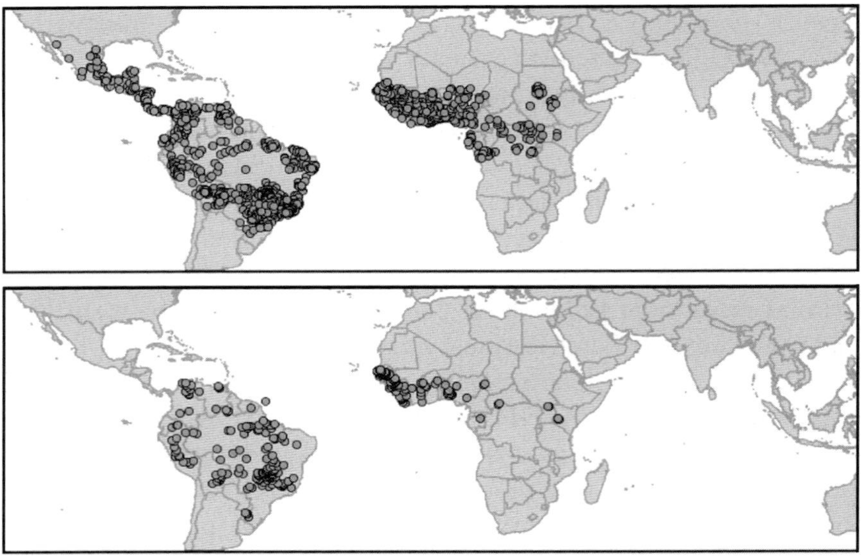

Figure 5 Yellow fever outbreak distribution for 1900–1959 (top) and 1960–2005 (bottom). The maps are displayed between 40°N and 40°S as these latitudes encompass all known areas of the disease.

3.3. Yellow Fever Data

At the beginning of the 20th century, a large number of yellow fever epidemics were recorded in both African and American cities, and these occurred against a background of annual cases. Documentation for this period is relatively complete because there was a concerted effort at the time to eradicate the disease.

In Africa yellow fever was mainly a problem of the sub-Saharan countries of West Africa, but reached as far east as central Sudan and Kenya (Soper, 1955; Haddow, 1965; Reiter *et al.*, 1998; Bell, 1999). There is little reason to think that the recorded distribution is an artefact of colonial data reporting. In the New World, a large number of outbreaks were reported in eastern Mexico and other Central American countries. At this time, yellow fever was an epidemic disease mainly of port cities, although a large number of inland reports came from south-eastern Brazil. These cases, and those from

Columbia, were recorded at very high spatial densities with many fewer reports from the Amazon Basin.

The period between about 1950 and the 1970s was one of complacency about the control of yellow fever, probably arising from the feeling that yellow fever vaccination had solved the problem. *Aedes aegypti* control was reduced and overall disease record keeping appears to have diminished.

For the period 1960–2005, only 110 yellow fever points were recorded in Africa and 171 in South America. In both regions, these records more or less fall within the same areas of risk shown for the first half of the last century (Figure 5b compared with Figure 5a), although there is a noticeable lack of new records in Central America and proportionately more cases within the Amazon basin.

3.4. Dengue Fever Data

For the first half of the last century and up until 1960, there was little control directed specifically at dengue, although some control was achieved through the collateral impacts of yellow fever vector eradication programmes (most notably in the Americas). Information for this period is difficult to obtain, although known outbreaks are recorded predominantly in coastal ports. Large movements of susceptible populations during the Second World War led to a rise in recorded dengue cases (Gubler, 1997), especially, for countries in South-East Asia (Gubler, 2004). Dengue point records since 1960 are shown in Figure 6 ($n = 897$). Dengue has now firmly established itself in three geographical areas; South-East Asia, Latin America and throughout the Pacific islands, and seems to be consolidating itself in

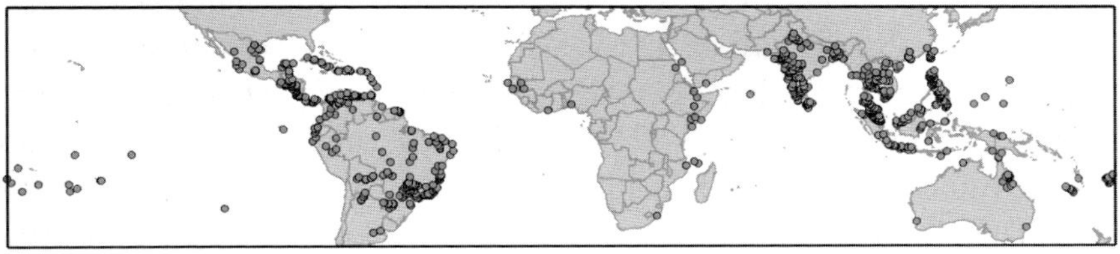

Figure 6 All-sera dengue outbreak distribution for 1960–2005. The map is displayed between 40°N and 40°S as these latitudes encompass all known areas of the disease.

these areas, benefiting from local, large human populations. Reported outbreaks are clustered around the Caribbean, in southern Brazil, western India and Bangladesh and in countries surrounding the South China Sea. Dengue has long been thought to be widespread in Africa but case reporting remains very poor. In the absence of laboratory test facilities, it is likely that dengue fever, as opposed to DHF, would often be mistaken for other diseases such as malaria.

For both yellow fever and dengue, all presence records from 1960 to 2005 were used to construct the risk maps. For each disease, it was also necessary to identify areas of absence and this was done by sampling at random regions no closer than 0.5° and no farther than 10° from any recorded presence site. A total of 3500 absence points for yellow fever and 9000 absence points for dengue were selected in this way.

3.5. Environmental Data from Satellites

The environmental data used to describe the distributions of yellow fever and dengue were derived from the Advanced Very High Resolution Radiometer (AVHRR) on board the National Oceanographic and Atmospheric Administration (NOAA) satellites and cover the period from 1982 to 1999. These data are described in detail in Hay *et al.* (this volume, pp. 37–77) and are provided on the DVD accompanying this volume. Briefly, monthly maximum value composited AVHRR Channel 3, the derived Land Surface Temperature (LST) and the Normalized Difference Vegetation Index (NDVI) data were temporal Fourier processed to extract annual, bi-annual and tri-annual seasonal signals, which were captured as separate images showing the amplitudes and phases or timing of the first peak of each of the three signals (Rogers, 2000). In addition, the signal means, maxima, minima and variances were also available, as was a single digital elevation surface (DEM) derived from the GTOPO30 coverage [http://edcdaac.usgs.gov/gtopo30/gtopo30.asp]. All AVHRR data were originally produced and made available at a spatial resolution of 8×8 km in the Goode's Interrupted Homolosine projection (James and Kalluri, 1994) and after Fourier processing were projected by bi-linear interpolation to latitude/longitude format at 0.10° spatial resolution. The DEM data at an original 30 arc second resolution (1/120th of a degree) were re-sampled to 0.1° resolution by averaging. Satellite and DEM data were later extracted for each of the disease presence and absence points and these data formed the training sets for model construction.

3.6. The Modelling Approach

The modelling approach followed that described by Rogers (this volume, pp. 1–35) and Rogers (2000). Briefly, non-linear discriminant analysis captured the covariance characteristics of sites of disease presence and absence and these were used to define the location within multivariate space of any point within the risk-mapped area. On this basis, the probability with which the point belonged to the category of disease presence or absence was calculated and this probability was entered into the final risk map. Disease presence and absence data were separately clustered before analysis into between one and eight clusters, but the final results presented here used only three clusters each in all the models. Clustering splits data that may be non-multivariate normal in their overall characteristics into groups that are more nearly multivariate normal, and thus make the data comply with the assumptions of the discriminant methods used. It is often the case that the resulting clusters occupy separate geographical regions or areas where different vector species are important, so that what is initially a mathematical requirement of the methods used can also be justified on ecological and biological grounds.

During discriminant analysis ten variables were selected for each model in a step-wise inclusion fashion, using the corrected Akaike Information Criterion (AICc) as the basis for choosing the next variable to add to the model (Rogers, this volume, pp. 1–35). The AICc is one of a number of criteria that might be used for variable selection, and is a measure of the Kullback–Leibler information or distance statistic that measures how well the current model fits the data (Burnham and Anderson, 2002); the smaller the value of the AICc, the more accurate the model.

As was the case with a set of Rift Valley Fever data (Rogers, this volume, pp. 1–35) derived using the same literature search criteria as were used here, the point records for yellow fever and dengue are likely to be incomplete to a lesser or greater degree. Rogers (this volume, pp. 1–35) describes two ways of extracting maximum information from such sparse datasets—bootstrap sampling and environmental envelope expansion—and the former was used here. For each disease, 100 bootstrap samples were taken, with replacement, from the training set (300 presence and 300 absence points for Yellow Fever and 900 of each for dengue) and a separate model was constructed for each bootstrap sample, producing one output risk map. The 100 risk maps were then averaged to produce the results given here. Each risk map shows the average predicted maximum likelihood posterior probabilities of environmental suitability for the disease in question, i.e. the probability with which each pixel belongs to the category of disease presence. Areas of high probability of disease presence are more likely to harbour the disease than are areas of lower probability, or are at greater risk of invasion by the disease if it is not already present. Conventionally, risk maps of this sort are thresholded at a probability of 0.5 to produce a binary presence/absence map, but the continuous scale more appropriately captures the variable risk of disease occurrence. Diseases may occur in areas of low predicted risk, but they should do so only infrequently, and they should not persist. The thresholded versions of the maps were used to calculate the kappa index (κ) of model fit (Congalton, 1991; Ma and Redmond, 1995), which is based on the matrix of observed and predicted presences and absences of each model's bootstrap sample of the training set data. Kappa varies from -1 (predictions completely opposite to observations) through 0 (model fit no better than random) to 1 (perfect fit) and Landis and Koch (1977) suggest the following ranges of agreement for the kappa statistic: poor, $\kappa < 0.4$; good, $0.4 < \kappa < 0.75$ and excellent, $\kappa > 0.75$.

4. RESULTS

4.1. Risk Maps for Yellow Fever and Dengue

The predicted distribution maps are shown in Figure 7 (yellow fever) and 8 (dengue). Figure 7 (Figure 7 is Plate 6.7 in the Separate Colour Plate Section) for yellow fever shows that the predicted high-risk areas are quite localised within the broad boundaries of the 2003 WHO map for this disease, and that there are predicted areas of high risk outside these boundaries, most notably in parts of the Minas Gerais region, south east of Brasilia in Brazil, the common border region between the Democratic Republic of Congo and western Zambia, the eastern border region of Zimbabwe and in Swaziland. Regions of high risk well outside these boundaries are also predicted in Malagasy, Thailand and parts of Malaysia and Indonesia.

Figure 8 (Figure 8 is Plate 6.8 in the Separate Colour Plate Section) for dengue shows a similar patchy distribution of high-risk areas within WHO's 2003 map for this disease, and high-risk areas outside these boundaries in the New World, most notably in southern Mexico. Given that some of the database points fall in these areas, it is clear that the WHO map should be extended to include more of this country. As expected, the predicted high-risk areas of Africa are much more extensive than the WHO map indicates, and are well outside any of the database records for this disease. This suggests an underreporting of dengue in Africa, for the reasons already mentioned. In India and South-East Asia the predictions more or less fill in the gaps between the database records. In Pakistan and India the predicted high-risk areas are well within the north-western boundary of the WHO map. Much of the northern part of the South-East Asian limits on the WHO map also seems to be at very low risk.

4.2. Overall Model Accuracy

The 100 bootstrap models for each disease were ranked in order of their AICc values, lowest (best fit) to highest (worst fit). Although there is no exact correspondence between the AICc values and the kappa index of agreement (since the former are based on probabilities and the latter on categorical assignment of those probabilities), there was, nevertheless, overall agreement between the figures. Thus, the mean kappa value for the top ten yellow fever models was 0.742 (s.d. = 0.023) and for the bottom ten was 0.644 (s.d. = 0.049). The equivalent figures for the dengue models were 0.700 (s.d. = 0.017) and 0.680 (s.d. = 0.011), respectively. These values indicate a good to excellent fit of the models to the point data.

4.3. Importance of Individual Variables

The summed Akaike weights of the models in which each variable occurred are given in Table 1 (yellow fever) and 2 (dengue). These sums are an indication of the importance of the individual variables regardless of the particular models in which, and other variables with which, they occurred (Rogers, this volume, pp. 1–35). Due to the 'best' model in each case being so much better than the second best,

Table 1 Yellow fever model predictor variables

Variable	Summed Akaike weight	Mean rank	*n*/100
wd1014vr	1.00000	3.64	77
wd1014p3	1.00000	5.81	76
wd1003p2	1.00000	9.90	31
wd1014a1	0.99906	6.09	61
wd1014p1	0.99906	7.16	51
wd1007p3	0.99906	9.58	45
wd1003a0	0.99906	10.03	14
wd1003mn	0.99906	10.06	20
wd1007mx	0.99906	10.21	22
wd1007mn	0.99906	10.40	15
wd1014a3	0.00094	7.14	81
wd1014a2	0.00094	7.47	69
wd1007a2	0.00094	7.68	55
wd1003a2	0.00094	8.90	45
wd1007p2	0.00094	8.90	44
wd1007a1	0.00094	9.61	21
wd1003vr	0.00094	10.23	17
wd1014mn	0.00000	10.39	13
wd1007a0	0.00000	10.61	13
wd1014a0	0.00000	10.79	5

Note: Summed Akaike weights (second column) for the top predictor variables (first column) of the 100 yellow fever bootstrap models used to produce Figure 7 (see text for details). The mean ranks (*i.e.* the order in which the variables were selected, where rank 1 = the first selected variable, rank 10 = the tenth selected variable and all non-selected variables are given a rank of 11) are given in the third column, and the number of times (out of 100 models) each variable was selected is given in the final column.

Key to variable names: *wd10* refers to AVHRR data at 0.10° resolution in the latitude/longitude format, *03* refers to the AVHRR channel 3 (MIR), *07* to LST and *14* to Normalized Difference Vegetation Index (NDVI) data; *a1*, *a2* and *a3* refer to the amplitudes of the annual, bi-annual and tri-annual cycles, respectively, of temporal Fourier processed imagery and *p1*, *p2* and *p3* to their corresponding phases (timing of the first peak); *mn* and *mx* refer to the minimum and maximum and *vr* to the variance.

these summed weights depend a great deal on whether the variables were chosen in the best model. For example, the third variable in the list for yellow fever is the phase of the bi-annual cycle of the AVHRR channel 3 (MIR) variable (Table 1). This was selected in only 9 of the top 25 models, while the two variables above it in the list were selected 19 and 18 times in the same models. The summed Akaike weights for all 3 variables are the same (1.0) because they were all selected for the top two models. The seven variables following these three all have the same Akaike weight (0.99906); they too were selected by the top model but not by the second- and third-best models. Vegetation index variables occupy four of the top five positions in Table 1, suggesting that yellow fever is particularly sensitive to the greenness or humidity of the environment. The fact that the NDVI variance is the most important variable also suggests that the variability of greenness or humidity is key to the distribution of this disease.

For dengue the situation with regard to the importance of the top few variables is the same (Table 2), but in this case LST variables occupy three of the top five slots. Once again, the most important variable appears to be the variance of this variable. In the case of dengue, the variability of environmental temperature rather than of moisture appears to be key to its distribution.

4.4. Variability of Bootstrap Results

Some ideas of the different sets of variables selected during bootstrapping, and also of differences between the two diseases, are given in Figure 9a (yellow fever) and b (dengue). Each row in each image in Figure 9 (Figure 9 is Plate 1.3 (middle and right) in the Separate Colour Plate Section) refers to one of the bootstrap models that are arranged in rank order, with 1 (lowest AICc value) at the top and 100 (highest AICc value) at the bottom. Each of the 31 columns on the right of the image indicates one of the satellite predictor variables available to describe the disease. The first column of these 31 columns is for the digital elevation layer (DEM), followed by three sets of 10 columns referring to the Fourier-processed AVHRR MIR, LST and NDVI imagery, respectively. Within each set, the Fourier layers are in

Table 2 Dengue model predictor variables

Variable	Summed Akaike weight	Mean rank	$n/100$
wd1007vr	1.00000	3.64	81
wd1007p1	1.00000	4.01	88
wd1014p3	1.00000	5.02	100
wd1014a0	1.00000	6.10	78
wd1007p3	1.00000	6.32	99
wd1003p3	1.00000	7.05	99
wd1014p2	1.00000	9.18	63
wd1003mx	0.99872	10.61	6
wd1007a2	0.99872	10.75	9
wd1007a0	0.99872	10.89	2
wd1003mn	0.00128	2.82	87
wd1003a0	0.00128	9.11	52
wd1014a1	0.00128	10.64	16
wd1014p1	0.00000	8.41	68
wd1003vr	0.00000	9.90	14
wd1003p2	0.00000	10.07	43
wd1007p2	0.00000	10.30	27
wd1003a1	0.00000	10.58	9
wd1014a3	0.00000	10.80	10
wd1014mn	0.00000	10.86	9

Note: Summed Akaike weights (second column) for the top predictor variables (first column) of the 100 dengue bootstrap models used to produce Figure 8 (see text for details). The mean ranks (*i.e.* the order in which the variables were selected, where rank 1 = the first selected variable, rank 10 = the tenth selected variable and all non-selected variables are given a rank of 11) are given in the third column, and the number of times (out of 100 models) each variable was selected is given in the final column.

Key to variable names: *wd10* refers to AVHRR data at 0.10° resolution in the latitude/longitude format, *03* refers to the AVHRR channel 3 (MIR), *07* to LST and *14* to Normalized Difference Vegetation Index (NDVI) data; *a1*, *a2* and *a3* refer to the amplitudes of the annual, bi-annual and tri-annual cycles, respectively, of temporal Fourier-processed imagery and *p1*, *p2* and *p3* to their corresponding phases (timing of the first peak); *mn* and *mx* refer to the minimum and maximum and *vr* to the variance.

the following order, from left to right; mean, phase of annual cycle, amplitude of annual cycle; phase of bi-annual cycle, amplitude of bi-annual cycle; phase of tri-annual cycle, amplitude of tri-annual cycle; maximum of fitted Fourier cycles (summed annual to tri-annual), minimum of fitted Fourier cycles and variance of the original signal. In any single model (row) the predictor variable selected first is coloured red, the second selected variable is coloured orange and so on according to the rainbow colour scale to the right of the image.

Variables not chosen in any model are not coloured at all in that row. Images like these are able to show visually both any consistent changes (if they occurred) in the suites of variables chosen by models of increasing overall accuracy and whether or not individual variables were consistently selected within those suites. For yellow fever and dengue (as was also the case with Rift Valley fever—Rogers, this volume, pp. 1–35), there do not appear to be any gradual changes in the suites of chosen variables, but (different) individual variables *are* consistently chosen.

In the case of yellow fever (Plate 1.3 (middle)), the predominantly red line down the right-most column of the image indicates variable 31 in the variable list, which is the NDVI variance. Not only is this variable overall the most important in the 100 models, it is also often selected first in the step-wise selection process; the annual amplitude of NDVI (variable 24 in the list) sometimes replaces it as the first-selected variable.

In the case of dengue (Plate 1.3 (right)) the predominantly red line down the middle of the image indicates variable 10 in the variable list, which is the minimum of AVHRR channel 3 (MIR). This variable is often selected first, but LST variance (number 21 in the sequence) is also often selected first, second or third. Crucially, LST variance was selected first in the best model; hence its highest overall summed Akaike weight (see above).

4.5. Populations at Risk

To be at all useful, risk maps must also contribute to initiatives aimed at public health intervention. To do this it is first necessary to relate the risk maps to the human populations in the 'at risk' regions. The GRUMP human population surface (Balk *et al.*, this volume, pp. 119–156) was aggregated to the same 0.1° resolution as the risk maps and the populations within each category of risk were read off from the resulting image. Figure 10 shows both the total human populations living within each risk category (Figure 10a) and the mean human population density per 0.10° grid square related to the same risk categories (Figure 10b). The total number of grid squares globally that fall

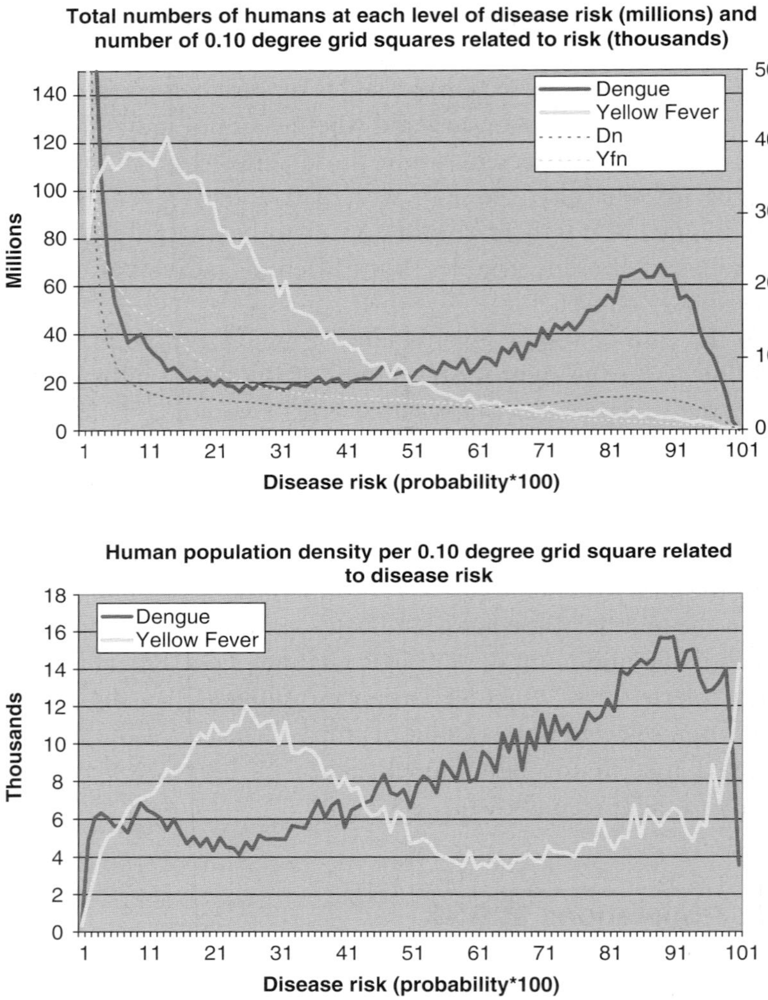

Figure 10 (a) Total numbers of humans within each category of yellow fever or dengue risk as shown in Figures 8 and 9 (thick lines, millions scale) and numbers of 0.10° grid squares within each risk category (dashed lines, thousands scale). (b) Mean human population density per 0.10° grid square for each category of yellow fever and dengue risk.

within each category of risk is indicated by the dashed lines in Figure 10a. The first point to be noted is that large portions of the globe—the colder, temperate regions—are free of any risk from either disease (note that the graph does not show the total human population at zero risk of either disease, because it is so large). Taking the 0.50 probability of risk as the cut-off between disease absence ($p<0.5$) and presence ($p \geqslant 0.5$) reveals that approximately 33.3% of the Earth's 6.05 billion people are at risk of dengue infection and 7.01% at risk of yellow fever infection (Figure 10a). Figure 10b shows that areas with the highest human population densities are associated with the higher risk categories for dengue and the lower risk categories for yellow fever. It seems that dengue is a disease associated with populous places of the globe, whereas yellow fever is associated with places where humans are relatively scarce. Curiously, the very highest risk categories of all are associated with low human population densities in the case of dengue (possible rural dengue in South-East Asia?), but high population densities in the case of yellow fever (possible urban yellow fever in South America?) (Figure 10b).

These results allow further interpretations of Figures 7 and 8. These figures were produced using only the environmental conditions associated with yellow fever or dengue infections in the past; the human population map was not an input data layer to the models. Figure 7 shows parts of Asia that are climatically suitable for yellow fever (which does not occur there); these include populous areas, and it is possible that the pressures of the human population exclude this disease. This could be for a number of reasons, including the exclusion by high population densities of humans of critical sylvatic maintenance hosts of yellow fever and the bridge mosquito vectors that transfer infections to the urban transmission cycle. Conversely, Figure 8 shows that many areas of Africa are climatically suitable for dengue, which is seldom reported from the continent. It was suggested previously in this review that this may be because of underdiagnosis of dengue in Africa, but a further possibility is that the low human population density across much of Africa (compared with India and South-East Asia) in some way precludes the occurrence of dengue. The essentially domestic vector, *Aedes aegypti*, may not exist at the levels required to transmit dengue, except in heavily populated places that provide just the right sorts of conditions for this vector species. If this is the case, any future increase of human populations in Africa, and their increasing urbanisation, may be associated with an increase in dengue across the continent, which is clearly already climatically suitable for this disease (Figure 8).

5. DISCUSSION

The models presented here are based purely on outbreak data. A large number of protection test surveys were undertaken as part of the yellow fever monitoring programmes (Beeuwkes *et al.*, 1930, 1934; Beeuwkes and Mahaffy, 1934; Sawyer and Whitman, 1936), but these data were not included in the models since some of them refer to immunity following vaccination rather than naturally acquired infection.

Future models of both diseases could be improved by incorporating information about the distributions of vector (Kumm, 1931; Whitman, 1951) and reservoir species (Balfour, 1915; Findlay *et al.*, 1936; de Thoisy *et al.*, 2004), either from point records or from models (Hopp and Foley, 2001) and the distribution of the human host populations (Balk *et al.*, this volume, pp. 119–156).

The restriction of yellow fever to Africa and South America has long been a puzzle (Bell, 1999; Monath, 2001). The present models indicate an area of high suitability for yellow fever in eastern Thailand and other areas of lower suitability for this disease in parts of Malaysia and Indonesia. Dengue occurs in all of these places, but the two diseases do not appear to co-occur. Explanations for this lack of co-existence include failed introduction prior to the modern transportation era (Gubler, 2002), the restriction of yellow fever outbreaks to communities which do not undertake international travel (Monath, 2001), cross protection by hyperendemic dengue (Theiler and Anderson, 1975) and low vector competence (Beaty and Aitken, 1979). The very different human population densities under which each disease appears to thrive (Figure 10) will tend to prevent their co-existence. Africa at the present time provides a test for some of these hypotheses. Parts of West Africa are predicted to be highly suitable for both diseases, while central parts of the Democratic Republic of the Congo (aka Zaire) are predicted suitable only for dengue (Figures 7 and 8). Does dengue occur only or mostly in the latter?

If dengue really only thrives in populous areas, will dengue increase as Africa's population increases?

6. CONCLUSION

Eradication attempts, when they fail, become merely temporary control measures and when these in turn end, or are underfunded, the controlled diseases can return. Devastating outbreaks of yellow fever in the 1970s and 1980s (Gubler, 2004), an expanding distribution of *Aedes aegypti* from their post-control levels (Gubler, 1998), the appearance of DHF and a resurgence of dengue from the 1970s, all demonstrate the resilience of yellow fever and dengue in the face of our attempts to control them globally. Figure 11 shows the annual occurrences of yellow fever outbreaks in South America and Africa since 1960. It suggests a period of relative quiescence in South America in the last 20 years, but a trend to increasing frequency of outbreaks in Africa over the same period of time. In the two decades before this period, the situation was, if anything, reversed. It is unlikely that any static risk map will capture this dynamic situation, and it is suggested that the risk-mapping approach presented here not

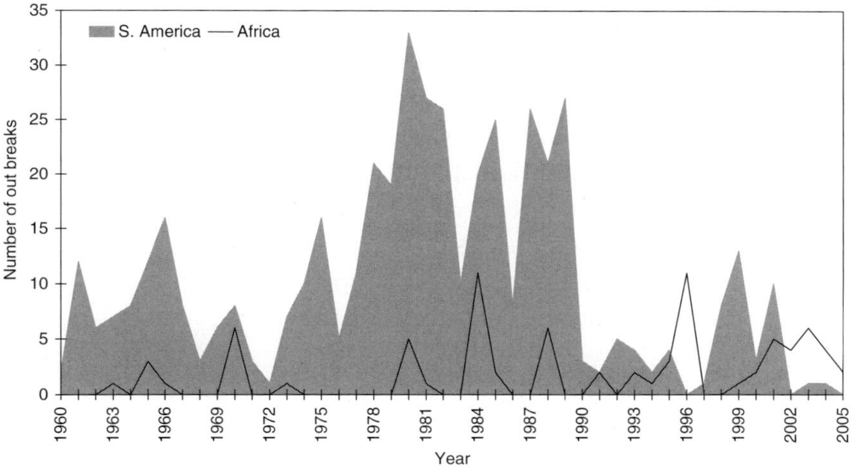

Figure 11 Reported number of yellow fever outbreaks per year per continent. *Source*: Weekly Epidemiological Record archives.

only delivers maps which more accurately reflect current disease situations, but can also become part of disease early warning systems, if the models are driven by new disease records and contemporary satellite data which are now freely available from a variety of sources (Hay *et al.*, this volume, pp. 37–77).

A major concern is that the rise of international sea and air traffic connects infected with suitable but presently uninfected regions of the globe and also connects remote infected regions in which different disease serotypes occur (Tatem *et al.*, this volume, pp. 293–343). Vector and disease spread to new regions seems almost inevitable, as are the consequences of co-circulation of different serotypes. These concerns have been in the minds of health officials for some time (Mhatre, 1934; Massad *et al.*, 2001), but have never been satisfactorily addressed. Given that we now have a semi-quantitative way of dealing with transportation network risks (Tatem *et al*, this volume, pp. 293–343) and with habitat suitability for each disease (this review), these concerns can now be quantified and perhaps prioritised.

The maps presented in this review are based solely on a pixel's environmental suitability for yellow fever and dengue, as judged by the locations of past cases of these diseases recorded in the literature, and the satellite measures of environmental conditions in these places. If, as Rogers points out (this volume, pp. 1–35), "*All maps are wrong; but some are useful*", it is now timely to see just how useful are satellite-informed, dynamic risk maps of vector-borne diseases for use in reconnaissance, surveillance and control. The relatively large discrepancies between the present 'standard' WHO maps for these two diseases on the one hand, and both the databases gleaned from the literature of the past 45 years and the predicted risk maps arising from them on the other, suggest that any effort put into improving existing maps should rapidly result in improvements in the maps, the data and therefore future generations of risk maps.

ACKNOWLEDGEMENTS

The authors are grateful to Dr. Carlos Guerra and Alya Dabbagh (WHO) for support in yellow fever data gathering, Dr. Andy Tatem for supplying Figure 2, Ben McCormick for discussions and Dr. Bethan Purse and the series editors for comments on the manuscript. AJG was funded by European Union grant GOCE-2003-010284 EDEN and the paper is catalogued by the EDEN steering committee as EDEN0010 [http://www.eden-fp6project.net/]. The contents of this publication are the sole responsibility of the authors and do not reflect the views of the European Union. AJW is funded by a University of Oxford Newton–Abraham Studentship in the Biological Sciences. SIH is funded by a Research Career Development Fellowship from the Wellcome Trust (no. 069045). SIH and DJR thank Monica Myers for continued support through the International Research Partnership for Infectious Diseases (INTREPID).

REFERENCES

Aitken, T.H., Tesh, R.B., Beaty, B.J. and Rosen, L. (1979). Transovarial transmission of yellow fever virus by mosquitoes (*Aedes aegypti*). *American Journal of Tropical Medicine and Hygiene* **28**, 119–121.

Armstrong, C. (1923). Dengue fever. *Public Health Reports* **38**, 1750–1784.

Balfour, A. (1915). Tropical problems in the New World. *Transactions of the Royal Society for Tropical Medicine and Hygiene* **8**, 75.

Barros, M.L. and Boecken, G. (1996). Jungle yellow fever in the central Amazon. *Lancet* **348**, 969–970.

Beaty, B.J. and Aitken, T.H.G. (1979). *In vitro* transmission of yellow fever virus by geographic strains of *Aedes aegypti*. *Mosquito News* **39**, 232–238.

Beeuwkes, H., Bauer, J.H. and Mahaffy, A.F. (1930). Yellow fever endemicity in West Africa, with special reference to protection tests. *The American Journal of Tropical Medicine* **10**, 305–333.

Beeuwkes, H. and Mahaffy, A.F. (1934). The past incidence and distribution of yellow fever in West Africa as indicated by protection test surveys. *Transactions of the Royal Society of Tropical Medicine and Hygiene* **28**, 39–76.

Beeuwkes, H., Mahaffy, A.F., Burke, A.W. and Paul, J.H. (1934). Yellow fever protection test surveys in the French Cameroons, French Equatorial Africa, the Belgian Congo and Angola. *Transactions of the Royal Society of Tropical Medicine and Hygiene* **28**, 233–258.

Bell, H. (1999). *Frontiers of Medicine in the Anglo-Egyptian Sudan, 1899–1940.* Oxford Historical Monographs. Oxford: Oxford University Press.

Bhamarapravati, N., Sabchareon, A., Yoksan, S., Forrat, R. and Lang, J. (2003). Progress in live attenuated, tetravalent dengue vaccine trials in Thailand. *Journal of Clinical Virology* **28**, S25.

Boyce, R. (1906). *Report to the government of British Honduras upon the outbreak of yellow fever in that colony in 1905 together with an account of the distribution of the stegomyia fasciata in Belize and the measures necessary to stamp out or prevent the re-occurrence of yellow fever*. London: Churchill.

Burke, D.S., Nisalak, A., Johnson, D.E. and Scott, R.M. (1988). A prospective study of dengue infections in Bangkok. *American Journal of Tropical Medicine and Hygiene* **38**, 172–180.

Burnham, K.P. and Anderson, D.R. (2002). *Model Selection and Multimodel Inference: A Practical Information Theoretic Approach*, 2nd ed. New York: Springer.

Bustamante, M.E. (1958). *La fiebre amarilla en Mexico y su origen en America*. Mexico City: Instituto de Enfermedades Tropicales.

Calisher, C. (2005). Persistant emergence of dengue. *Emerging Infectious Diseases* **11**, 738–739.

CDC. (2001). *Information on Aedes albopictus*, http://www.cdc.gov/ncidod/dvbid/arbor/albopic_new.htm/ Atlanta: CDC.

Chen, W.J., Chen, S.L., Chien, L.J., Chen, C.C., King, C.C., Harn, M.R., Hwang, K.P. and Fang, J.H. (1996). Silent transmission of the dengue virus in southern Taiwan. *American Journal of Tropical Medicine and Hygiene* **55**, 12–16.

Christophers, S.R. (1960). *Aëdes aegypti (L.): The Yellow Fever Mosquito: Its Life History, Bionomics and Structure*. Cambridge: Cambridge University Press.

Clark, D.V., Mammen, M.P., Jr., Nisalak, A., Puthimethee, V. and Endy, T.P. (2005). Economic impact of dengue fever/dengue hemorrhagic fever in Thailand at the family and population levels. *American Journal of Tropical Medicine and Hygiene* **72**, 786–791.

Cliff, A., Haggett, P. and Smallman-Raynor, M. (2004). *World Atlas of Epidemic Diseases*. London: Arnold.

Congalton, R.G. (1991). A review of assessing the accuracy of classifications of remotely sensed data. *Remote Sensing and Environment* **37**, 35–46.

de Thoisy, B., Dussart, P. and Kazanji, M. (2004). Wild terrestrial rainforest mammals as potential reservoirs for flaviviruses (yellow fever, dengue 2 and St. Louis encephalitis viruses) in French Guiana. *Transactions of the Royal Society for Tropical Medicine and Hygiene* **98**, 409–412.

Deubel, V., Digoutte, J.P., Monath, T.P. and Girard, M. (1986). Genetic heterogeneity of yellow fever virus strains from Africa and the Americas. *Journal of General Virology* **67**, 209–213.

Diallo, M., Ba, Y., Sall, A.A., Diop, O.M., Ndione, J.A., Mondo, M., Girault, L. and Mathiot, C. (2003). Amplification of the sylvatic cycle of dengue virus type 2, Senegal, 1999–2000: entomologic findings and epidemiologic considerations. *Emerging Infectious Diseases* **9**, 362–367.

Division of Epidemiological Surveillance and Health Situation and Trend Assessment. (1992). *Global Health Situation and Projections: Estimates.* Geneva: WHO.

Doherty, R.L., Westaway, E.G. and Whitehead, R.H. (1967). Further studies of the aetiology of an epidemic of dengue in Queensland, 1954–1955. *Medical Journal of Australia* **2**, 1078–1080.

Durieux, C. (1956). Mass yellow fever vaccination in French Africa South of Sahara. In: *Yellow Fever Vaccination* (K.C. Smithburn, C. Duriex, R. Koerber, H.A. Penna, G.W.A. Dick, G. Courtois, C. de Sousa Manso, G. Stuart, P.H. Bonnel, eds), Vol. Monograph series no 30, pp. 115–121. Geneva: WHO.

Effler, P.V., Pang, L., Kitsutani, P., Vorndam, V., Nakata, M., Ayers, T., Elm, J., Tom, T., Reiter, P., Rigau-Perez, J.G., Hayes, J.M., Mills, K., Napier, M., Clark, G.G. and Gubler, D.J. (2005). Dengue fever, Hawaii, 2001–2002. *Emerging Infectious Diseases* **11**, 742–749.

Findlay, G.M., Stefanopoulo, G.J., Davey, T.H. and Mahaffy, A.F. (1936). Yellow fever immune bodies in the blood of African animals. *Transactions of the Royal Society for Tropical Medicine and Hygiene* **29**, 419–424.

Fontenille, D., Diallo, M., Mondo, M., Ndiaye, M. and Thonnon, J. (1997). First evidence of natural vertical transmission of yellow fever virus in *Aedes aegypti*, its epidemic vector. *Transactions of the Royal Society for Tropical Medicine and Hygiene* **91**, 533–535.

Gallup, J. L. & Sachs, J. D. (2000). *The economic burden of malaria.* CID Working paper No. 52, p. 22.

General Government of French West Africa. (1929). *Proceedings of the African Conference of the Yellow Fever*. Dakar, April, 1928.

Goh, K.T. (1995). Changing epidemiology of dengue in Singapore. *Lancet* **346**, 1098.

Goh, K.T. (1997). Dengue—a re-emerging infectious disease in Singapore. *Annals of the Academy for Medicine Singapore* **26**, 664–670.

Gordon-Smith, C.E., Turner, L.H. and Armitage, P. (1962). Vaccination in Malaya by subcutaneous injection and multiple puncture. *Bulletin of the World Health Organization* **27**, 717–727.

Gorgas, W.C. (1915). *Sanitation in Panama*. New York: Appleton & Co.

Gratz, N.G. (2004). Critical review of the vector status of *Aedes albopictus*. *Medical and Veterinary Entomology* **18**, 215–227.

Great Britain Colonial Office Yellow Fever Commission. (1915a). *Reports on questions connected with the investigation of non-malarial fevers in West Africa*, Vol. 1. Liverpool: University Press of Liverpool.

Great Britain Colonial Office Yellow Fever Commission. (1915b). *Reports on questions connected with the investigation of non-malarial fevers in West Africa: supplemental information*, Vol. 4. Liverpool: University Press of Liverpool.

Great Britain Colonial Office Yellow Fever Commission. (1915c). *Reports on questions connected with the investigation of non-malarial fevers in West Africa: supplemental information*, Vol. 2. Liverpool: University Press of Liverpool.

Gubler, D.J. (1997). Dengue and dengue hemorrhagic fever: its history and resurgence as a global public health problem. In: *Dengue and Dengue Hemorrhagic Fever* (D.J. Gubler and G. Kuno, eds), pp. 1–22. London: CAB International.

Gubler, D.J. (1998). Dengue and dengue hemorrhagic fever. *Clinical Microbiology Reviews* **11**, 480–496.

Gubler, D.J. (2002). The global emergence/resurgence of arboviral diseases as public health problems. *Archives of Medical Research* **33**, 330–342.

Gubler, D.J. (2003). *Aedes albopictus* in Africa. *The Lancet* **3**, 751–752.

Gubler, D.J. (2004). The changing epidemiology of yellow fever and dengue 1900 to 2003: full circle? *Comparative Immunology, Microbiology and Infectious Diseases* **27**, 319–330.

Gubler, D.J. and Kuno, G. (1997). *Dengue and Dengue Hemorrhagic Fever*. London: CAB International.

Guzman, M.G., Kouri, G.P., Bravo, J., Soler, M., Vazquez, S. and Morier, L. (1990). Dengue haemorrhagic fever in Cuba. A retrospective seroepidemiologic survey. *American Journal of Tropical Medicine and Hygiene* **42**, 179–184.

Haddow, A.J. (1965). Yellow fever in central Uganda, 1964. *Part I. Historical introduction. Transactions of the Royal Society of Tropical Medicine and Hygiene* **59**, 436–441.

Hales, S., de Wet, N., Maindonald, J. and Woodward, A. (2002). Potential effect of population and climate changes on global distribution of dengue fever: an empirical model. *Lancet* **360**, 830–834.

Halstead, S.B. (1970). Observations related to pathogenesis of dengue hemorrhagic fever. VI. *Hypothesis and discussion. Yale Journal of Biological Medicine* **42**, 350–360.

Halstead, S.B. (1979). *In vivo* enhancement of dengue virus infection in rhesus monkeys by passively transferred antibody. *Journal of Infectious Diseases* **140**, 527–533.

Halstead, S.B. (1988). Dengue hemorrhagic fever. In: *Handbook of Viral and Rickettsial Hemorrhagic Fevers* (J.H.S. Gear, ed.), Vol. 1, pp. 85–94. Boca Raton, FL: CRC Press.

Halstead, S.B. (1997). Epidemiology of dengue and dengue hemorrhagic fever. In: *Dengue and Dengue Hemorrhagic Fever* (D.J. Gubler and G. Kuno, eds), p. 478. Cambridge: CAB International.

Halstead, S.B. (2005). More dengue, more questions. *Emerging Infectious Diseases* **11**, 740–741.

Halstead, S.B. and Deen, J. (2002). The future of dengue vaccines. *Lancet* **360**, 1243–1245.

Halstead, S.B. and O'Rourke, E.J. (1977). Dengue viruses and mononuclear phagocytes. I. Infection enhancement by non-neutralising antibody. *Journal of Experimental Medicine* **146**, 201–217.

Halstead, S.B., Porterfield, J.S. and O'Rourke, E.J. (1980). Enhancement of dengue virus infection in monocytes by flavivirus antisera. *American Journal of Tropical Medicine and Hygiene* **29**, 638–642.

Hare, F.E. (1898). The 1897 epidemic of dengue in Northern Queensland. *Australasian Medical Gazette* **17**, 98–107.

Hay, S.I., Guerra, C.A., Tatem, A.J., Atkinson, P.M. and Snow, R.W. (2005). Urbanization, malaria transmission and disease burden in Africa. *Nature Reviews Microbiology* **3**, 81–90.

Holmes, E.C., Bartley, L.M. and Garnett, G.P. (1998). The emergence of dengue: past, present and future. In: *Emerging Infections* (R.M. Krause, ed.), pp. 301–325. San Diego: Academic Press.

Hopp, M.J. and Foley, J.A. (2001). Global-scale relationships between climate and the dengue fever vector. *Aedes aegypti. Climatic Change* **48**, 441–463.

Innis, B.L., Eckels, K.H., Kraiselburd, E., Dubois, D.R., Meadors, G.F., Gubler, D.J., Burke, D.S. and Bancroft, W.H. (1988). Virulence of a live dengue virus vaccine candidate: a possible new marker of dengue virus attenuation. *Journal of Infectious Diseases* **158**, 876–880.

James, M.E. and Kalluri, S.N.V. (1994). The pathfinder AVHRR land data set—an improved coarse resolution data set for terrestrial monitoring. *International Journal of Remote Sensing* **15**, 3347–3363.

Jorge, R. 1938. *Fievre Jaune*. Lisbon.

Kilpatrick, J.W., Tonn, R.J. and Jatanasen, S. (1970). Evaluation of ultra-low-volume insecticide dispensing systems for use in single-engined aircraft and their effectiveness against *Aedes aegypti* populations in South-East Asia. *Bulletin of the World Health Organization* **42**, 1–14.

Kliks, S.C., Nisalak, A., Brandt, W.E., Wahl, L. and Burke, D.S. (1989). Antibody-dependent enhancement of dengue virus growth in human monocytes as a risk factor for dengue hemorrhagic fever. *American Journal of Tropical Medicine and Hygiene* **40**, 444–451.

Knudsen, A.B. (1977). The silent jungle transmission cycle of dengue virus and its tenable relationship to endemic dengue in Malaysia. *The Malaysian Nature Journal* **31**, 41–47.

Knudsen, A.B. (1995). Global distribution and continuing spread of *Aedes albopictus*. *Parasitologia* **37**, 91–97.

Kochel, T.J., Watts, D.M., Halstead, S.B., Hayes, C.G., Espinoza, A., Felices, V., Caceda, R., Bautista, C.T., Montoya, Y., Douglas, S. and Russell, K.L. (2002). Effect of dengue-1 antibodies on American dengue-2 viral infection and dengue hemorrhagic fever. *Lancet* **360**, 310–312.

Kouri, G.P., Guzman, J.R., Bravo, J.R. and Triana, C. (1989). Dengue hemorrhagic fever/dengue shock syndrome: lessons from the Cuban epidemic, 1981. *Bulletin of the World Health Organization* **67**, 375–380.

Kumm, H.W. (1931). The geographical distribution of the yellow fever vectors: a compilation of material recorded in the literature, unpublished communications and certain collections made by the author in Nigeria, West Africa. *American Journal of Hygiene*. **Monograph Series 12**, 110.

Landis, J.R. and Koch, G.C. (1977). The measurement of observer agreement for categorical data. *Biometrics* **33**, 159–174.

Lopez-Correa, R.H., Moore, C.G., Sather, G.E., Morens, D.M., Chiriboga, J., Banegura, F. and Woodall, J.P. (1979). The 1977 dengue epidemic in Puerto Rico: epidemiologic and clinical observations. *Pan American Health Organization Scientific Publication* **375**, 60–67.

Lounibos, L.P. (2002). Invasions by insect vectors of human disease. *Annual Review of Entomology* **47**, 233–266.

Ma, Z. and Redmond, R.L. (1995). Tau coefficients for accuracy of assessment of classification of remote sensing data. *Photogrammetric Engineering and Remote Sensing* **61**, 435–439.

Maguire, T., Miles, J.A., MacNamara, F.N., Wilkinson, P.J., Austin, F.J. and Mataika, J.U. (1974). Mosquito-borne infection in Fiji, 1971–73 Dengue epidemic. *Journal of Hygiene* **73**, 263–270.

Massad, E., Coutinho, F.A.B., Burattini, M.N. and Lopez, L.F. (2001). The risk of yellow fever in a dengue-infested area. *Transactions of the Royal Society of Tropical Medicine and Hygiene* **95**, 370–374.

McCullough, D. (1977). *The Path between the Seas: The Creation of the Panama Canal 1870–1914*. New York: Simon and Schuster.

Medlock, J.M., Snow, K.R. and Leach, S. (2005). Potential transmission of West Nile virus in the British Isles: an ecological review of candidate mosquito bridge vectors. *Medical and Veterinary Entomology* **19**, 2–21.

Meltzer, M.I., Rigau-Pérez, J.G., Clark, G.G., Reiter, P. and Gubler, D.J. (1998). Using disability-adjusted life years to assess the economic impact of dengue in Puerto Rico, 1984–1994. *American Journal of Tropical Medicine and Hygiene* **59**, 265–271.

Mendes Luz, P., Torres Codeço, C., Massad, E. and José Struchiner, C. (2003). Uncertainties regarding dengue modeling in Rio de Janeiro, Brazil. *Memorias do Instituto Oswaldo Cruz* **98**, 871–878.

Mhatre, R.K. (1934). *A Survey of Aedes Mosquitos in Bombay and the Measures Suggested for Their Control*. Bombay: British India Press.

Monath, T.P. (2001). Yellow fever: an update. *Lancet Infectious Diseases* **1**, 11–20.

Mongkolsapaya, J., Dejnirattisai, W., Xu, X.-n., Vasanawathana, S., Tangthawornchaikul, N., Chairunsri, A., Sawasdivorn, S., Duangchinda, T., Dong, T., Rowland-Jones, S., Yenchitsomanus, P.-t., McMichael, A.J., Malasit, P. and Screaton, G. (2003). Original antigenic sin and apoptosis in the pathogenesis of dengue hemorrhagic fever. *Nature Medicine* **9**, 921–927.

Moore, C.G. (1999). *Aedes albopictus* in the United States: current status and prospects for further spread. *Journal of the American Mosquito Control Association* **15**, 221–227.

Moore, C.G. and Mitchell, C.J. (1997). *Aedes albopictus* in the United States: ten-year presence and public health implications. *Emerging Infectious Diseases* **3**, 329–334.

Mutebi, J.-P., Wang, H., Li, L., Bryant, J.E. and Barrett, A.D.T. (2001). Phylogenetic and evolutionary relationships among yellow fever isolates in Africa. *Journal of Virology* **75**, 6999–7008.

Nisalak, A., Endy, T.P., Nimmannitya, S., Kalayanrooj, S., Thisayakorn, U., Scott, R.M., Burke, D.S., Hoke, C.H., Innis, B.L. and Vaughn, D.W. (2002). Serotype-specific dengue virus circulation and dengue disease in Bangkok, Thailand from 1973 to 1999. *American Journal of Tropical Medicine and Hygiene* **68**, 191–202.

Noguchi, H., Muller, H. R., Torres, O., Silva, F., Martins, M.D., Vianna, G. and Bião, M. (1924). *Experimental studies of yellow fever in northern Brazil*. Monographs of the Rockefeller Institute for Medical Research. 20. New York.

Pearson, E.F. and Miles, W. (1980). Disinfection of mail in the United States. *Bulletin of the History of Medicine* **54**, 111–124.

Pinheiro, F.P. (1989). Dengue in the Americas, 1980–1987. *Epidemiological Bulletin of PAHO* **10**, 1–8.

Porter, K.R., Beckett, C.G., Kosasih, H., Tan, R.I., Alisjahbana, B., Rudiman, P.I.F., Widjaja, S., Listiyaningsih, E., Ma'Roef, C.N., McArdle, J., Parwati, I., Sudjana, P., Jusuf, H., Yuwono, D. and Wuryadi, S. (2005). Epidemiology of dengue and dengue hemorrhagic fever in a cohort of adults living in Bandung, west Java, Indonesia. *American Journal of Tropical Medicine and Hygiene* **72**, 60–66.

Reinert, J.F., Harbach, R.E. and Kitching, I.J. (2004). Phylogeny and classification of *Aedini* (Diptera: Culicidae), based on morphological characters of all life stages. *Zoological Journal of the Linnean Society* **142**, 289–368.

Reiter, P., Cordellier, R., Ouma, J.O., Cropp, C.B., Savage, H.M., Sanders, E.J., Marfin, A.A., Tukei, P.M., Agata, N.N., Gitau, L.G., Rapuoda, B.A. and Gubler, D.J. (1998). First recorded outbreak of yellow fever in Kenya, 1992–1993. II. Entomologic investigations. *American Journal of Tropical Medicine and Hygiene* **59**, 650–656.

Reiter, P. and Gubler, D.J. (1997). Surveillance and control of urban dengue vectors. In: *Dengue and Dengue Hemorrhagic Fever* (D.J. Gubler and G. Kuno, eds), pp. 425–462. London: CAB International.

Rodriguez-Roche, R., Alvarez, M., Holmes, E.C., Bernardo, L., Kouri, G., Gould, E.A., Halstead, S. and Guzman, M.G. (2005). Dengue virus type 3, Cuba, 2000–2002. *Emerging Infectious Diseases* **11**, 773–774.

Rogers, D.J. (2000). Satellites, space, time and the African trypanosomiases. *Advances in parasitology* **47**, 130–171.
Rosen, L. (1987). Sexual transmission of dengue viruses by *Aedes albopictus*. *American Journal of Tropical Medicine and Hygiene* **37**, 398–402.
Rosen, L., Roseboom, L.E., Gubler, D.J., Lien, J.C. and Chaniotis, B.N. (1985). Comparative susceptibility of mosquito species and strains to oral and parenteral infection with dengue and Japanese encephalitis viruses. *American Journal of Tropical Medicine and Hygiene* **34**, 603–615.
Rothman, A.L. and Ennis, F.A. (1999). Immunopathogenesis of dengue hemorrhagic fever. *Virology* **257**, 1–6.
Rush, A. B. (1789). An account of the bilious remitting fever, as it appeared in Philadelphia in the summer and autumn of the year 1780. In: *Medical Enquiries and Observations*. Philadelphia, Pritchard & Hall, pp. 104–117.
Saluzzo, J.F., Cornet, M., Adam, C., Eyraud, M. and Digoutte, J.P. (1986). Dengue-2 in eastern Senegal—serological survey in simian and human populations (1974 to 1985). *Bulletin de la Societe de Pathologie Exotique* **79**, 313–322.
Sangkawibha, N., Rojanasuphot, S., Ahandrik, S., Viriyapongse, S., Jatanasen, S., Salitul, V., Phanthumachinda, B. and Halstead, S.B. (1984). Risk factors in dengue shock syndrome: a prospective epidemiologic study in Rayong, Thailand. I. The 1980 outbreak. *American Journal of Epidemiology* **120**, 653–669.
Sawyer, W. A. (1934). The present geographic distribution of yellow fever and its significance. 66–92.
Sawyer, W.A. and Whitman, L. (1936). The yellow fever immunity survey of north, east and south Africa. *Transactions of the Royal Society of Tropical Medicine and Hygiene* **29**, 397–412.
Scott, T.W., Chow, E., Strickman, D., Kittapayong, P., Wirtz, R.A., Lorenz, L.H. and Edman, J.D. (1993). Blood-feeding patterns of *Aedes aegypti* (Diptera: Culicidae) collected in a rural Thai village. *Journal of Medical Entomology* **30**, 922–927.
Sheppard, P.M., MacDonald, W.W., Tonn, R.J. and Grab, B. (1969). The dynamics of an adult population of *Aedes aegypti* in relation to dengue haemorrhagic fever in Bangkok. *Journal of Animal Ecology* **38**, 661–702.
Siler, J.F., Hall, M.W. and Hitchens, A.P. (1926). Dengue: the history, epidemiology, mechanism of transmission, etiology, clinical manifestations, immunity and prevention. *Philippine Journal of Science* **29**, 1–304.
Soper, F.L. (1934). Some notes on the epidemiology of yellow fever in Brazil. *Revista de Hygiene e Saude Publica* **8**, 73–94.
Soper, F.L. (1935a). Recent extensions of knowledge of yellow fever. *Quarterly Bulletin of the Health Organisation of the League of Nations* **5**, 1–50.
Soper, F. L., (1935b). *Rural and jungle yellow fever: a new public health problem in Columbia*. Lecture given before the Faculty of Medicine of Bogota.

Soper, F.L. (1955). *The unfinished business of Yellow Fever. A speech given at a Symposium in Commemoration of Carlos Juan Finlay*. Jefferson Medical College of Philadelphia.

Southwood, T.R.E., Murdie, G., Yasuno, M., Tonn, R.J. and Reader, P.M. (1972). Studies on the life budget of *Aedes aegypti* in Wat Samphaya, Bangkok, Thailand. *Bulletin of the World Health Organization* **46**, 211–226.

Spiegel, J., Yassi, A. and Tate, R. (2002). Dengue in Cuba: mobilisation against *Aedes aegypti. Lancet Infectious Diseases* **2**, 207–208.

Teixeira, M.D.G., Barreto, M.L., Costa, M.D.C.N., Ferreira, L.D.A., Vasconcelos, P.F.C. and Cairncross, S. (2002). Dynamics of dengue virus circulation: a silent epidemic in a complex urban area. *Tropical Medicine and International Health* **7**, 757–762.

Theiler, M. and Anderson, C.R. (1975). The relative resistance of dengue-immune monkeys to yellow fever virus. *American Journal of Tropical Medicine and Hygiene* **24**, 115–117.

Trpis, M. and Hausermann, W. (1986). Dispersal and other population parameters of *Aedes aegypti* in an African village and their possible significance in epidemiology of vector-borne diseases. *American Journal of Tropical Medicine and Hygiene* **35**, 1263–1279.

Vainio, J. and Cutts, F. (1998). *Yellow Fever*. Geneva: *WHO Division of Emerging and other Communicable Diseases Surveilance and Control.*

van Rooyen, C.E. and Rhodes, A.J. (1948). *Virus Diseases of Man*. New York: Thomas Nelson & Sons.

Vasconcelos, P.F.C., Costa, Z.G., da Rossa, E.S.T., Luna, A., Rodrigues, S.G., Barros, V.L.R.S., Dias, J.P., Monteiro, H.A.O., Oliva, O.F.P., Vasconcelos, H.B., Oliveira, R.C., Sousa, M.R.S., Da Silva, J.B., Croz, A.C.R., Martins, E.C. and Da Rossa, J.F.S.T. (2001). Epidemic of jungle yellow fever in Brazil, 2000: implications of climatic alterations in disease spread. *Journal of Medical Virology* **65**, 598–604.

Vaughn, D.W. (2000). Invited commentary: lessons from Cuba. *American Journal of Epidemiology* **152**, 800–803.

Von Allmen, S.D., Lopez-Correa, R.H., Woodall, J.P., Morens, D.M., Chiriboga, J. and Castavelez, A. (1979). Epidemic dengue fever in Puerto Rico, 1977: a cost analysis. *American Journal of Tropical Medicine and Hygiene* **28**, 1040–1044.

Whitman, L. (1951). The arthropod vectors of yellow fever. In: *Yellow fever* (G.K. Strode, ed.), p. 263. New York: McGraw-Hill.

WHO. (1971). *WHO expert committee on yellow fever 3rd report.*

WHO. (1986). *Prevention and Control of Yellow Fever in Africa*. Geneva: WHO.

WHO. (2001). *Yellow Fever*, Vol. 2005, pp. WHO fact sheet 100. Geneva: WHO.

WHO. (2002). *Dengue and Dengue Hemorrhagic Fever*, Vol. 2004, pp. WHO fact sheet 117. Geneva: WHO.

Wilder-Smith, A., Foo, W., Earnest, A., Sremulanathan, S. and Paton, N.I. (2004). Seroepidemiology of dengue in the adult population of Singapore. *Tropical Medicine and International Health* **9**, 305–308.

Zivna, I., Green, S., Vaughn, D.W., Kalayanrooj, S., Stephens, H.A.F., Chandanayingyong, D., Nisalak, A., Ennis, F.A. and Rothman, A.L. (2002). T cell responses to an HLA-B*07-restricted epitope on the dengue NS3 protein correlate with disease severity. *Journal of Immunology* **168**, 5959–5965.

Global Epidemiology, Ecology and Control of Soil-Transmitted Helminth Infections

S. Brooker[1], A.C.A. Clements[1,2] and D.A.P. Bundy[3]

[1]*Department of Infectious and Tropical Disease, London School of Hygiene and Tropical Medicine, Keppel Street, London WC1E 7HT, UK*
[2]*Schistosomiasis Control Initiative, Department of Infectious Disease Epidemiology, Faculty of Medicine, Imperial College, Norfolk Place, London W2 1PG, UK*
[3]*Human Development Division, The World Bank, Washington DC 20433, USA*

Abstract . . . 221
1. Introduction . . . 222
2. Transmission Dynamics and the Environment . . . 224
3. Ecological Correlates . . . 227
4. Predicting Distributions . . . 231
5. Urbanization . . . 232
6. Global Control Strategies . . . 234
7. Control Applications of GIS/RS . . . 237
8. Global Distributions . . . 243
9. Predicted Numbers of Infections . . . 246
10. The Future . . . 250
Acknowledgements . . . 252
References . . . 252

ABSTRACT

Soil-transmitted helminth (STH) infections are among the most prevalent of chronic human infections worldwide. Based on the demonstrable impact on child development, there is a global commitment to finance and implement control strategies with a focus on school-based

ADVANCES IN PARASITOLOGY VOL 62
ISSN: 0065-308X $35.00
DOI: 10.1016/S0065-308X(05)62007-6

chemotherapy programmes. The major obstacle to the implementation of cost-effective control is the lack of accurate descriptions of the geographical distribution of infection. In recent years, considerable progress has been made in the use of geographical information systems (GIS) and remote sensing (RS) to better understand helminth ecology and epidemiology, and to develop low-cost ways to identify target populations for treatment. This review explores how this information has been used practically to guide large-scale control programmes. The use of satellite-derived environmental data has yielded new insights into the ecology of infection at a geographical scale that has proven impossible to address using more traditional approaches, and has in turn allowed spatial distributions of infection prevalence to be predicted robustly by statistical approaches. GIS/RS have increasingly been used in the context of large-scale helminth control programmes, including not only STH infections but also those focusing on schistosomiasis, filariasis and onchocerciasis. The experience indicates that GIS/RS provides a cost-effective approach to designing and monitoring programmes at realistic scales. Importantly, the use of this approach has begun to transition from being a specialist approach of international vertical programmes to becoming a routine tool in developing public sector control programmes. GIS/RS is used here to describe the global distribution of STH infections and to estimate the number of infections in school-age children in sub-Saharan Africa (89.9 million) and the annual cost of providing a single anthelmintic treatment using a school-based approach (US$5.0–7.6 million). These are the first estimates at a continental scale to explicitly include the fine spatial distribution of infection prevalence and population, and suggest that traditional methods have overestimated the situation. The results suggest that continent-wide control of parasites is, from a financial perspective, an attainable goal.

1. INTRODUCTION

There are four main nematode species of human soil-transmitted helminth (STH) infections, also known as geohelminths: *Ascaris lumbricoides* (roundworm), *Trichuris trichiura* (whipworm), *Ancylostoma duodenale* and *Necator americanus* (hookworms). These infections are most prevalent in tropical and sub-tropical regions of the developing world where adequate water and sanitation are lacking, with recent estimates suggesting that *A. lumbricoides* infects 1221 million people, *T. trichiura* 795 million and hookworms 740 million (de Silva *et al.*, 2003). The greatest numbers of STH infections occur in sub-Saharan Africa (SSA), East Asia, China, India and South America.

Chronic and intense STH infections can contribute to malnutrition and iron-deficiency anaemia, and also can adversely affect physical and mental growth in childhood (Drake *et al.*, 2000; Stephenson *et al.*, 2000; Hotez *et al.*, 2004). In recognition of the global health importance of STH infections, there is a renewed global commitment to finance and implement control strategies to reduce the disease burden of STH and other helminths, including schistosomiasis (Fenwick *et al.*, 2003), filariasis and onchocerciasis (Molyneux *et al.*, 2003). The development of effective helminth control is possible because of the availability of proven, cost-effective and logistically feasible intervention strategies. In the case of STH infections, regular periodic chemotherapy, using benzimidazole anthlemintics (BZAs), of school-aged children delivered through the school system is the main intervention strategy (Aswashi *et al.*, 2003; Hotez *et al.*, 2006; Bundy *et al.*, 2006). Understanding where at-risk populations live is fundamental for appropriate resource allocation and cost-effective control. In particular, it allows for reliable estimation of the overall drug needs of programmes and efficient geographical targeting of control efforts (Brooker and Michael, 2000). The precise global distribution of STH infection and how many people are infected and at risk of morbidity however remains poorly defined. This limits how national governments and international organizations define and target resources to combat the disease burden due to STH infection.

A previous review in this series highlighted the potential use of geographical information systems (GIS) and remote sensing (RS) to better understand helminth distributions and their ecological correlates, but also to serve as geographic decision-making tools for identifying areas of particular risk as well as for the design, implementation and monitoring of control programmes (Brooker and Michael, 2000). As an increasing number of large-scale control efforts are underway, it is timely to assess how the potential of GIS and RS to guide control has been realized in practice. This article begins by describing the scientific basis of how environmental factors affect the biology and transmission dynamics of STH infection. We then show how satellite data can be used to establish and predict species-specific distributions, and how these tools can help shed additional light on the ecology and epidemiology of infection. Next, we describe how these tools have been effectively used within the context of large-scale control programmes. Finally, we adopt a data-driven approach to map the contemporary global distributions of STH infection, and relate these to global human population-distribution data to derive regional and national estimates of population at risk by parasite species. Although focusing on STH infections, examples will also be presented for other helminthiases, including schistosomiasis, filariasis and onchocerciasis.

2. TRANSMISSION DYNAMICS AND THE ENVIRONMENT

To understand and ultimately predict the global distribution of STH infections, it is essential to appreciate their biology, ecology and transmission dynamics. The life cycles of STH infection follow a general pattern. The adult parasite stages inhabit some part of the host intestine (*A. lumbricoides* and hookworm in the small intestine; *T. trichiura* in the colon), reproduce sexually and produce eggs, which are passed in human faeces and deposited in the external environment. Adult worms survive for several years and produce large numbers of eggs after 4–6 weeks (Table 1). Eggs can remain viable in the soil for several months (*A. lumbricoides* and *T. trichiura*) and larvae several weeks (hookworms), dependent on prevailing environmental conditions. *A. duodenale* larvae can undergo hypobiosis (arrested development at a specific point in the nematode life cycle) in the human body under certain environmental conditions for several months. Infection occurs through accidental ingestion of eggs (*A. lumbricoides* and *T. trichiura*) or penetration of the skin (by hookworm larvae).

As is common for infectious diseases, the transmission of STH infections can be summarized by the basic reproductive number (R_0).

Table 1 Population parameters, development rates and life expectancies of parasites and free-living STH infective stages

	A. lumbricoides	*T. trichiura*	Hookworm
Infective stage	Ova	Ova	Larvae
Egg production (eggs/female worm/day)[a]	10 000–200 000	2000–20 000	3000–20 000
Life expectancy of free-living infective stages[a]	28–84 days	10–30 days	3–10 days
Adult life span[b]	1–2 years	1–2 years	3–4 years
Pre-patency (adult development to sexual maturity)[b]	50–80 days	50–84 days	28–50 days
Larvae development time to infective stage[c]	8–37 days	20–100 days	2–14 days
Max. temp. of viable development[c]	35–39°C	37–39°C	40°C
Basic reproductive number[b]	1–5	4–6	2–3

[a]Data taken from Anderson (1982), Bundy and Cooper (1989) and Crompton (2001).

[b]Data taken from Anderson and May (1991).

[c]Data on *A. lumbricoides* (Seamster (1950); *T. suis* (Beer, 1976); hookworm (Nwosu, 1978; Smith and Schad, 1989)).

This is defined as the average number of female offspring produced by one adult female parasite that attain reproductive maturity, in the absence of density dependent constraints (Anderson and May, 1991). R_0 values of between 1 and 6 are estimated, with rates intrinsically highest for *T. trichiura* and lowest for hookworm. In practice, epidemiological studies fail to differentiate between the main hookworm species, *A. duodenale* and *N. americanus*, which will have different epidemiological and ecological characteristics.

Increases in R_0 give rise to increases in infection prevalence (percentage of individuals infected) and infection intensity (number of worms per human host). The dynamic processes involved in STH transmission, such as free-living infective stage development and survival, depend on the prevailing environmental conditions (Pavlovsky, 1966; Anderson, 1982). For example, as indicated in Figure 1, free-living infective stages present in the environment develop and die at temperature-dependent rates. Maximum survival rates of hookworm larvae, as indicated by proportion of larvae surviving, occur at 20–30 °C (Figure 1a). Experimental studies suggest that maximum development

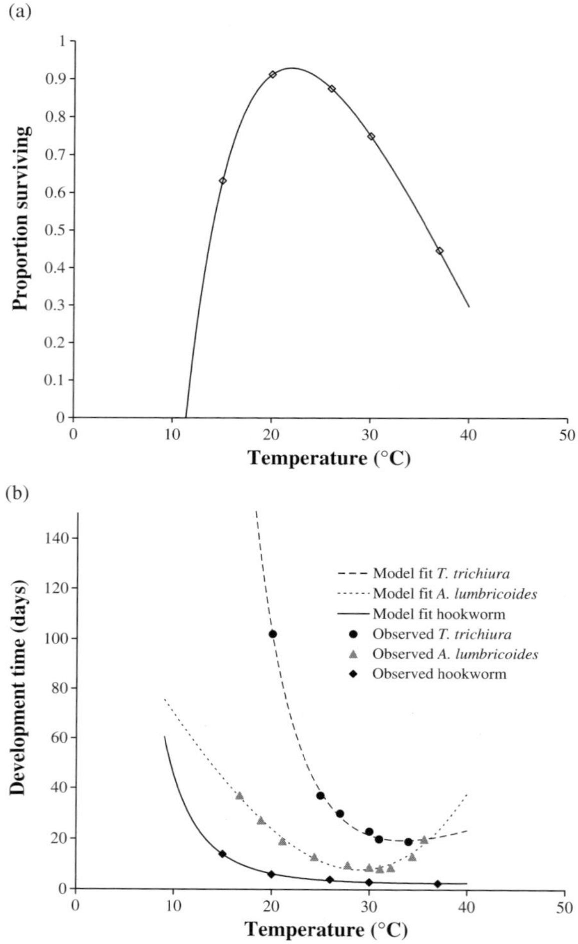

Figure 1 Relationship between temperature and (a) hookworm survival and (b) development duration. Points indicate experimental data (Seamster, 1950; Beer, 1973; Nwosu, 1978; Udonsi and Atata, 1987) and lines are fits derived from fractional polynomials analyses. Regression details: parasite survival $y = 0.884 + (-22.88x^{-0.5}) + (-7.73\ln(x))$; *A. lumbricoides* duration $y = 8.601 + (-63.718x^{-2}) + 2.526x^{-3}$; *T. trichiura* duration $y = 26.079 + 41209.68x^{-2} + (-1715.02x^{-2})$; hookworm duration $y = 3.701 + 40.88x^{-2} + (-46.18x^{-2})$.

rates of free-living infective stages occur at temperatures between 28 and 32°C, with development of *A. lumbricoides* and *T. trichiura* arresting below 5 and above 38°C (Beer, 1976; Seamster, 1950), and development of hookworm larvae ceasing at 40°C (Udonsi and Atata, 1987; Smith and Schad, 1989) (Figure 1b). It is suggested that *A. lumbricoides* eggs are more resistant to extreme temperatures than *T. trichiura* eggs (Bundy and Cooper, 1989).

Soil moisture and relative atmospheric humidity are also known to influence the development and survival of ova and larvae: higher humidity is associated with faster development of ova; and at low humidity (<50%) the ova of *A. lumbricoides* and *T. trichiura* do not embryonate (Otto, 1929; Spindler, 1929). Field studies show that the abundance of hookworm larvae is related to atmospheric humidity (Nwosu and Anya, 1980; Udonsi *et al.*, 1980).

These differing rates of development and survival will influence parasite establishment in the human host and hence the infection levels. Thus, a climate-induced increase in the rate of establishment, while holding parasite mortality constant, causes the parasite equilibrium to rise (Bundy and Medley, 1992). Although seasonal dynamics in transmission may occur, such fluctuations may be of little significance to the overall parasite equilibrium within communities. This is because the life span of adult worms is typically much longer (1–10 years) than the periods in the year during which R_0 is less than unity, and R_0 on average will be greater than 1, maintaining overall endemicity (Anderson, 1982). For all these reasons, spatial variability in long-term synoptic environmental factors will have a greater influence on transmission success and patterns of STH infection than seasonal variability in a location.

3. ECOLOGICAL CORRELATES

Given the importance of environmental factors on transmission processes, it is unsurprising that statistical relationships between environmental factors and spatial patterns of infection can be observed. Several studies, some dating back to the early twentieth century, report ecological associations between STH distributions and temperature, rainfall and altitude (reviewed in Brooker and Michael, 2000). As outlined in Hay *et al.*, (this volume, pp. 37–77), the ability to investigate ecological correlates of infection has been enhanced by satellite imagery, which can provide data from which information about temperature, humidity and vegetation conditions can be derived, and by the use of GIS to overlay multiple layers of data. These tools have been used successfully to describe the environmental factors associated with patterns of STH infection in selected geographical locations, and helped identify the relative importance of different environmental factors in determining geographic distributions (Appleton and Gouws, 1996; Appleton *et al.*, 1999; Brooker *et al.*, 2002a,b, 2003, 2004b; Saathoff *et al.*, 2005).

Here we extend these analyses and investigate the spatial ecology of STH infection across SSA, where surprisingly little is still known about the distributions of STH infection and their underlying environmental determinants. Although intensity of STH infection is a key determinant of transmission dynamics and morbidity (Anderson and May, 1991), prevalence of infection, based on microscopic examination of STH eggs in stool samples, remains the most widely used indicator of infection status and the need for control. We derive estimates of STH prevalence from dedicated surveys, conducted among schoolchildren after 1985 and geo-referenced using global positioning systems. Detailed data are available for Cameroon, Chad, Eritrea, Guinea, South Africa, Tanzania, Uganda, and Zambia, which show that prevalence of *A. lumbricoides* and *T. trichiura* is greatest in equatorial, central and west Africa and southeast South Africa, whereas hookworm has a wider distribution across the continent (Figure 2 is Plate 7.2 in the Separate Color Plate Section). By relating these distributions to satellite-derived environmental data, we can investigate their large-scale ecological correlates.

For each species, a clear relationship exists between prevalence of infection and remotely sensed Land Surface Temperature (LST) (Cracknell, 1997): prevalence of *A. lumbricoides* and *T. trichiura* is generally $<5\%$ in areas where LST exceeds 38–40°C, whereas hookworm infection remains highly prevalent throughout the upper end of the thermal range (Figure 3). This is an intriguing observation since experimental studies suggest that each STH species has similar thermal

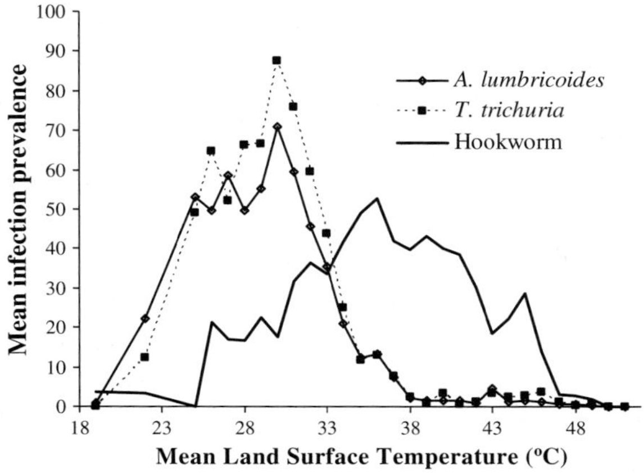

Figure 3 Relationship between mean LST, estimated from satellite data, and prevalence of STH infection. Estimates are derived for each survey location and data for locations experiencing the same temperature are averaged for presentation. (see legend of Figure 2 for details.)

thresholds (Table 1). The apparent ability of hookworm to survive hotter conditions may be explained in part by the ability of mobile larvae to migrate to more suitable thermal and moisture conditions. In particular, whereas hookworm larvae stages have some limited motility and can move downward into the soil, thereby avoiding desiccation, the ova of *A. lumbricoides* and *T. trichiura* are non-motile, and high surface temperatures will result in ova dying from desiccation (Beaver, 1953).

Some features of helminth life cycles, such as hypobiosis, equip the parasite to survive periods that are unsuitable for transmission. However, hypobiosis only occurs for *A. duodenale* in humans (Schad *et al.*, 1973) and not for *N. americanus*, the hookworm species predominant in SSA (Kilama, 1990) and is therefore excluded here as a possible explanation for hookworm's apparent wider thermal tolerance. Other species-differences in life history traits may play an important role. It is suggested that the probability of hookworm larvae surviving to infect hosts is enhanced by their more rapid development in the soil (3–10 days) in comparison with *A. lumbricoides* larvae (28–84 days) and *T. trichiura* larvae (10–30 days) (Table 1). This hypothesis is supported by an observed relationship between prevalence and the number of days in the year that below the thermal threshold (Figure 4). The prevalence of *A. lumbricoides* and *T. trichiura* was generally low in locations where temperatures fall below the thermal threshold for less than 35–40 days, and increased with increasing number of days. Hookworms, however, required a much smaller (8 day) window of thermal suitability for transmission and so were able to persist even when the period available for development was of the order of 10 days.

Finally, a potentially very important factor is the longevity of the adult worm, since the location inside the host is essentially a refuge

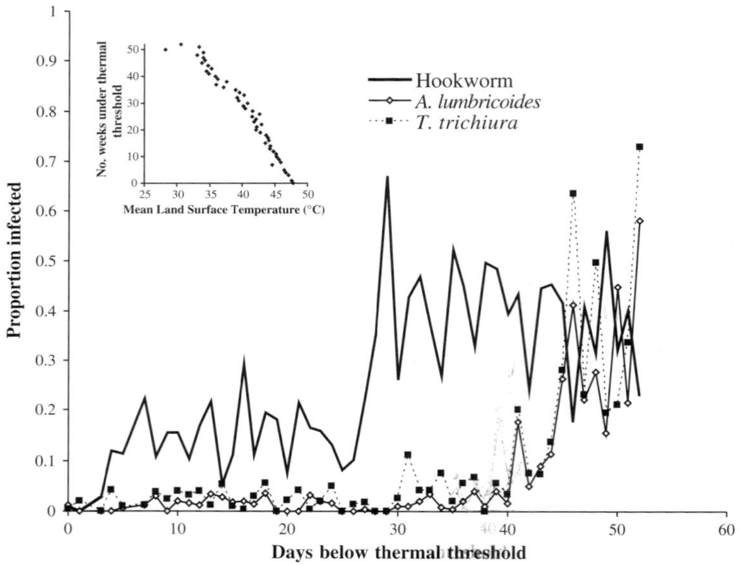

Figure 4 Relationship between the proportion of humans infected with worms and the number of days during which temperature was less than 40 °C, which is suitable for survival of free-living STH infective stages: based on observed data for 601 locations from nationwide surveys in Cameroon (Ratard *et al.*, 1991, 1992), Chad (Brooker *et al.*, 2002a) and Uganda (Brooker *et al.*, 2004b). Data were collected in cross-sectional school surveys using similar diagnostic technique (Kato–Katz method) and sampling designs (stratified, random), and encompass a broad range of infection rates. Insert: Relationship between mean LST and number of weeks temperature falls below 40 °C ($y = -3.006x + 147.8$, $r = 0.94$, $p < 0.001$). Estimates were derived for each survey location and locations with the same number of days under the thermal threshold were averaged for presentation.

from the high external temperatures. Hookworms have a longer adult stage life expectancy (3–4 years) than either *A. lumbricoides* or *T. trichiura* (1–2 years). This implies that hookworms can find refuge from external temperatures for more than twice the length of time of the other species, and are effectively protected from extreme temperatures over a 3–4 year period in appropriate conditions for development. This greatly increases the chances of hookworm transmission stages being deposited and developing in suitable thermal conditions.

4. PREDICTING DISTRIBUTIONS

These analyses have enabled the first continent-wide predictions of infection patterns on the basis of satellite-derived environmental covariates as well as providing insight into the ecology of infection. Predictions of prevalence were based on binomial logistic regression analysis, where, for each location, the response variable contained the total number of positive responses and the total number examined, and the independent variables were satellite-derived mean LST and Normalized Difference Vegetation Index (NDVI) for 1982–1999 (Hay *et al.*, this volume, pp. 37–77) and elevation (http://edcwww.cr.usgs.gov/landdaac/gtopo30/). Models were then cross-validated using a jack-knife procedure (King *et al.*, 2004) and predicted values were compared to observed values using Receiver Operating Characteristics (ROC) analysis (Brooker *et al.*, 2002b). The models for *A. lumbricoides* and *T. trichiura* provided impressive descriptive accuracies of low transmission (prevalence $<5\%$) areas and of areas which would be the target of mass treatment programmes (prevalence $>50\%$) (Figure 2). By contrast, the hookworm model has only moderate accuracy, a probable reflection of the apparent wider distribution of hookworm. The models indicate that prevalence of *A. lumbricoides* and *T. trichiura* is greatest in equatorial, central and west Africa, eastern Madagascar and southeast Africa, whereas hookworm is more widely distributed across the continent. A limitation of this analysis is the use of infection prevalence rather than infection intensity, which has greater relevance to transmission dynamics and morbidity, but which few studies quantify. There remains the need to investigate geographical heterogeneity in infection intensity.

Satellite data can therefore help define the large-scale distributions of STH infection, which are demonstrated to be influenced by heterogeneities in climate. At smaller spatial scales other factors, including variability in human behaviour, including personal hygiene, as well as differences in sanitation and socio-economic status, have to be considered. For example, in climatically unsuitable areas, microhabitats, influenced by local housing and sanitation, may provide suitable transmission foci, and *vice versa*. Such microhabitats are commonly found in urban areas.

5. URBANIZATION

In common with many other parasitic infections, STH infections flourish in impoverished areas characterized by inadequate sanitation and overcrowding. It is commonly assumed that *A. lumbricoides* and *T. trichiura* are more prevalent in urban areas whereas hookworm is more often found in rural areas (Crompton and Savioli, 1993). However, comparable data for STH infections in urban and rural settings are remarkably few and those that do exist indicate a more complicated picture. Studies which surveyed similar age groups and socio-economic areas indicate that the prevalence of *A. lumbricoides* and *T. trichiura* differ between urban and rural communities, but in no systematic manner (Table 2). By contrast, hookworm appears to be equally prevalent in both urban and rural settings (Table 2). The precise reasons for the urban–rural dichotomies for *A. lumbricoides* and *T. trichiura* are as yet unclear. Differences in prevalence of *A. lumbricoides* and *T. trichiura* in urban and rural areas may reflect differences in sanitation or population density; socio-economic differences will also play an important role. It is clear that further work is needed to resolve these issues.

By 2007, it is predicted that more than half of the global human population will be urban citizens, most of them living in the rapidly growing cities of Africa, Asia and Latin America (United Nations, 2003). Urbanization often accompanies social and economic

Table 2 Prevalence of STH infections among schoolchildren in urban and rural communities in developing countries. Included studies sought to restrict investigation to areas of similar socio-economic characteristics

Setting	Sample size	*A. lumbricoides*		*T. trichiura*		Hookworm		Reference
		Urban	Rural	Urban	Rural	Urban	Rural	
Blantyre, Malawi	553 children (3–14 years old)	15.4	0.7	—	—	0.4	2.1	Phiri *et al.* (2000)
Pemba, Tanzania	256 children (3–14 years old)	60.6	63.6	100	100	97.6	94.6	Albonico *et al.* (1997)
Buea, Cameroon	211 children (8–15 years old)	33.9	56.4	32.3	59	0	5.1	Ndenecho *et al.* (2002)
Rolandia, Brazil	236 children (5–15 years old)	6.1	1.3	0.7	0.1	4.3	4.4	Giraldi *et al.* (2001)
Malaysia	3073 children (< 15 years old)	51.7	21.2	65.3	29.1	5.7	5.9	Kan *et al.* (1989)
Penang, Malaysia	192 children (7–12 years old)	37.4	33.4	100	92	18.7	19.7	Rahman (1998)

development, with better opportunities for education, adequate living standards and higher incomes. However, overcrowding and lack of adequate water and sanitation in urban slum communities may increase transmission of STH infections. Investigation of the impact of increased urbanization on STH infections together with assessment of the effectiveness of urban helminth control measures in low-income settings is clearly warranted because increased urbanization may promote the transmission of STH infections, especially *A. lumbricoides* and *T. trichiura*.

6. GLOBAL CONTROL STRATEGIES

Recommended drugs for use in public health programmes to control STH infection are the BZAs, albendazole or mebendazole; older drugs including pyrantel pamoate and levamisole are also occasionally used in some developing countries (WHO (World Health Organization), 2002; Utzinger and Keiser, 2004). In areas where STH infections co-occur with schistosomiasis, BZAs are co-administered with praziquantel (PQZ), the major drug used for the treatment of schistosomiasis (Fenwick *et al.*, 2003).

Current efforts to control STH infection, as well as schistosomiasis, focus on the school-age population. It is estimated that between 25% and 35% of school-aged children are infected with one or more of the major species of worms (Bundy, 1997; de Silva *et al.*, 2003). The most intense worm infections and related illnesses occur at school age (Partnership for Child Development (PCD), 1998, 1999).

Infection can result in significant consequences for health and development, affecting growth, promoting anaemia and causing some overt clinical disease, much of which is rapidly reversed by treatment (Warren *et al.*, 1993; Hotez *et al.*, 2006). In addition to these impacts on health and physical development, infected schoolchildren perform poorly in tests of cognitive function; when treated, immediate educational and cognitive benefits are apparent only for children with heavy worm burdens or with concurrent nutritional deficits (Bundy *et al.*, 2006). Treatment alone can neither reverse the cumulative effects of lifelong infection nor compensate for years of missed learning, but studies suggest that children are more ready to learn after treatment for worm infections and may be able to catch up if this learning potential is exploited effectively in the classroom. In Kenya, treatment reduced absenteeism by one quarter, with the largest gains for the youngest children who suffered the most ill health (Miguel and Kremer, 2004).

For these reasons, school-age children are the natural targets for treatment, and school-based treatment delivery programmes offer major cost advantages because of the use of the existing school infrastructure and the fact that schoolchildren are accessible through schools. An important element of the approach is to minimize the need for clinical diagnosis, which is often more expensive than the treatment itself, and to focus on mass delivery of services. Evidence suggests that mass delivery of deworming is preferable on efficacy, economic and equity grounds to approaches that require diagnostic screening (Warren *et al.*, 1993). School-based deworming also has major externalities for untreated children and the whole community by reducing disease transmission in the community as a whole (Bundy *et al.*, 1990).

Recognizing the centrality of school-age children to the response to helminth infection, in 2001, the 54th World Health Assembly of the WHO passed a resolution to provide regular deworming treatment to 75% of school-age children at risk (an estimated target population of 398 million) by 2010. School health and nutrition programmes provide the vehicle for delivering regular but infrequent (every 6 months or more) anthelmintic treatment to schoolchildren. Operational research has demonstrated how interventions can be implemented and evaluated at the country level, for example enabling mass deworming of schoolchildren (Bundy and Guyatt, 1996; PCD, 1998, 1999; Guyatt *et al.*, 2001).

A major step forward in international coordination and cohesion was achieved when a framework to Focus Resources on Effective School Health (FRESH) was launched at the World Education Forum in Dakar in April 2000 (World Bank, 2000). Among the early partners in this effort were UNESCO, UNICEF, the World Food Programme (WFP), the WHO and the World Bank, with the Education Development Centre, Education International and the PCD. The FRESH framework provides a consensus approach of agreed good practice for the effective implementation of health and nutrition services within school health programmes. The framework proposes four core components that should be considered in designing an effective school health and nutrition programme and suggests that the programme will be most equitable and cost-effective if all of these components are made available, together, in all schools. The four components also provide the appropriate mix of interventions for responding to helminth infection globally: (1) *Policy*: health- and nutrition-related school policies that promote the nutrition and health of staff and children (and promote the role of teachers in delivering anthelmintic treatment); (2) *School environment*: access to safe water and provision of effective sanitation facilities (which helps break the helminth transmission cycle); (3) *Education*: skills-based education, including life skills that addresses health and hygiene issues and promotion of positive behaviours (including promoting handwashing and other hygienic behaviours that protect against helminth infection); and (4) *Services*: simple, safe and familiar health and nutrition services that can be delivered cost-effectively in schools (such as deworming).

This common focus has encouraged concerted action by the participating agencies and has increased significantly the number of countries implementing school health reforms. The simplicity of the approach, combined with the enhanced resources available from donor coordination, has helped ensure that these programmes can go to scale. For example, annual external support from the World Bank for these actions approaches US$90 million, targeting some 100 million schoolchildren.

The FRESH framework does not prescribe the design of school-based deworming programmes, and in practice these are highly variable and country specific (Bundy *et al.*, 2006). In low-income countries a public sector model is commonly used, involving the Ministry of Health in supervising the activity, and the Ministry of Education in implementing the intervention through teachers. In middle-income countries, including Indonesia and, historically, Japan and South Korea, a private-sector model involving non-governmental organizations delivering treatment that is paid for by the community has proven sustainable and effective.

Whatever the design, identifying which schools and communities require treatment is an essential part of the process, and requires a key role for GIS.

7. CONTROL APPLICATIONS OF GIS/RS

As recently as 5 years ago, applications of GIS and RS in helminthology had only been attempted for schistosomiasis and filariasis (reviewed in Brooker and Michael, 2000; Brooker, 2002). Since then, studies have investigated spatial patterns of STH infection (Brooker *et al.*, 2002b, 2003, 2004b; Saathoff *et al.*, 2005), *Loa loa* (Thomson *et al.*, 2004) and onchocerciasis (Carabin *et al.*, 2003). These studies have focused on the use of RS data to identify ecological correlates of infection and develop statistical models of disease risk. While these applications are attractive research objectives, the challenge remains to apply these geographic tools in the context of large-scale control programmes. Here we examine how GIS and RS have contributed to the design and implementation of helminth control programmes.

Human onchocerciasis or river blindness results from infection with a parasitic filarial worm, *Onchocerca volvulus*, which is transmitted by female *Simulium* blackflies. Blackflies breed in areas close to fast-flowing and well-oxygenated rivers and seldom travel more than 15 km in search of a bloodmeal. This means that high-prevalence communities are located close to breeding sites. Rapid Epidemiological Mapping of Onchocerciasis (REMO), developed by TDR (Tropical Disease Research)/WHO, has been a key geographic tool for the control of onchocerciasis (Ngoumou *et al.*, 1994; Katabarwa *et al.*, 1999). With REMO, it is possible to assess quickly and cheaply, which communities are at high risk of onchocerciasis and where they are located. REMO uses geographical information, particularly the locations of river basins, to identify communities that are likely to be at high risk. A subsample of these communities is then rapidly assessed by screening individuals for onchocercal nodules. This enables communities to be classified into three categories: priority areas which require community-directed treatment with ivermectin; areas which do not require treatment; and possible endemic areas but which require further investigation. Results of REMO have been effectively incorporated into a GIS to visualize priority areas for mass distribution of ivermectin and estimate the number of individuals to be treated (Noma *et al.*, 2002). This has helped the African Programme for Onchocerciasis Control (APOC) to prioritize allocation of resources according to need. The robustness of REMO following several rounds of interventions remains, however, to be fully investigated since there has been little validation of the approach since its initial development.

Severe adverse (and sometimes fatal) encephalopathic reactions following treatment with ivermectin have been reported in individuals co-infected with *O. volvulus* and *Loa loa* (loiasis), and as such, there is an operational necessity to identify areas with a high prevalence of *L. loa* (Addiss *et al.*, 2003). Thomson *et al.* (2000, 2004) developed a spatial model that predicted the prevalence of *L. loa* microfilaraemia on the basis of satellite-derived environmental data in Cameroon, with applications for defining areas at-risk of post-ivermectin *Loa*-related severe adverse reactions. Like onchocerciasis, treatment strategies have been defined according to levels of endemicity of onchocerciasis and loiasis. In areas where both diseases are highly endemic, detailed measures, such as training of medical staff, provision of medical supplies and heightened surveillance of treated individuals, are required.

A rapid mapping method has also been developed for lymphatic filariasis. This disease is caused by the filarial parasite *Wuchereria bancrofti* and is being targeted for eradication by the Global Alliance for the Elimination of Lymphatic Filariasis (www.filariasis.org). As with other control programmes, delimitation of endemic localities is an essential prerequisite for planning control elimination programmes, based on treatment with diethylcarbamazine (DEC) plus albendazole or ivermectin plus albendazole. To address this, a method for the Rapid Geographical Assessment of Bancroftian Filariasis (RAGFIL) has been developed by TDR/WHO. This is based on the use of a spatial sampling grid with either 25 km or 50 km between sampled communities, rapid prevalence assessments using immunochromatographic card tests (ICT) for the detection of circulating antigen from adult *W. bancrofti* filarial antigenaemia, and geostatistical methods for predicting the distribution of filariasis throughout the target area (WHO, 1998). Using this method, Gyapong *et al.* (2002) predicted prevalence in four countries in West Africa, enabling control planning to be initiated. Other analyses suggest, however, that endemic foci can persist within the interstices of the proposed grid and that smaller grids are required (Srividya *et al.*, 2002; Alexander *et al.*, 2003).

The Schistosomiasis Control Initiative (SCI) is currently supporting six countries in SSA to implement national control programmes for schistosomiasis and STH infections, including Burkina Faso, Mali, Niger, Tanzania, Uganda and Zambia (www.schisto.org). In Uganda, where *Schistosoma mansoni* is widespread, GIS and RS have been employed to classify the country according to different treatment strategies. Regular chemotherapy with PQZ and albendazole is being provided to schoolchildren and other high-risk groups (Kabatereine *et al.*, 2005). Following WHO guidelines, the programme is classifying communities according to three strategies: (1) in communities with a high prevalence ($\geq 50\%$) schoolchildren are treated every year and high-risk groups, such as fishermen, are treated; (2) in communities with a moderate prevalence ($\geq 20\%$ and $<50\%$) schoolchildren are treated once every 2 years; and (3) in communities with a low prevalence ($<20\%$) chemotherapy is made available in health facilities for treatment of suspected cases. With the use of GIS coupled with satellite and climatic data, geographical analysis found that no transmission of *S. mansoni* typically occurs in areas of Uganda where total annual rainfall was <850 mm or altitude was >1400 m (Kabatereine *et al.*, 2004). These areas were subsequently set aside without the need for further surveys (Brooker *et al.*, 2004a) (Figure 5). It was also shown that prevalence consistently exceeded 50% in areas within 5 km of Lakes Victoria and Albert, and thus in these areas warranted mass treatment without the need for further surveys. Prospective surveys have validated these predicted classifications (Kabatereine *et al.*, unpublished results).

Outside these two ecological areas, where smaller rivers and water bodies are numerous, it is suggested that individual communities are surveyed using standard parasitological methods (Brooker *et al.*, 2004a). Rapid mapping of communities is based on the Lot Quality Assurance Sampling (LQAS) method (Brooker *et al.*, 2005). Developed in industry for quality control, LQAS makes it possible to use small sample sizes when conducting surveys among populations (lots), best used in situations where classification of a population is useful and where the emphasis is on decision making (e.g. whether or not to intervene in a particular community) rather than estimation of prevalence and intensity of infection. Field testing showed that a LQAS sampling plan where only 15 children are sampled and a decision to intervene is made if seven children are found to be infected had excellent diagnostic performance, while economic analysis demonstrated that screening using LQAS was more cost-effective than mass treating

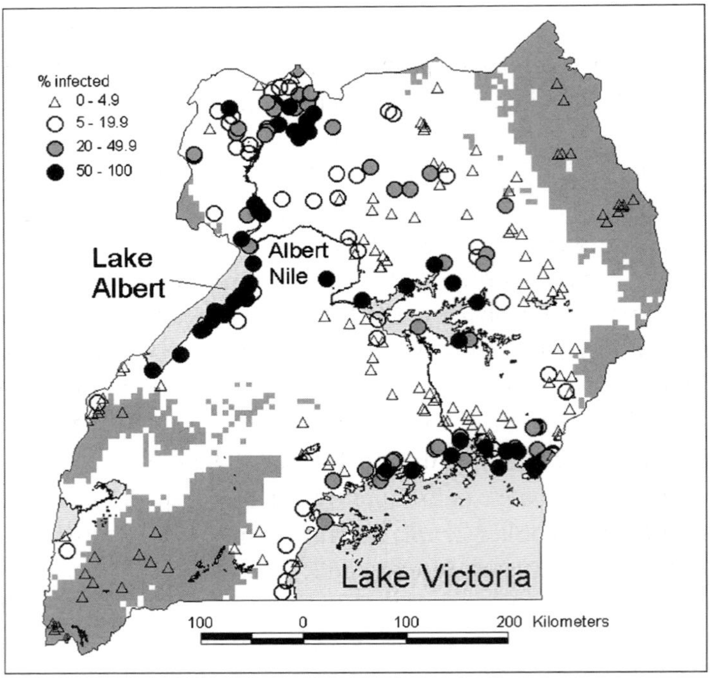

Figure 5 Distribution of *S. mansoni* in Uganda and classification of the country according to treatment category: (1) mass treatment without further surveys, <5 km from Lake Victoria and Lake Albert (not shown); (2) non-treatment areas, altitude >1400 m or annual rainfall <850 mm (grey areas); (3) and areas requiring further investigation using LQAS (white areas). For further details see Kabatereine *et al*., (2004) and Brooker *et al*., (2004a, 2005).

all schools in a single sub-county (Brooker *et al.*, 2005). Based on these findings, LQAS is employed in Uganda to quantify the fine-scale distribution of *S. mansoni* in areas of potential risk. Thus, GIS/RS coupled with field data can target schistosomiasis control from the national to the local level.

Predictive risk-mapping has also been employed to determine target areas for mass treatment in Tanzania (Clements *et al.*, 2006). First, simulation studies were conducted to determine the number of individuals and schools that were required to give a desired level of precision in risk estimates. Parasitological surveys were then conducted in 143 schools in northwestern Tanzania. The proportions of children found to have *S. haematobium* and *S. mansoni* infections were determined for each school and these were used as the outcome variables in separate spatially explicit binomial logistic regression models. The models were developed in a Bayesian framework and included environmental covariates derived from RS and a geostatistical component, using a powered exponential function to describe spatial correlation in the datasets. Models were externally validated against an independently collected dataset from one district within the study area and were used to make spatial predictions for *S. haematobium* and *S. mansoni* risk at prediction co-ordinates, defined by a grid of equally spaced locations and predictions were interpolated to produce a continuous risk surface for the study area as shown in Figure 6a and 6b. (Figure 6a and b are Plates 7.6a and b in the Separate Color Plate Section).

Subsequently, prediction surfaces for *S. haematobium* and *S. mansoni* were combined to make a single intervention map (as the treatment programme and its use of PQZ, makes no distinction between the two schistosome species), consisting of contours that equated to a prevalence of *S. haematobium* or *S. mansoni* of 10% and 50% (Figure 6c is Plate 7.6 in the Separate Color Plate Section). The lower contour (10%) separates areas that would and would not be targeted by the mass treatment campaign and the upper contour (50%) separates areas where mass treatment would only be conducted in school-age children and areas where both school-age children and other high-risk groups in the community would be targeted. Estimates of uncertainty in the spatial predictions help identify the area where further data collection is required.

The experience of SCI in Tanzania and Uganda amply demonstrates the usefulness of GIS/RS as geographic decision-making tools for implementing helminth control at both national scales and local scales. Geographical distributions are continually updated as new epidemiological data are collected, and as intervention reduces the prevalence of infection. Analysis of the cost-effectiveness of the tools, which is germane to their long-term and sustainable use, is currently underway.

The above examples have shown how research and international programmes have led the way in developing the use of GIS/RS for directing control programmes. An important emerging trend is that national governments are beginning to use this approach for designing and developing sustainable national programmes. GIS/RS has been employed by national governments to plan and conduct nationwide rapid epidemiological assessments of STH and schistosomiasis in Chad (Brooker *et al.*, 2002a) and Eritrea (PCD, 2003), and to design and implement national parasite control programmes, in both cases as part of national development programmes with World Bank assistance. In Chad, RS data were used to define seven ecological zones, which were combined with population data in a GIS to define the sample protocol, whereby 20 schools, in different ecological zones were surveyed. This approach substantially reduced the cost of the sample survey, while preserving its utility and effectiveness. The analysis showed that the patterns of hookworm and *S. haematobium* had a close association with the RS/GIS defined ecological zones, and significant relationships with environmental variables; it was correctly predicted that *A. lumbricoides* and *T. trichiura* would not occur in the country. The results from the survey helped the government plan the country's school-based control programme, and resulted in significant cost savings for the programme since it identified the need to target far fewer schools than had first been anticipated. A similar geographical approach was adopted in planning the school health programme in Eritrea (PCD, 2003). Again the sampling methodology proved substantially less expensive, and more practical, than traditional approaches developed without the benefit of GIS/RS. The national survey revealed that infection was highly focal, and that deworming interventions could be precisely targeted, with significant savings in financial and technical resources.

8. GLOBAL DISTRIBUTIONS

The above examples show that, used appropriately, GIS/RS can provide a practical and low-cost tool for designing and implementing sustainable helminth control programmes. GIS has the potential to promote evidence-based priority setting and careful targeting of finite financial resources, resulting in savings in resources and enhancement of sustainability. For this to become a reality, a critical first step is defining the geographical distribution of infection globally.

As with all diseases, there is long history of attempts to define global distributions of STH infections, and provide estimates of numbers infected. A first seminal effort was provided by Stoll (1947), which has been frequently updated (Peters, 1978; Crompton and Tulley, 1987; Bundy and Cooper, 1989; Chan *et al.*, 1994; Brooker *et al.*, 2000; Bundy *et al.*, 2004), with the most recent estimates provided by de Silva *et al.* (2003).

Since Stoll provided his estimates nearly 60 years ago, efforts to control STH infections and morbidity have varied across different regions of the world. The analysis by de Silva *et al.*, (2003) showed that in certain regions there has been a reduction in STH infection prevalence. In the Americas, for example, there has been a precipitous decline in prevalence and in absolute numbers since the 1960s, a change largely attributable to national treatment programmes coincidental with social and economic development that have brought about improved access to clean water and proper sanitation (Pan-American Health Organization (PAHO), 2000; Ehrenberg *et al.*, 2003). Well-documented declines in prevalence exist for Brazil and Mexico, the two most populous countries in the region (Tay *et al.*, 1976, 1995; Vinha, 1971; PAHO, 2000). Because of the success of control we do not consider the geographical distribution of infection in the region any further although we recognize that there may be small foci of high prevalence which require control activities.

In Asia, several countries, notably Japan, South Korea and Taiwan, have achieved sustained and successful control of STH infections over the last 40 years (WHO, 1996). More recently, a national control programme in Sri Lanka has reduced prevalence to $<5\%$. Control programmes have also been launched in several other Asian countries including China, Indonesia, Malaysia, Nepal, The Philippines and Thailand. Despite these control efforts, data available for south and southeast Asia suggest the need for effective control (Figure 7). STH

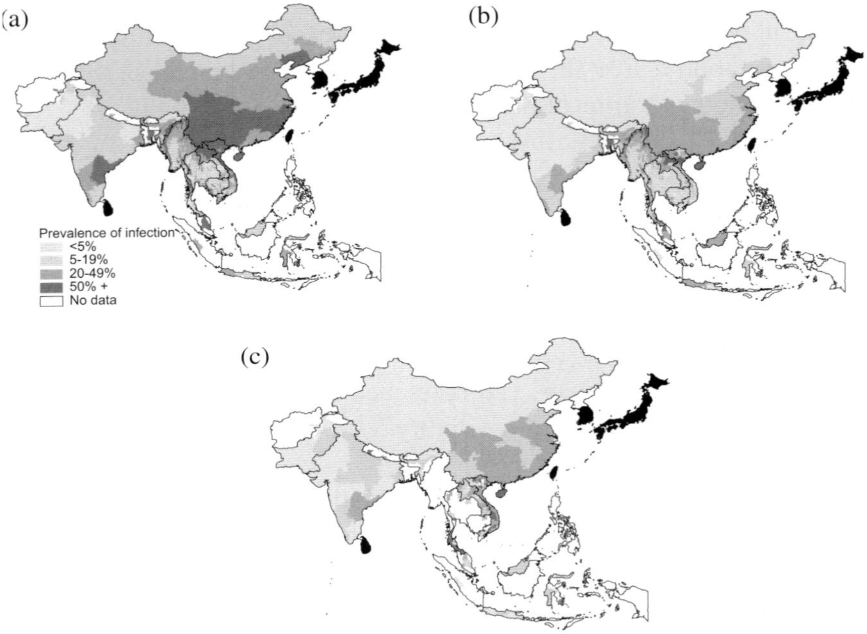

Figure 7 Prevalence of STH infection by province in Asia. (a) *A. lumbricoides*, (b) *T. trichiura* and (c) hookworm. Horizontal hatched areas indicate areas where sustained control has resulted in prevalence levels of <5%; white areas indicate a lack of data. Data were derived from published surveys or reviews: Afghanistan (Albis Gabrielli, unpublished data), Bangladesh (Hall and Nahar, 1994; Mascie-Taylor *et al.*, 1999), Bhutan (Allen *et al.*, 2004), Cambodia (Sinuon *et al.*, 2003; Urbani *et al.*, 2001), China (Xu *et al.*, 1995), India (de Silva *et al.*, 2003), Indonesia (Margono, 2001), Lao PDR (Rim *et al.*, 2003), Malaysia (Singh and Cox-Singh, 2001) Myanmar (Montresor *et al.*, 2004), Pakistan (Government of Pakistan, 1988), Thailand (Anantaphruti et al., 2000, 2002, 2004; Chongsuvivatwong *et al.*, 1994; Kasuya *et al.*, 1989; Nacher *et al.*, 2002; Waikagul *et al.*, 2002), Pacific Islands (Hughes *et al.*, 2004); Vietnam (Anon, 1995; van der Hoek *et al.*, 2003). In Cambodia and Myanmar, where empirical data are lacking, prevalence of *A. lumbricoides* and *T. trichiura* was estimated from RS-based prediction models (Brooker *et al.*, 2003).

infection remains prevalent in China and India, countries that account for a third of the world's population. The highest prevalence rates are observed for southern China and the northern regions of southeast Asia, and the lowest rates in northern China, northern India and Pakistan. In northern China, *A. lumbricoides* is more prevalent than either *T. trichiura* or hookworm. STH species are also widespread in the Pacific Islands (Hughes *et al.*, 2004).

We suggest that such distribution patterns reflect environmental suitability of STH transmission, especially thermal constraints. Here, this hypothesis is explored by investigating the relationship between prevalence of infection and satellite data for each administrative unit in the Asia region. LST was expressed as the median value for each of a number of 5 °C intervals (bins) spanning the full range of temperatures across the region (Figure 8). Median LST is estimated to be <20°C in northern China and >40°C in northern India and Pakistan, where STH infection prevalence is lowest. It appears that *A. lumbricoides* is more widespread in Asia and is able to survive colder temperatures than either *T. trichiura* or hookworm. It can thus be interpreted that these represent the lower thermal limits of STH transmission. Insufficient data were available to adequately explore the upper thermal limits of transmission in the region, as was possible for SSA (Figure 3). The current regional analysis for Asia supports previous studies, which show that heterogeneities in STH infection prevalences are correlated with temperature and humidity in China (Xu *et al.*, 1995; Lai and His, 1996) and southeast Asia (Brooker *et al.*, 2003). The latter study used satellite data to predict prevalence in areas lacking detailed data (Brooker *et al.*, 2003).

While there have been declines in the prevalence of STH infection in the Americas and parts of Asia, estimated prevalence rates for SSA are equivalent to those first estimated by Norman Stoll more than 60 years ago (de Silva *et al.*, 2003). In fact, the prevalence and distribution of STH infection remain largely unknown, because the region has weak disease surveillance systems. The helminthological data that do exist in the formal and 'grey' literature have been collated into a GIS database (Brooker *et al.*, 2000). However, this continental database includes data for only 15% of districts in SSA, with most data on hookworm infection. Furthermore, for reasons outlined

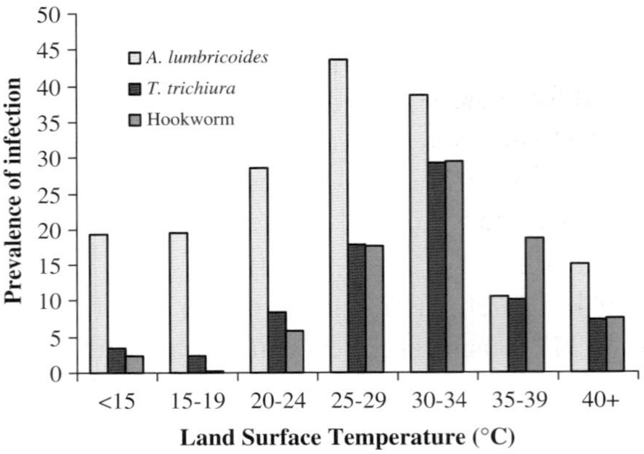

Figure 8 The relationship between prevalence of STH infection in Asia and satellite-derived mean LST for 1982–1999, obtained from NOAA's AVHRR. Prevalence is expressed as median prevalence for each temperature category and the median temperature was calculated for each geographical region; it is recognized that this approach masks the heterogeneity in STH prevalence and temperature within regions, but epidemiological data are not available at finer spatial resolution.

previously (Stoll, 1947; Brooker and Michael, 2000), the inherent variability in the data, because of differences in the timing, methodology and study population, makes it difficult to use these data to reliably define the distribution of STH species. Instead, we have utilized the robust predictions of STH prevalence developed here (Figure 2). These predictions provide perhaps the most detailed description of STH infection in SSA to date.

9. PREDICTED NUMBERS OF INFECTIONS

As indicated, school-based deworming represents the most cost-effective and feasible intervention strategy for STH infections and the greatest need for control exists in SSA. For these reasons, we derive country-specific estimates of school-aged children infected in the region. Consistent with other analyses in this edition, population was estimated from the Gridded Population of the World version 3

(GPW3) global human population distribution for 2005 (Balk *et al.*, this volume, pp. 119–156). Country-specific medium variant population growth rates and proportions of the population aged 5–14 years from the United Nations Population Division – World Population Prospects database (http://esa.un.org/unpp) were used to project this age cohort of the population total to 2005. These population data were spatially related to our species-specific risk models (Figure 2), which were re-sampled at 1 km, the spatial resolution of GPW3. Since the models were developed using prevalence estimates among school-aged children it is assumed that models define prevalence for this age group. For each district, population totals and predicted infection prevalence were extracted and used to estimate numbers of school-aged children infected with each species. Since no clear patterns in urban–rural differences in species prevalence exist (Table 2) no adjustment was made for the effect of urbanization.

Infections from *A. lumbricoides, T. trichiura* and hookworm are often found in the same communities and individuals and BZAs, such as albendazole and mebendazole, are broadly effective against all species. In order to estimate the cost of interventions against multiple STH infections, an estimate of the numbers of infections of any species, either single- or multiple- species infections, is required. To quantify the numbers of multiple-species infections we use a validated probabilistic model to predict the prevalence of multiple-species infections for each district for which only overall prevalence data exist (Booth and Bundy, 1995), thereby allowing the number of children infected with any STH species to be estimated.

It is estimated that over 30.7 million African school-aged children are infected with *A. lumbricoides*, 36.5 million with *T. trichiura* and 50.0 million with hookworm. Since many children have multiple infections, it is estimated that 89.9 million are infected with one or more STH species. Forty four per cent of the infections are concentrated in just four countries: in descending order of magnitude these are Nigeria, the Democratic Republic of Congo, South Africa and Tanzania (Table 3). Previous estimates have suggested larger numbers of infections; the most recent estimate suggesting that 53 million school-aged children (5–15 years) are infected with *A. lumbricoides*, 50 million with *T. trichiura* and 47 million with hookworm (de Silva *et al.*, 2003). Previous

Table 3 Estimated numbers of STH infection among school-aged children in SSA by country, 2005

Country	School-aged population (1000s)	Estimated numbers of infections (1000s)				Estimated numbers requiring mass treatment based on 50% threshold (1000s)	Total annual per treatment cost (US$ 1000s)[b]
		A. lumbricoides	*T. trichiura*	Hookworm	All STH species		
Angola	3666	610	712	1048	1896	2031	122–183
Benin	1737	256	352	399	831	836	50–75
Botswana	422	8	10	121	113	—	—
Burkina Faso	3222	30	35	759	671	—	—
Burundi	1768	309	341	583	956	753	45–68
Cameroon	4091	1227	1490	1059	2728	3017	181–272
Cape Verde	111	7	13	44	50	22	1–2
CAR[a]	1014	200	227	256	555	640	38–58
Chad	2186	25	27	529	474	—	—
Comoros	146	72	81	45	131	147	9–13
DRC[a]	14151	4111	4710	3790	9643	13410	805–1207
Republic of Congo	839	242	310	201	574	828	50–75
Cote d' Ivoire	4399	1400	1783	044	3189	3967	238–357
Djibouti	170	1	1	41	14	—	—
Equatorial Guinea	125	70	81	37	118	126	8–11
Eritrea	1020	11	14	324	281	4	0.3–0.4
Ethiopia	17424	1653	1641	6085	7362	4476	269–403
Gabon	332	155	191	9	295	331	20–30
Gambia	353	15	21	99	112	41	2–4
Ghana	5326	1330	1699	1300	3284	3551	213–320
Guinea	2226	505	622	584	1325	1450	87–131
Guinea Bissau	330	47	66	84	161	105	6–9
Kenya	8385	1134	1158	2682	3921	3661	220–330

Lesotho	549	175	217	206	423	553	33–50
Liberia	807	401	485	185	734	813	49–73
Madagascar	4413	1538	1739	1224	3146	3312	199–298
Malawi	3113	471	548	857	1525	1308	79–118
Mali	3160	41	49	750	688	—	—
Mauritania	337	5	11	75	76	—	—
Mauritius	292	168	190	84	277	295	18–27
Mozambique	5006	995	1188	1216	2767	3319	199–299
Namibia	473	7	9	140	126	—	—
Niger	3034	14	9	725	613	—	—
Nigeria	31742	4690	6346	8332	15193	14286	857–1286
Rwanda	2105	474	526	714	1272	1326	80–119
Réunion	192	116	128	61	185	194	12–17
Senegal	2574	424	495	589	1023	560	34–50
Seychelles	22	7	8	16	19	22	1–2
Sierra Leone	1224	475	590	319	1004	1234	74–111
Somalia	2458	136	155	799	803	74	16–25
South Africa	11787	3436	4123	3358	7779	7770	466–699
Sudan	8571	86	104	2140	1897	—	—
Swaziland	256	84	99	66	189	258	15–23
São Tomé & Príncipe	37	26	29	9	37	37	2–3
Tanzania	9628	1385	1542	2839	4591	3906	234–352
Togo	1250	149	213	294	550	348	21–31
Uganda	6540	1324	1388	1977	3630	3498	210–315
Zambia	2853	382	423	840	1334	1044	63–94
Zimbabwe	3434	289	322	1002	1312	302	18–27
Total	179308	30710	36530	50010	89900	84056	5043–7565

[a]CAR, Central African Republic; DRC, Democratic Republic of Congo.
[b]Range based on delivery costs of US$0.03–0.04 per child and drug costs US$0.03–0.05 per child.

estimates at a regional or continental scale have all been based on the approach first established by Stoll in 1947: the prevalence data from the few studies available within a country are used to estimate a mean prevalence for the country as a whole, and then expressed as a product of the estimated number of school-age children in the country. The estimates we present here are the first at this scale to explicitly include the fine spatial variation in distribution of both infection and population, and indicate that the earlier methodology tended to overestimate. It is also worth emphasizing that this level of precision has been achieved affordably only because of the use of GIS/RS: using traditional survey methods to obtain these data would be prohibitively expensive.

The estimates provide a basis for quantifying the financial resources required to support school-based deworming in SSA. It is assumed that mass treatment is only provided in districts where prevalence of any STH infection exceeds 50%, and in these districts treatment is provided to all school-age children irrespective of infection status, according to WHO guidelines on mass treatment (WHO, 2002). Cost analyses estimate that the school-based delivery costs of albendazole or mebendazole, anthelmintics that are effective against all the common STH species, are in the range between US$0.03 and 0.04 per capita (PCD, 1999), while the drug costs are in the range between US$ 0.03 and 0.05 per capita (Guyatt, 2003). Under these assumptions, using the school-based approach to provide a single annual treatment to all school-age children in all districts in Africa where mass treatment is justified would cost US$5.0–7.6 million.

These cost estimates are for maintaining a programme and do not include the higher costs of start-up. Nevertheless they suggest that continent-wide control of parasites is, from a financial perspective, an attainable goal.

10. THE FUTURE

GIS/RS is a powerful tool that has evolved from supporting sophisticated epidemiological research, through a role in directing public health interventions, to now showing a real potential for assisting the design of sustainable development programmes. The initial epidemiological research was critical to this evolution, because it has provided the evidence for the link between transmission dynamics and the environmental factors that can be detected by RS. Furthermore, as illustrated here, it can provide important new insights into patterns of transmission at a geographical scale that has proven impossible to address using more traditional approaches.

The present analyses have shown that, with the available evidence, GIS/RS can predict patterns of STH transmission, and can do so at a scale that is relevant to the design of national control programmes. Most importantly, this approach is able to achieve this even for areas for which limited empirical data are available and at low cost relative to traditional survey technologies. This approach has the potential to facilitate the design of national control programmes, and the key question now is how best to realize this potential.

The experience of international vertical programmes aimed at controlling lymphatic filariasis, schistosomiasis and onchocerciasis provides some insights, but it is perhaps the examples from Chad, Eritrea, Tanzania and Uganda of national government efforts that more clearly point the way forward. GIS/RS has emerged as a tried and tested and increasingly acceptable technology that can provide national governments with a relatively low-cost approach to surveying and programme design, and which can significantly reduce the cost of practical programmes through more precise geographical targeting and simplifying the processes of monitoring and evaluation. It is important to recognize that GIS/RS can reduce both the upstream (e.g. survey and design) and downstream (e.g. targeting, monitoring and evaluation) costs of programmes, while at the same time enhancing programme effectiveness.

As more use is made of GIS/RS in control, there is a need to assess their cost and cost-effectiveness in targeting control efforts. At present, most funding for their use comes from international agencies and northern researchers. Determining the long-term sustainability of the use of GIS/RS in disease control and how they influence allocation of resources is crucial. It is also important to assess the inherent uncertainties in spatial data (epidemiological and environmental) within a GIS and the propagation of uncertainty in predictions and estimations.

Such research is underway for ecological modelling but similar research is clearly needed for epidemiological applications.

Notwithstanding these issues, the clear message is that this technology should be made available to policy makers and planners at the national level. In terms of helminth control an essential component is to enhance access to the predicted geographical patterns of transmission, such as those described here. As a contribution to improving access, all the maps presented in this text are made available at www.schoolsandhealth.org, and are available separately to individual countries in CD format by request from the same site. In this same spirit we welcome the release of the population and environmental data with this volume.

ACKNOWLEDGEMENTS

We thank the following individuals who have generously contributed data to this review: Henrietta Allen, Chris Appleton, Narcis Kabatereine, Nicholas Lwambo and Antonio Montresor. We are also grateful to Paul Coleman for his contribution to the STH ecological analysis and to Christine Budke, Maiza Campos Ponce, Bob Snow and Anthony Wilson for their comments on this manuscript. SB is supported by a Wellcome Trust Advanced Training Fellowship (#073656) and ACAC is funded by the Schistosomiasis Control Initiative, which receives support from the Bill and Melinda Gates Foundation.

REFERENCES

Addiss, D.G., Rheingans, R., Twum-Danso, N.A. and Richards, F.O. (2003). A framework for decision-making for mass distribution of Mectizan in areas endemic of *Loa loa*. *Filaria Journal* **2**, S9.

Albonico, M., Chwaya, H.M., Montresor, A., Stolfzfus, R.J., Tielsch, J.M., Alawi, K.S. and Savioli, L. (1997). Parasitic infections in Pemba Island school children. *East African Medical Journal* **74**, 294–298.

Alexander, N.D., Moyeed, R.A., Hyun, P.J., Dimber, Z.B., Bockarie, M.J., Stander, J., Grenfell, B.T., Kazura, J.W. and Alpers, M.P. (2003). Spatial variation of Anopheles-transmitted *Wuchereria bancrofti* and *Plasmodium falciparum* infection densities in Papua New Guinea. *Filaria Journal* **2**, 14.

Allen, H., Sithey, G., Padmasiri, E.A. and Montresor, A. (2004). Epidemiology of soil-transmitted helminths in the western region of Bhutan. *Southeast Asian Journal of Tropical Medicine and Public Health* **35**, 1–3.

Anantaphruti, M.T., Jongsuksuntigul, P., Imsomboon, T., Nagai, N., Muennoo, C., Saguankiat, S., Pubampen, S. and Kojima, S. (2002). School-based helminthiases control: I. a baseline study of soil-transmitted helminthiases in Nakhon Si Thammarat Province, Thailand. *Southeast Asian Journal of Tropical Medicine and Public Health* **33**, 113–119.

Anantaphruti, M.T., Nuamtanong, S., Muennoo, C., Sanguankiat, S. and Pubampen, S. (2000). *Strongyloides stercoralis* infection and chronological changes of other soil-transmitted helminthiases in an endemic area of southern Thailand. *Southeast Asian Journal of Tropical Medicine and Public Health* **31**, 378–382.

Anantaphruti, M.T., Waikagul, J., Maipanich, W., Nuamtanong, S. and Pubampen, S. (2004). Soil-transmitted helminthiases and health behaviors among schoolchildren and community members in a west-central border area of Thailand. *Southeast Asian Journal of Tropical Medicine and Public Health* **35**, 260–266.

Anderson, R.M. (1982). The population dynamics and control of hookworm and roundworm infection. In: *Population Dynamics of Infectious Diseases: Theory and Applications* (R.M. Anderson, ed.), pp. 67–109. London: Chapman and Hall.

Anderson, R.M. and May, R.M. (1991). *Infectious Diseases of Humans: Dynamics and Control*. Oxford: Oxford University Press.

Anon. (1995). *Report of the National Anemia and Nutrition Risk Factor Survey Vietnam 1995*. Hanoi: United Nations Children's Fund (UNICEF).

Appleton, C.C. and Gouws, E. (1996). The distribution of common intestinal nematodes along an altitudinal transect in Kwa-Zulu Natal, South Africa. *Annals of Tropical Medicine and Parasitology* **90**, 181–188.

Appleton, C.C., Maurihungirire, M. and Gouws, E. (1999). The distribution of helminth infections along the coastal plain of Kwazulu-Natal province, South Africa. *Annals of Tropical Medicine and Parasitology* **93**, 859–868.

Aswashi, S., Bundy, D.A.P. and Savioli, L. (2003). Helminthic infections. *British Medical Journal* **23**, 431–433.

Beaver, P.C. (1953). Persistence of hookworm larvae in soil. *American Journal of Tropical Medicine and Hygiene* **2**, 102–108.

Beer, R.J. (1976). The relationship between *Trichuris trichiura* (Linnaeus 1758) of man and *Trichuris suis* (Schrank 1788) of the pig. *Research in Veterinary Science* **20**, 47–54.

Booth, M. and Bundy, D.A.P. (1995). Estimating the number of multiple-species geohelminth infections in human communities. *Parasitology* **111**, 645–653.

Brooker, S. (2002). Schistosomes, snails and satellites. *Acta Tropica* **82**, 209–216.

Brooker, S., Beasley, N.M.R., Ndinaromtan, M., Madjiouroum, E.M., Baboguel, M., Djenguinabe, E., Hay, S.I. and Bundy, D.A.P. (2002a). Use of remote sensing and a geographical information system in a national helminth control programme in Chad. *Bulletin of the World Health Organization* **80**, 783–789.

Brooker, S., Hay, S.I., Tchuem Tchuenté, L.A. and Ratard, R. (2002b). Using NOAA-AVHRR data to model helminth distributions for planning disease control in Cameroon, West Africa. *Photogrammetric Engineering and Remote Sensing* **68**, 175–179.

Brooker, S., Kabatereine, N.B., Clements, A.C.A. and Stothard, J.R. (2004a). Schistosomiasis control. *Lancet* **363**, 658–659.

Brooker, S., Kabatereine, N.B., Myatt, M., Stothard, J.R. and Fenwick, A. (2005). Rapid assessment of *Schistosoma mansoni*: the validity, applicability and cost-effectiveness of the Lot Quality Assurance Sampling method in Uganda. *Tropical Medicine and International Health* **10**, 647–658.

Brooker, S., Kabatereine, N.B., Tukahebwa, E.M. and Kazibwe, F. (2004b). Spatial analysis of the distribution of intestinal nematode infections in Uganda. *Epidemiology and Infection* **132**, 1065–1071.

Brooker, S. and Michael, E. (2000). The potential of geographical information systems and remote sensing in the epidemiology and control of human helminth infections. *Advances in Parasitology* **47**, 245–288.

Brooker, S., Pratap, S., Waikagul, J., Suvanee, S., Kojima, S., Takeuchi, T., Luong, T.V. and Looareesuwan, S. (2003). Mapping soil-transmitted helminth infections in Southeast Asia and implications for parasite control. *Southeast Asian Journal of Tropical Medicine and Public Health* **34**, 24–35.

Brooker, S., Rowlands, M., Haller, L., Savioli, L. and Bundy, D.A.P. (2000). Towards an atlas of human helminth infection in sub-Saharan Africa: the use of geographical information systems (GIS). *Parasitology Today* **16**, 303–307.

Bundy, D., Shaeffer, S., Jukes, M., Beegle, K., Gillespie, A., Drake, L., Seung-hee Frances Lee, B., Hoffman, A.-M., Jones, J., Mitchell, A., Wright, C., Barcelona, D., Camara, B., Golmar, C., Savioli, L., Takeuchi, T. and Sembene, M. (2006). School based health and nutrition programs, 2nd edition. In: *Disease Control Priorities for Developing Countries* (D. Jamison, M. Claeson, J. Breman and A. Meacham, eds). Oxford, New York: Oxford University Press.

Bundy, D.A.P. (1997). This wormy world then and now. *Parasitology Today* **13**, 407–408.

Bundy, D.A.P., Chan, M.S., Medley, G.F., Jamison, D., Savioli, L. 2004. Intestinal nematode infections (Chapter 9). In: Murray, C.J.L., Lopez, A.D., Mathers, C.D., (Eds.), Global Epidemiology of Infectious Disease, Global Epidemiology of Infectious Disease Volume IV. Geneva, WHO, pp. 243–300. http://whqlibdoc.who.int/publications/2004/9241592303.pdf

Bundy, D.A.P. and Cooper, E.S. (1989). *Trichuris* and trichuriasis in humans. *Advances in Parasitology* **28**, 107–173.

Bundy, D.A.P. and Guyatt, H.L. (1996). Schools for health: focus on health, education and the school-aged child. *Parasitology Today* **12**, 1–16.

Bundy, D.A.P. and Medley, G.F. (1992). Immuno-epidemiology of human geohelminthiasis: ecological and immunological determinants of worm burden. *Parasitology* **104**, S105–S119.

Bundy, D.A.P., Wong, M.S., Lewis, L.L. and Horton, J. (1990). Control of geohelminths by delivery of targeted chemotherapy through schools. *Transactions of the Royal Society of Tropical Medicine and Hygiene* **84**, 115–120.

Carabin, H., Escalona, M., Marshall, C., Vivas-Martinez, S., Botto, C., Joseph, L. and Basanez, M.G. (2003). Prediction of community prevalence of human onchocerciasis in the Amazonian onchocerciasis focus: Bayesian approach. *Bulletin of the World Health Organization* **81**, 482–490.

Chan, M.S., Medley, G.F., Jamison, D. and Bundy, D.A.P. (1994). The evaluation of potential global morbidity attributable to intestinal nematode infections. *Parasitology* **109**, 373–387.

Chongsuvivatwong, V., Pas-Ong, S., Ngoathammatasna, W., McNeil, D., Vithsupakorn, K., Bridhikitti, V., Jongsuksuntigul, P. and Jeradit, C. (1994). Evaluation of hookworm control program in southern Thailand. *Southeast Asian Journal of Tropical Medicine and Public Health* **25**, 745–751.

Clements, A. C. A., Lwambo, N. J. S., Blair, L., Nyandindi, U., Kataano, G., Kinung'hi, S., Webster, J. p., Fenwick, A. and Brooker, S. (2006). Bayesian spatial analysis and risk-mapping: tools to enhance planning of a schistosomiasis control programme in Tanzania. *Tropical Medicine and International Health* (in press).

Cracknell, A. P. (1997). Earth surfaces temperatures. In: The Advanced Very *High Resolution Radiometer*. pp. 181–231. London: Taylor & Francis.

Crompton, D.W.T. (2001). *Ascaris* and ascariasis. *Advances in Parasitology* **48**, 285–375.

Crompton, D.W.T. and Savioli, L. (1993). Intestinal parasitic infections and urbanization. *Bulletin of the World Health Organization* **71**, 1–7.

Crompton, D.W.T. and Tulley, J.J. (1987). How much Ascariasis is there in Africa? *Parasitology Today* **3**, 123–127.

de Silva, N.R., Brooker, S., Hotez, P.J., Montresor, A., Engels, D. and Savioli, L. (2003). Soil-transmitted helminth infections: updating the global picture. *Trends in Parasitology* **19**, 547–551.

Drake, L.J., Jukes, M.C.H., Sternberg, R.J. and Bundy, D.A.P. (2000). Geohelminth infections (Ascariasis, Trichuriasis and Hookworm): cognitive

and developmental impacts. *Seminars in Pediatric Infectious Diseases* **11**, 245–251.
Ehrenberg, J.P., de Mérida, A.M. and Sentz, J. (2003). An epidemiological overview of geohelminth and schistosomiasis in the Caribbean. Washington DC: Pan-American Health Organization (PAHO).
Fenwick, A., Savioli, L., Engels, D., Bergquist, R.N. and Todd, M.H. (2003). Drugs for the control of parasitic diseases: current status and development in schistosomiasis. *Trends in Parasitology* **19**, 509–515.
Giraldi, N., Vidotto, O., Navarro, I.T. and Garcia, J.L. (2001). Enteroparasites prevalence among daycare and elementary school children of municipal schools, Rolandia, PR, Brazil. *Revista da Sociedade Brasileira de Medicina Tropical* **34**, 385–387.
Government of Pakistan. (1988). *National Nutrition Survey 1985–87*. Islamabad: National Institute of Health, Ministry of Health.
Guyatt, H.L. (2003). The cost of delivering and sustaining a control programme for schistosomiasis and soil-transmitted helminthiasis. *Acta Tropica* **86**, 267–274.
Guyatt, H.L., Brooker, S., Hall, A., Kihamia, C.K. and Bundy, D.A.P. (2001). Evaluation of efficacy of school-based anthelmintic treatments against anaemia in children in the United Republic of Tanzania. *Bulletin of the World Health Organization* **79**, 695–703.
Gyapong, J.O., Kyelem, D., Kleinschmidt, I., Agbo, K., Ahouandogbo, F., Gaba, J., Owusu-Banahene, G., Sanou, S., Sodahlon, Y.K., Biswas, G., Kale, O.O., Molyneux, D.H., Roungou, J.B., Thomson, M.C. and Remme, J. (2002). The use of spatial analysis in mapping the distribution of bancroftian filariasis in four West African countries. *Annals of Tropical Medicine and Parasitology* **96**, 695–705.
Hotez, P.J., Brooker, S., Bethony, J.M., Bottazzi, M.E., Loukas, A. and Xiao, S. (2004). Current concepts: hookworm infection. *New England Journal of Medicine* **351**, 799–807.
Hotez, P. J., Bundy, D. A. P., Beegle, K., Brooker, S., de Silva, N., Montresor, A., Engels, D., Drake, L., Chitsulo, L., Michaud, C., Bethony, J. M., Oliveira, R., Xiao, S. H., Fenwick, A. and Savioli, L. (2006). Helminth infections. In: *Disease Control Priorities in Developing Countries, 2nd Edition.* (D. Jamison, M. Claeson, J. Breman and A. Meacham, eds). New York: Oxford University Press.
Hughes, R.G., Sharp, D.S., Hughes, M.C., Akau'ola, S., Heinsbroek, P., Velayudhan, R., Schulz, D., Palmer, K., Cavalli-Sforza, T. and Galea, G. (2004). Environmental influences on helminthiasis and nutritional status among Pacific schoolchildren. *International Journal of Environmental Health Research* **14**, 163–177.
Kabatereine, N.B., Brooker, S., Tukahebwa, E.M., Kazibwe, F. and Onapa, A. (2004). Epidemiology and geography of *Schistosoma mansoni* in

Uganda: implications for planning control. *Tropical Medicine and International Health* **9**, 372–380.

Kabatereine, N. B., Tukahebwa, E. M., Kazibwe, F., Namwangye, H., Zaramba, S., Brooker, S., Stothard, J. R., Kamenka, C., Whawell, S., Webster, J. P. and Fenwick, A. (2006). Progress towards country-wide control of schistosomiasis and soil-transmitted helminthiasis in Uganda. *Transactions of the Royal Society of Tropical Medicine and Hygiene* (in press).

Kan, S.P., Guyatt, H.L. and Bundy, D.A.P. (1989). Geohelminth infection of children from rural plantations and urban slums in Malaysia. *Transactions of the Royal Society of Tropical Medicine and Hygiene* **83**, 817–820.

Kasuya, S., Khamboonruang, C., Amano, K., Murase, T., Araki, H., Kato, Y., Kumada, Y., Koyama, A., Higuchi, M. and Nakamura, J. (1989). Intestinal parasitic infections among schoolchildren in Chiang Mai, northern Thailand: an analysis of the present situation. *Journal of Tropical Medicine and Hygiene* **92**, 360–364.

Katabarwa, M., Onapa, A.W. and Nakileza, B. (1999). Rapid epidemiological mapping of onchocerciasis in areas of Uganda where *Simulium neavei* sl is the vector. *East Africa Medical Journal* **76**, 440–446.

Kilama, W.L. (1990). Hookworm infection and disease in Africa and the Middle East. In: *Hookworm Disease. Current Status and New Directions* (G.A. Schad and K.S. Warren, eds), pp. 17–32. London: Taylor & Francis.

King, R.J., Campbell-Lendrum, D.H. and Davies, C.R. (2004). Predicting geographic variation in cutaneous leishmaniasis, Colombia. *Emerging Infectious Diseases* **10**, 598–607.

Lai, D. and His, B.P. (1996). Soil-transmitted helminthiasis in China: a spatial statistical analysis. *Southeast Asian Journal of Tropical Medicine and Public Health* **27**, 754–759.

Margono, S.S. (2001). Large-scale control against intestinal helminthic infections in Japan, with special reference to the activities of Japan Association of Parasite Control. In: *Collected Papers on the Control of Soil-transmitted Helminthiases Vol VII* (S. Hayashi, ed.), pp. 169–172. Tokyo: Asian Parasite Control Organization.

Mascie-Taylor, C.G.N., Alam, M., Montanari, R.M., Karim, R., Ahmed, T., Karim, E. and Akhtar, S. (1999). A study of the cost effectiveness of selective health interventions for the control of intestinal parasites in rural Bangladesh. *Journal of Parasitology* **85**, 6–11.

Miguel, E. and Kremer, M. (2004). Worms: identifying impacts on education and health in the presence of treatment externalities. *Econometrica* **72**, 159–217.

Molyneux, D.H., Bradley, M., Hoerauf, A., Kyelem, D. and Taylor, M.J. (2003). Mass drug administration for lymphatic filariasis and onchocerciasis. *Trends in Parasitology* **19**, 516–522.

Montresor, A., Zin, T.T., Padmasiri, E., Allen, H. and Savioli, L. (2004). Soil-transmitted helminthiasis in Myanmar and approximate costs for countrywide control. *Tropical Medicine and International Health* **9**, 1012–1015.

Nacher, M., Singhasivanon, P., Yimsamran, S., Manibunyong, W., Thanyavanich, N., Wuthisen, R. and Looareesuwan, S. (2002). Intestinal helminth infections are associated with increased incidence of *Plasmodium falciparum* malaria in Thailand. *Southeast Asian Journal of Tropical Medicine and Public Health* **88**, 55–58.

Ndenecho, L., Ndamukong, K.J. and Matute, M.M. (2002). Soil transmitted nematodes in children in Buea Health District of Cameroon. *East Africa Medical Journal* **79**, 442–445.

Ngoumou, P., Walsh, J.F. and Mace, J.M. (1994). A rapid mapping technique for the prevalence and distribution of onchocerciasis: a Cameroon case study. *Annals of Tropical Medicine and Parasitology* **88**, 463–474.

Noma, M., Nwoke, B.E.B., Nutall, I., Tambala, P.A., Enyong, P., Namsenmo, A., Remme, J., Amazigo, U.V., Kale, O.O. and Seketeli, A. (2002). Rapid epidemiology mapping of onchocerciasis (REMO): its application by the African Programme for Onchocerciasis Control (APOC). *Annals of Tropical Medicine and Parasitology* **96**, S29–S39.

Nwosu, A.B.C. (1978). Investigations into the free-living phase of the cat hookworm life-cycle. *Zeitschrift für Parasitenkunde* **56**, 243–249.

Nwosu, A.B.C. and Anya, A.O. (1980). Seasonality in human hookworm infection in an endemic area of Nigeria, and its relationship to rainfall. *Tropenmedizin Und Parasitologie* **31**, 201–208.

Otto, G.F. (1929). A study of the moisture requirements of the eggs of the horse, the dog, human and pig ascarids. *American Journal of Hygiene* **10**, 497–520.

Pan-American Health Organization (PAHO). (2000). *Meeting on the Control of Intestinal Helminthiasis in the Context of AIEPI*. Washington DC: PAHO.

Partnership for Child Development (PCD). (1998). The health and nutritional status of school children in Africa: evidence from school-based health programmes in Ghana and Tanzania. *Transactions of the Royal Society of Tropical Medicine and Hygiene* **92**, 254–261.

Partnership for Child Development (PCD). (1999). The cost of large-scale school health programmes which deliver anthelmintics to children in Ghana and Tanzania. *Acta Tropica* **73**, 183–204.

Partnership for Child Development (PCD). (2003). *A Situation Analysis of the Health of School Children in Eritrea*. London: Imperial College.

Pavlovsky, E.P. (1966). *Natural Nidality of Transmissible Diseases—with Special Reference to the Landscape Epidemiology of Zooanthroponoses*. Urbana: University of Illinois Press.

Peters, W. (1978). The relevance of parasitology to human welfare today. *Symposia of the British Society for Parasitology* **16**, 25–40.

Phiri, K., Whitty, C.J., Graham, S.M. and Ssembatya-Lule, G. (2000). Urban/rural differences in prevalence and risk factors for intestinal helminth infection in southern Malawi. *Annals of Tropical Medicine and Parasitology* **94**, 381–387.

Rahman, W.A. (1998). Helminthic infections of urban and rural schoolchildren in Penang Island, Malaysia: implications for control. *Asian Journal of Tropical Medicine and Public Health* **29**, 596–598.

Ratard, R.C., Kouemeni, L.E., Bessala, M.M.E., Ndamkou, C.N., Sama, M.T. and Cline, B.L. (1991). Ascariasis and trichuriasis in Cameroon. *Transactions of the Royal Society of Tropical Medicine and Hygiene* **85**, 84–88.

Ratard, R.C., Kouemeni, L.E., Ekani Bessala, M.K. and Ndamkou, C.N. (1992). Distribution of hookworm infection in Cameroon. *Annals of Tropical Medicine and Parasitology* **86**, 413–418.

Rim, H.J., Chai, J.Y., Min, D.Y., Cho, S.Y., Eom, K.S., Hong, S.J., Sohn, W.M., Yong, T.S., Deodato, G., Standgaard, H., Phommasack, B., Yun, C.H. and Hoang, E.H. (2003). Prevalence of intestinal parasite infections on a national scale among primary schoolchildren in Laos. *Parasitological Research* **91**, 267–272.

Saathoff, E., Olsen, A., Sharp, B., Kvalsvig, J.D., Appleton, C.C. and Kleinschmidt, I. (2005). Ecologic covariates of hookworm infection and reinfection in rural KwaZulu-Natal/South Africa: a geographic information system-based study. *American Journal of Hygiene and Tropical Medicine* **72**, 384–391.

Schad, G.A., Chowdhury, A.B., Dean, C.G., Kochar, V., Nawalinski, T.A., Thomas, J. and Tonascia, J.A. (1973). Arrested development in human hookworm infections: an adaptation to seasonality unfavourable external environment. *Science* **180**, 502–504.

Seamster, A.P. (1950). Developmental studies concerning the eggs of *Ascaris lumbricoides* var. *suum. The American Midland Naturalist* **43**, 450–468.

Singh, B. and Cox-Singh, J. (2001). Parasites that cause problems in Malaysia: soil-transmitted helminths and malaria parasites. *Trends in Parasitology* **17**, 597–600.

Sinuon, M., Anantaphruti, M.T. and Socheat, D. (2003). Intestinal helminthic infections in schoolchildren in Cambodia. *Southeast Asian Journal of Tropical Medicine and Public Health* **34**, 254–258.

Smith, G. and Schad, G.A. (1989). *Ancylostoma duodenale and Necator americanus*: effect of temperature on egg development and mortality. *Parasitology* **99**, 127–132.

Spindler, L.A. (1929). The relation of moisture to the distribution of human trichuris and ascaris. *American Journal of Hygiene* **10**, 476–496.

Srividya, A., Michael, E., Palaniyandi, M., Pani, S.P. and Das, P.K. (2002). A geostatistical analysis of the geographic distribution of lymphatic filariasis prevalence in southern India. *American Journal of Tropical Medicine Hygiene* **67**, 480–489.

Stephenson, L.S., Latham, M.C. and Ottesen, E.A. (2000). Malnutrition and parasitic helminth infections. *Parasitology* **121**, S23–S38.

Stoll, N.R. (1947). This wormy world. *Journal of Parasitology* **33**, 1–18.

Tay, J., Ruiz, A., Sanchez Vega, J.T., Romero-Cabello, R., Robert, L. and Becerril, M.A. (1995). Intestinal helminthiasis in the Mexican Republic. *Boletín Chileno de Parasitología* **50**, 10–16.

Tay, J., Salazar-Schettino, P.M., de Haro Arteaga, I. and Bucio Torres, M.I. (1976). Incidence of intestinal helminthiasis in Mexico. *Revista de Investigacíon en Salud Pública* **36**, 241–280.

Thomson, M.C., Obsomer, V., Dunne, M., Connor, S.J. and Molyneux, D.H. (2000). Satellite mapping of *Loa loa* prevalence in relation to ivermectin use in west and central Africa. *Lancet* **356**, 1077–1078.

Thomson, M.C., Obsomer, V., Kamgno, J., Gardon, J., Wanji, S., Takougang, I., Enyong, P., Remme, J.H., Molyneux, D.H. and Boussinesq, M. (2004). Mapping the distribution of *Loa loa* in Cameroon in support of the African Programme for Onchocerciasis Control. *Filaria Journal* **3**, 7.

Udonsi, J.K. and Atata, G. (1987). *Necator americanus*: temperature, pH, light, and larval development, longevity, and desiccation tolerance. *Experimental Parasitology* **63**, 136–142.

Udonsi, J.K., Nwosu, A.B.C. and Anya, A.O. (1980). *Necator americanus*: population structure, distribution, and fluctuations in population densities of infective larvae in contaminated farmlands. *Zeitschrift für Parasitenkunde* **63**, 251–259.

United Nations. (2003). *World Urbanization Prospects: The 2003 Revision*. Department of Economic and Social Affairs, Population Division. New York: United Nations.

Urbani, C., Odermatt, P., Socheat, D., Sinoun, M., Hoyer, S. and Hatz, C. (2001). Control of soil-transmitted helminth infections in schoolchildren in Cambodia. Implications for an integrated approach to helminth control. In: *Controlling Disease Caused by Helminth Infections* (D.W.T. Crompton, M.C. Nesheim and L. Savioli, eds). Geneva: World Health Oraganization (WHO).

Utzinger, J. and Keiser, J. (2004). Schistosomiasis and soil-transmitted helminthiasis: common drugs for treatment and control. *Expert Opinion in Pharmacotherapy* **5**, 263–286.

van der Hoek, W., De, N.V., Konradsen, F., Cam, P.D., Hoa, N.T., Toan, N.D. and Cong le, D. (2003). Current status of soil-transmitted helminths in Vietnam, *Southeast Asian Journal of Tropical Medicine and Public Health* **34**, 1–11.

Vinha, C. (1971). Occurrence, in Brazil, of helminthiasis transmitted by the soil. Croposcopic routine of the ex-Departmento Nacional de Endemias Rurais. *Revista Brasileira de Malariologia e Doencas Tropicais* **23**, 3–17.

Waikagul, J., Krudsood, S., Radomyos, P., Radomyos, B., Chalemrut, K., Jonsuksuntigul, P., Kojima, S., Looareesuwan, S. and Thaineau, W. (2002). A cross-sectional study of intestinal parasitic infections among schoolchildren in Nan Province, Northern Thailand. *Southeast Asian Journal of Tropical Medicine and Public Health* **33**, 218–223.

Warren, K.S., Bundy, D.A.P., Anderson, R.M., Davis, A.R., Henderson, D.A., Jamison, D.T., Prescott, N. and Senft, A. (1993). Helminth infections. In: *Disease Control Priorities in Developing Countries* (D.T. Jamison, W.H. Mosley, A.R. Measham and J.L. Bobadilla, eds), pp. 131–160. Oxford: Oxford University Press.

WHO. (1996). Report of the WHO Informal Consultation on the use of chemotherapy for the control of morbidity due to soil-transmitted nematodes in humans. WHO/ CTD/ SIP/ 96.2. Geneva: World Health Organization (WHO).

WHO. (1998). *Research on Rapid Geographical Assessment of Bancroftian Filariasis*. TDR/TDF/COMDT/98.2. Geneva: World Health Organization (WHO).

WHO. (2002). *Prevention and Control of Schistosomiasis and Soil-Transmitted Helminthiasis*. WHO Technical Series Report 912. Geneva: World Health Organization (WHO).

World Bank. (2000). FRESH: Focusing Resources on Education and School Health. A FRESH start to improving the quality and equity of education. Washington DC: UNESCO, UNICEF, WHO and World Bank. http://www.schoolsandhealth.org/FRESH.htm

Xu, L.-Q., Yu, S.-H., Jiang, Z.-X., Yang, J.-L., Lai, C.-Q., Zhang, X.J. and Zheng, C.-Q. (1995). Soil-transmitted helminthiases: nationwide survey in China. *Bulletin of the World Health Organization* **73**, 507–513.

Tick-borne Disease Systems: Mapping Geographic and Phylogenetic Space

S.E. Randolph[1] and D.J. Rogers[2]

[1]*Oxford Tick Research Group, Tinbergen Building, Department of Zoology, South Parks Road, Oxford OX1 3PS, UK*
[2]*TALA Research Group, Tinbergen Building, Department of Zoology, University of Oxford, South Parks Road, Oxford OX1 3PS, UK*

Abstract . . . 263
1. Predicting Changing Risk of Infection on Evolutionary Time Scales . . . 264
   1.1. Evolutionary Emergence of Vector-Borne Pathogens . . . 265
2. The Evolutionary Time Scale for Vector-Borne Flaviviruses . . . 267
3. Correlates of Phylogenetic Patterns . . . 269
   3.1. Biotic Selectors . . . 269
   3.2. Geography and Phylogeny . . . 270
   3.3. Biotic Liberators, Abiotic Constraints . . . 272
4. Testing the Role of Climate in the Evolution of Tick-borne Flaviviruses . . . 276
   4.1. Constructing Phenetic Eco-Climatic Trees for Viruses . . . 277
   4.2. Data Quantity and Quality . . . 279
   4.3. Composite Predictive Map for Six Viruses . . . 281
   4.4. Congruence between Phylogenetic and Eco-Climatic Trees? . . . 283
Acknowledgements . . . 285
References . . . 285

ABSTRACT

Evidence is presented that the evolution of the tick-borne flaviviruses is driven by biotic factors, principally the exploitation of new hosts as transmission routes. Because vector-borne diseases are limited by

ADVANCES IN PARASITOLOGY VOL 62
ISSN: 0065-308X $35.00
DOI: 10.1016/S0065-308X(05)62008-8

climatic conditions, however, abiotic factors have the potential to direct and constrain the evolutionary pathways. This idea is explored by testing the hypothesis that closely related viruses occupy more similar eco-climatic spaces than do more distantly related viruses. A statistical comparison of the conventional phylogenetic tree derived from molecular distances and a novel phenetic tree derived from distances between the climatic spaces within which each virus circulates, indicates that these trees match each other more closely than would be expected at random. This suggests that these viruses are indeed limited in the degree to which they can evolve into new environmental conditions.

1. PREDICTING CHANGING RISK OF INFECTION ON EVOLUTIONARY TIME SCALES

In this review, we shall apply the same methods used in the global mapping of infectious diseases to exploring the role of climatic factors in the evolution of vector-borne diseases. Natural selection acts on the limiting components of biological systems; those features that are limiting are likely to show evolutionary change. Vector-borne pathogen systems are climate-limited, principally through the impact of climate on the distribution, abundance and population dynamics of the arthropod vectors. This leads to the prediction that climate has been a significant factor in the evolution of these pathogens, and, by implication, will be important in future events on both evolutionary and ecological time scales. This is not to deny the important role of biotic interactions in the evolution of these pathogens. Evidence from many sources indicates that biotic interactions are important in selecting for novel mechanisms in vector–pathogen and pathogen–host interactions to overcome the many intrinsic biological barriers, effectively the innate immunity of host and vector, inherent in indirect transmission cycles. New pathogen strains evolve to exploit enhanced or even new transmission routes. This raises the questions, which new strains may establish persistent transmission cycles and where might this be? This is determined largely by extrinsic multi-factorial environmental factors, together with intrinsic factors such as host-acquired immunity, upon which depends the quantitative balance between all the rates of the transmission processes: transmission may be biologically possible but may not always occur effectively enough to allow persistent cycles of transmission. The same small handful of basic processes (birth, development, feeding and death of the vectors, for example) generate a vast array of patterns simply by virtue of their differing rates.

It is because rates of processes may change more rapidly than the underlying mechanisms that the risk of infection varies so markedly through space and time. It is important to distinguish non-evolutionary emergence of pathogens from evolutionary emergence, the subject of this review.

1.1. Evolutionary Emergence of Vector-Borne Pathogens

The ever-changing nature of threats posed by arthropod-borne microbes involves two separate phenomena. Most commonly, the incidence of infection with existing types of microbes changes, and where the incidence in humans exceeds zero for the first time, new foci of human disease may develop. Many vector-borne diseases are zoonoses, where the pathogen circulates naturally between vectors and wildlife hosts; humans are incidental hosts, typically unable to transmit the pathogen back to biting vectors and so contributing nothing to pathogen persistence. In these cases, a substantial 'iceberg' of enzootic cycles may exist hidden beneath the surface (i.e. undetected without specific prospective searches for infection in vectors or vertebrates) until it 'emerges' through some quantitative change in the dynamic vector–host–pathogen interactions. The recent cases of western tick-borne encephalitis (WTBE) in southeast Norway provide an excellent example. The first clinical case was recognized in 1998 (Ormaasen *et al.*, 2001), with four additional cases reported from 1998 to 2001, all from the small coastal island of Tromøya in Aust-Agder County (Skarpass *et al.*, 2002). Yet, between 1992 and 2000, 16.4% of 317 serum samples collected from dogs in Arendal, on the mainland just west of Tromøya, were positive for antibodies to WTBE virus (Csango *et al.*, 2004). This animal survey suggests persistent enzootic cycles, with infected ticks biting dogs but not yet sufficiently abundant to cause infections in humans (upon which ticks feed much less frequently). Meanwhile, in Tromøya, the annual number of human cases increased to four in 2004. Clearly, this is an emergence of existing microbes within an existing transmission system. With a clear explanation and quantification of the driving forces of existent vector-borne disease systems, as derived from methods described in this volume, it is possible to predict this sort of emergence (non-evolutionary) in response to known or forecast changes in the critical abiotic and biotic factors.

Less common is the emergence of new types of microbes. This involves three essential steps, evolution, establishment within host populations and geographic spread, which differ in their predictability. It is not possible to predict the first step, arising initially from genetic changes that affect biotic interactions. It is, however, possible to predict that vector-borne microbes may be more constrained than directly transmitted microbes in their tolerance to mutations and recombinations, given that they must remain adaptive to two very different sorts of host environments, invertebrate and vertebrate.

The probability of establishment may be predicted only with sufficiently detailed knowledge of the biology of the new host relationships and transmission conditions. More common are post-hoc explanations, which may themselves contribute to future predictions of similar systems. For example, we now understand why we escaped so lightly from the first appearance of the SARS virus, despite its rapid global distribution; fortunately and somewhat atypically, the incubation period (time to infectivity) was longer than the latent period (time to clinical symptoms), allowing isolation of patients before they became infectious (but exposing healthcare workers to highest risk) (Anderson *et al.*, 2004). In effect, this SARS virus did not establish itself within the human population, even though repeated sporadic entry into human populations remains a distinct possibility.

Third, the extent of spread from the point of origin can be predicted by risk mapping, the subject of this volume. As soon as there are sufficient initial observed and accurately geo-referenced foci, the critical limiting factors, both abiotic and biotic, can be identified statistically.

Any place with matching conditions elsewhere in the world, would, in theory, be at potential risk of invasion by the new microbe. Nowhere in the UK, for example, matches the climatic conditions seen within known foci of West Nile virus (WNV) in Europe, suggesting that the UK is not at risk of invasion (D.J. Rogers, unpublished). Mutations of the virus, however, might easily change the precise nature (qualitative or quantitative) of those limiting factors. Statistical, pattern-matching methods, by definition, assume that the correlations upon which the model is built will apply in different times and different places, and cannot take account of dynamic biological (evolutionary) change. In addition to the extent, the speed of any spread could be predicted from estimates of the force of transmission, specifically the basic reproductive number, R_0, based on the number of secondary cases that arise from each primary case introduced into a fully susceptible population. A higher R_0 value also implies greater robustness in the system that permits persistence under a wider range of conditions. In this way, the speed of spread may be functionally related to the extent of spread, as illustrated by the genetically distinct type of WNV (Davis *et al.*, 2003) that swept throughout North America within 5 years of its arrival, apparently from Israel (Giladi *et al.*, 2001; Lanciotti *et al.*, 1999).

Thus, while biotic factors *drive* the first and second steps above, we hypothesize that the final step in the evolution of new epidemiological patterns is *directed* and *constrained* by abiotic factors. We have tested this by examining one clade within the vector-borne RNA flaviviruses.

2. THE EVOLUTIONARY TIME SCALE FOR VECTOR-BORNE FLAVIVIRUSES

Before trying to relate evolutionary history to environmental conditions, it is necessary to establish that the time frames for the observed patterns of evolution and the environment are compatible. Modern climatic conditions can tell us nothing about the abiotic constraints on the processes occurring millions of years ago that have generated palaeontological patterns. Microbes, however, are capable of rapid evolutionary change through their very large population sizes arising from exceptionally high rates of replication from single copies. The simpler the microbe, the more these features apply, making RNA viruses among the most rapidly evolving biological entities known (Moya *et al.*, 2004). Natural selection for newly adaptive strains, however, operates on the complete life cycle, or transmission cycle for parasitic microbes, so that more complex indirect transmission routes not only impose strong selection forces, but also prolong generation times, and therefore slow down the rate of evolutionary change.

The force of this brake appears to differ with the contrasting biology of the two major classes of vectors, insects and ticks. For insect vectors, the relatively short blood-feeding interval, a few days, must be added to the microbe's development cycle within its vector (the so-called extrinsic incubation period), which may take 1–2 weeks. This allows a complete transmission cycle within a very few weeks. At the other extreme, hard ticks of the family Ixodidae feed only once per life stage, as larvae, nymphs and adults. The tick's inter-stadial development period varies from weeks to many months depending on the geographically and seasonally variable climate, after which there may be a considerable period before the new tick stage feeds again. This introduces a long delay into the transmission process, during which a high percentage of ticks die, especially in dry summers (Randolph, 2004b). This protracted mode of transmission by ticks (with each complete cycle measured in months or even years) is likely to act as a brake on evolutionary change (Gritsun *et al.*, 1995; Zanotto *et al.*, 1995). Indeed, based on rates of non-synonymous nucleotide substitution among the RNA flaviviruses, the tick-borne species show an apparent rate of evolution estimated to be about 56% of the rate seen in the insect-borne species (Zanotto *et al.*, 1996). The patterns of cladogenesis also differ. Tick-borne flaviviruses appear to have diverged more or less continually over the last 2000 years, with only one, Louping ill (LI) virus, having arisen <800 years ago and diversifying within the last 300 years (Gould *et al.*, 2001; McGuire *et al.*, 1998; Zanotto *et al.*, 1996). In contrast, although the major groups of mosquito-borne flaviviruses diverged several thousand years ago, periods of intense cladogenesis in the dengue and Japanese encephalitis (JE) serotypes have occurred during the last two centuries (Zanotto *et al.*, 1996).

These best estimates of the dates of divergence within phylogenetic trees of vector-borne RNA viruses indicate an evolutionary time scale of the order of hundreds of years. This significantly shortens the conventional gap between evolutionary and ecological time scales, making it more reasonable to use modern abiotic conditions to nvestigate evolutionary constraints. Although climate has varied over the past one or two millennia, these changes through time are small compared with spatial variation across continents, with which the analysis addressed in this review is concerned. It also makes evolutionary change a matter of more immediate relevance to changing epidemiological risk. If we can *explain* the driving forces and constraints that have shaped the evolutionary pathways of pathogen strains, that are now so well *described* in the new and improving phylogenies based on molecular analyses, we shall be better equipped to predict the likely establishment and spread, if not the first appearance, of new strains as they appear in the future.

3. CORRELATES OF PHYLOGENETIC PATTERNS

3.1. Biotic Selectors

The evolutionary history of the flaviviruses has been deduced from their molecular phylogeny and has been well described in recent years (Gould *et al.*, 2001, 1997; McGuire *et al.*, 1998; Zanotto *et al.*, 1995, 1996). Division into major clades coincides with different principal vectors (insects, ticks or no known vector), with further sub-division according to vertebrate hosts (Gaunt *et al.*, 2001). Among the insect-borne species, neurotropic viruses associated with encephalitic disease in humans (e.g. JE, WNV) persist in cycles between *Culex* mosquitoes and birds, while non-neurotropic viruses associated with haemorrhagic disease in humans (e.g. Dengue, Yellow Fever) cycle between *Aedes* mosquitoes and primates. Although mosquitoes show species-specific patterns of feeding behaviour, this is not absolute, and each genus of mosquitoes may feed on and infect both birds and mammals (Gaunt *et al.*, 2001). Nevertheless, complete transmission cycles appear to be largely confined to one class of vertebrates or the other (*Culex*-birds or *Aedes*-primates) (Gaunt *et al.*, 2001), indicating the dominant role of the vertebrate host in driving the selection process underlying this pattern of clades.

Gould *et al.* (2001) make a clear case for the role of vertebrate hosts in selecting for different genotypes of tick-borne viruses. Early in the evolution of the genus (but apparently not until after the most recent glaciation) a major split occurred into two clades of viruses, the first being transmitted via seabirds (the TYU serogroup) and the second transmitted principally via mammals (the TBE complex viruses) (Figure 1). Furthermore, a group of closely related encephalomyelitis viruses all associated with sheep diverged from the other TBE complex viruses that are associated mostly with forest rodents.

3.2. Geography and Phylogeny

The following account is taken from Gould *et al.* (2001). One of the earliest nodes of the tick-borne clades includes an African virus (KAD), suggesting that during the past 5000 years, infected ticks have been dispersed by seabirds from an African (or Asian) origin, to northwest France (MEA), to the eastern coast of Russia (TYU) and to the Great Barrier Reef (SRE). Corresponding to the migratory behaviour of seabirds, this most ancient lineage of tick-borne viruses shows the greatest geographical and genetic separation (*c.* 60% amino acid identity compared with 70% between most of the TBE complex viruses). At the deeper nodes of the tree, the earliest TBE complex viruses are also found at great geographic distance from each other, possibly suggesting similar bird-based introductions to far-flung places (e.g. GGY virus, found in ticks under rocks and debris used by penguins and other seabirds on Macquarie Island, several hundred miles off southern Australia).

The most recent TBE complex viruses, however, show an asymmetric topology on all phylogenetic trees so far constructed, indicating a continuous rather than interrupted evolution. Moreover, the correlation between the genetic and geographic distances between these viruses is interpreted as indicating an evolutionary cline from east to west across the northern hemisphere (Zanotto *et al.*, 1995).

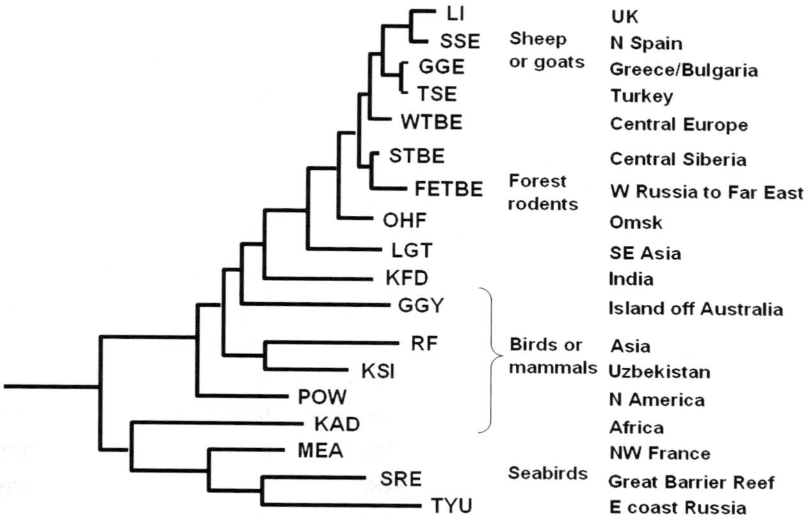

Figure 1 Phylogenetic tree of the tick-borne clade of flaviviruses: consensus tree based on the 1st and 2nd codon positions for 41 E genes and the NS5 gene sequence, taken from Gould *et al.* (2001) and Gaunt *et al.* (2001). The source of the gene sequences used to construct this tree is given in the original publications. The genus *Flavivirus* contains about 70 distinct antigenically related flaviviruses. These are positive-stranded RNA viruses that consist of three structural proteins (C capsid, M membrane and E envelope) and seven non-structural (NS) proteins. TYU, Tyuleniy; SRE, Saumarez Reef; MEA, Meaban; KAD, Kadam; POW, Powassan; KSI, Karshi; RF, Royal Farm; GGY, Gadget's Gully; KFD, Kyasanur Forest disease; LGT, Langat; OHF, Omsk haemorrhagic fever; FETBE, Far Eastern tick-borne encephalitis; STBE, Siberian tick-borne encephalitis; WTBE, Western tick-borne encephalitis; TSE, Turkish sheep encephalitis; GGE, Greek goat encephalitis; SSE, Spanish sheep encephalitis; LI, Louping ill. Principal vertebrate host types and the geographical distribution are shown.

Thus, within the mammal-associated lineage, the viruses of Southeast Asia (KFD and LGT) are the ancestors of the viruses of the northern forests of central Russia, Far East and Siberia (OHF, FETBE and STBE), which in turn are the ancestors of the viruses of eastern and central Europe (WTBE, TSE, GGE), culminating in a virus found only in northern Spain (SSE) and the most westerly virus (LI) found in the British Isles. It is only these last four viruses that are transmitted via sheep, and LI is evidently also transmissible via red grouse (*Lagopus scoticus*) (Hudson *et al.*, 1995) and mountain hares (*Lepus timidus*) (Jones *et al.*, 1997), but not rodents (Gilbert *et al.*, 2000). Despite being found in isolated foci within upland sheep-grazing regions of Turkey (TSE), Bulgaria and Greece (GGE), Spain (SSE) and Ireland, UK and Norway (LI), all are antigenically very closely related, but distinguishable by a unique tripeptide sequence in the envelope (E) gene (Gao *et al.*, 1993; Gould *et al.*, 2001).

The existence of this cline immediately raises the question: what directed its development? It is not sufficient to point to the extremely limited mobility of ticks. Ticks rely for their displacement on the movement of their hosts, which is also fairly limited in the case of rodents, but not so for birds and ruminants (deer as well as livestock). Clearly, through movement of hosts, either natural or due to man's intervention, TBE complex viruses have been able to colonize a large proportion of the northern hemisphere during the past few thousand years. Yet, each virus is characterized by a geographically coherent, focal distribution within the much more continuous ranges of both the competent vector tick species and rodent or ovine (sheep and goats) transmission hosts. The implication is that the chance of each specific virus (within an infected tick or host) successfully invading a new region is rare, and that the distribution of viruses is limited less by their ability to reach a new place than by their ability to survive there. At least for WTBE virus, this limitation is now known to be due to rather specific climatic conditions required for persistent enzootic cycles, which in turn are determined by the biological basis of transmission.

3.3. Biotic Liberators, Abiotic Constraints

3.3.1. Fragile Cycles of WTBE Virus Associated with Rodents

Under laboratory conditions, WTBE virus is transmissible by a wide range of Eurasian tick species (Korenberg and Kovaleskii, 1994) and even an African species (Nuttall and Labuda, 1994), and also via a range of natural rodent host species (Labuda *et al.*, 1993a). In natural conditions, however, only the specific ecological association of

I. ricinus with field mice (*Apodemus flavicollis*) contributes significantly to WTBE virus maintenance (Labuda and Randolph, 1999). The vertebrate specificity is due to a particular route of transmission, now recognized as crucial because of its quantitative impact. A complete transmission cycle, from tick to tick via vertebrates, occurs most efficiently between co-feeding ticks in the absence of a systemic viraemia (Labuda *et al.*, 1993a). Only certain vertebrate species (notably *Apodemus* mice) allow the necessary virus replication within particular cells of the skin's immune system (Labuda *et al.*, 1996), enhanced by pharmacologically active substances in tick saliva (Nuttall, 1998; Wikel, 1996). Epidemiologically, the quantitative significance of this transmission pathway is that non-systemic infections are less virulent than fully developed systemic viraemias, and so allow *Apodemus* to survive long enough to permit ticks to complete their prolonged blood meals (Labuda *et al.*, 1993b). The consequent $>50\%$ increase in the force of transmission relative to the systemic pathway is crucial to virus survival given the inherent fragility of this system (Randolph *et al.*, 1996), but it is not sufficient.

In either case, non-viraemic or viraemic, the period of vertebrate infectivity is limited to a few days. Specific seasonal climatic conditions are therefore required to ensure that larvae and nymphs actively seek their hosts in seasonal synchrony (Randolph *et al.*, 2000), so that large numbers of larvae co-feed with any one infected nymph. Only this condition, plus the coincident aggregated distributions of larvae and nymphs concentrated principally on sexually active male rodents, ensures enough amplification of infected ticks (Randolph *et al.*, 1999) to offset the high percentage mortality that inevitably occurs between each life stage (Randolph, 1998). The geographical distribution of these specific seasonal climatic conditions dictates the focal distribution of TBEV within a subset of the extensive range of its vectors and hosts. Based on correlations between the patterns of remotely sensed climatic conditions and TBE foci, the critical climatic factors that determine where these biological requirements are satisfied and therefore where WTBE foci exist, have been identified (Randolph, 2000). First, high humidity at ground level during the summer allows good tick survival, active host-seeking activity and therefore adequate tick populations. These conditions are typically found in deciduous woodlands (Daniel and Kolar, 1990) and also on some upland moorlands, where large host species such as deer or sheep, that feed the reproductive adult ticks, are also found. Secondly, a particular seasonal profile of land surface temperature is statistically associated with both synchronous feeding by larvae and nymphs and the presence of WTBE human infections (Randolph *et al.*, 2000), although the precise biological processes that link these temperature and tick patterns have not yet been fully uncovered.

Among the potential tick vectors in Europe, only *I. ricinus* has the correct host relationships and the appropriate natural life cycle in certain places to support WTBE virus transmission cycles. This is a good example of the quantitative ecological hurdles referred to at the start of this review, in this case imposing biotic specificity.

3.3.2. Geographically Limited TBE Complex Viruses Associated with Sheep

So far, WTBE is the only tick-borne flavivirus for which this degree of quantitative explanation of natural enzootic cycles has been achieved. Intensive field investigations and mathematical modelling of the biotic elements of the LI system indicate that persistent cycles of LI virus depend on specific combinations of vertebrate hosts, including sheep, grouse and mountain hares (Hudson, 1992; Hudson *et al.*, 1995; Jones *et al.*, 1997; Norman *et al.*, 1999), that have not yet been fully identified in all situations (Laurenson *et al.*, 2003). Abiotic factors also appear to be important; recent analysis of climatic correlates of LI foci indicates the need for specific temperature conditions that confine LI virus to sub-sets of the distribution of the vector tick, *I. ricinus* (Benjamin McCormick, pers. comm.). Very little is known about the natural host relationships of TSE, GGE and SSE viruses beyond their ability to be transmitted via sheep or goats by *I. ricinus* and perhaps also by other tick species.

Gritsun *et al.* (1995) suggest that the mutations in the variable regions of the E gene probably reflect adaptation to different wildlife host species. Escape from the ancestral rodent transmission pathway by LI, TSE and GGE viruses may have had significant evolutionary consequences. Unlike the sheep-transmitted viruses, WTBE virus may be funnelled through vectors of greater than average genetic homogeneity at each transmission cycle, because a large proportion of each cluster of larval ticks co-feeding on any one mouse is likely to come from a single egg batch (larvae quest close to where they emerge). WTBE virus also has a considerably narrower effective host range than LI virus. Both these factors might explain the greater genetic and antigenic homogeneity of WTBE across its wide geographical range (2% amino acid diversity in continental Europe) compared with the higher strain diversity (3–4%) of LI over its much smaller range within the UK (Gould *et al.*, 2001; Guirakhoo *et al.*, 1987; Heinz and Kunz, 1981, 1982).

The same mutations might also have significant epidemiological consequences. Larger vertebrate host species feed both nymphs and adults of *I. ricinus*, allowing virus transmission between these two tick stages, which much more commonly (always?) show synchronous feeding seasons. These systems, therefore, might be expected to be more robust than the WTBE virus system. Any excess slack in the system would permit invasion into less than optimum biotic or abiotic contexts. Yet, despite the widespread, overlapping or contiguous distributions of rodents and sheep (and goats) throughout Europe, each virus is confined to its own geographically separated, well-defined range. For example, 'hundreds of thousands of sheep' migrate from Bosnia to an area of a natural WTBE focus in Croatia for winter grazing, and take local ticks back with them when they return to Bosnia in the spring, but neither WTBE virus nor any sheep-transmitted viral type has been introduced to Bosnia (Borcic *et al.*, 1999). Isolates of TSE have been recorded only from six villages close to Gebze and Kirklareli in northwest Turkey (Hartley *et al.*, 1969; Whitby *et al.*, 1993), and of GGE from northeast Greece (Papadopoulos, 1980) and Plodiv district of central Bulgaria (Pavlov, 1968). Their full distributions are unknown, but appear to be somewhat limited. Interestingly (but of no known significance), in the Bulgarian focus, which is several hundred kilometres southwest of WTBE foci in Romania, GGE was isolated from sheep, ticks and shrews (*Sorex araneus*).

Once TSE or GGE had established itself in livestock hosts, transportation by humans could easily permit dissemination to other regions where distinct strains have evolved. Yet, this appears to have occurred very infrequently: once from Turkey, Greece and Bulgaria to the Basque highlands of Spain (SSE), thence to Ireland (LI/MA54) and thence to parts of Wales (LI/I), northern England (LI/917 *et al.*), Scotland (LI/31 *et al.*) and finally to southwest England (LI/A and LI/DEV4) (McGuire *et al.*, 1998). Indeed, most LI virus dispersal in the British Isles has been related to the history of sheep farming during the last 300 years (Gould *et al.*, 2001; McGuire *et al.*, 1998). Sheep were imported into Norway from Britain in the 19th century, and LI was isolated from goats in Norway in 1978 and from sheep in 1982 (Ulvund *et al.*, 1983). LI, however, did not spread beyond a limited southwest coastal region closest to Scotland, despite extensive tick populations and other tick-transmitted infections of sheep and wild ruminants along most of the coastal regions of southern Norway as far north as Nordland county (Stuen, 2003).

Why has each virus type remained so isolated? What has prevented its spreading out of the area of its presumed introduction, infecting sheep over their much wider distributions? It appears that each virus may be 'trapped', presumably by a set of peculiar abiotic environmental conditions acting on any one part of the pathogen–vector–host interaction, just as has been identified for WTBE virus. In that case, while biotic elements have selected for new strains, abiotic factors seem to have imposed constraints, preventing spread throughout Europe despite the ubiquity of competent vectors and hosts.

4. TESTING THE ROLE OF CLIMATE IN THE EVOLUTION OF TICK-BORNE FLAVIVIRUSES

If climate dictates the distribution of these viruses, it is reasonable to hypothesize that climate has directed and constrained their evolution. One way to investigate this is to test whether the eco-climatic spaces occupied by closely related viruses are more or less similar than those occupied by more distantly related ones. If they prove to be very similar, one could conclude that climate has been a significant evolutionary constraint, whereas if they are significantly different, one could conclude that viruses have been free to leap from one eco-space to another with no constraint imposed by climate *per se* despite the evident geographic constraints.

We are applying the methods outlined in other reviews in this volume to test for matches between virus phylogeny and the environmental conditions in which each virus circulates by seeking correlations between two very different sorts of trees. First are the familiar molecular phylogenetic trees, constructed on the basis of genetic differences as a direct measure of evolutionary distance between related species. These describe evolutionary history. Second are eco-climatic trees, constructed from the statistical distances between the eco-climatic spaces in which each virus circulates. These trees are purely descriptive, containing no information on evolutionary history.

4.1. Constructing Phenetic Eco-Climatic Trees for Viruses

The basis for these eco-climatic trees is the ability to define the conditions within which each virus can survive (i.e. exists), plotted in multi-variate space. With the advent of Geographical Information Systems, this procedure has now become commonplace for a wide variety of organisms, although the quality of the input observations and predictor variables, and of the statistical analyses and their interpretation, varies considerably. Based on correlations between environmental and distributional patterns, over the past decade the distributions of single species of organisms or single diseases have been routinely mapped with high accuracy. Good accuracy statistics, however, do not always guarantee good visual matches between the observed and predicted maps of presence and absence. Both statistical and visual matches must be good for the predictions to be convincing. It is the use to which these maps are put that is now the challenge. Most simply they paint a fuller picture of a pre-existing but incomplete description of the distribution, providing a more accurate blueprint upon which to base better public health decisions or control strategies. More interestingly, the variables identified as limiting an organism's presence may help to quantify the complex of biological processes that determine the distributional patterns, and predict any possible changes in these patterns under dynamic eco-climatic conditions.

For the present purpose, it is a specific product of the statistical method itself, in this case discriminant analysis (Green, 1978; Rogers *et al.*, 1996), that is of use for creating phenetic trees. The particular virtue of this method is that it provides a measure of distance: the Mahalanobis distance is the distance between the geometric centres of multi-variate eco-climatic spaces, adjusted for the co-variance between the variables that define that space (Figure 2). It is a way of reducing the many differences in eco-climatic conditions to a single measure of separation between presence and absence, or between the presence of pairs of species. The ecologist's Mahalanobis distance is thus comparable to the evolutionary biologist's molecular distance. Using forward stepwise selection of variables, discriminant analysis

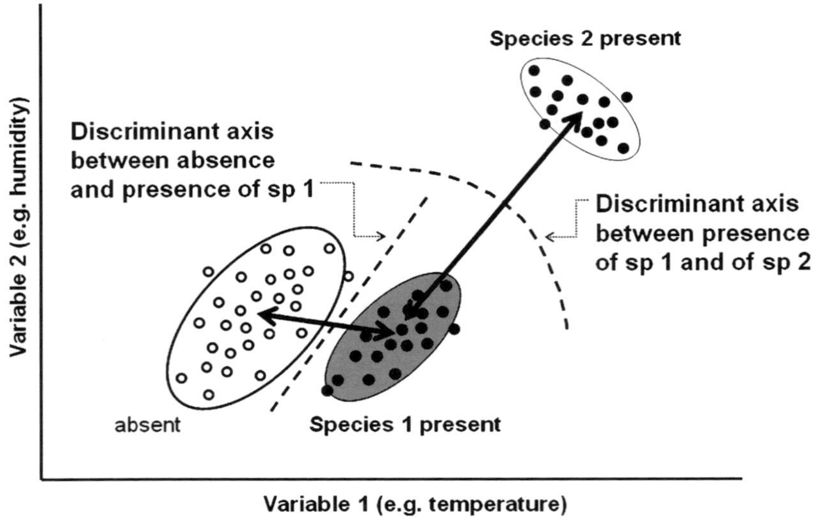

Figure 2 An illustration of the principle of linear discriminant analysis used to distinguish between the abiotic conditions in which an organism is present or absent, and the conditions occupied by two different species. Mahalanobis distances are represented by the heavy double-ended arrows. (Reprinted from Randolph (2004a), with permission of Cambridge University Press.)

also assigns relative importance to the variables that define the species distribution, thereby yielding both discrete and distance information.

Furthermore, creating a map for all the tick-borne flaviviruses is itself a novel challenge here. To create phenetic eco-climatic trees, the Mahalanobis distances between species must be derived from the same set of variables. One way of doing this would be to make predictive maps of each virus individually in the conventional manner and derive a consensus list of the 10 most commonly selected predictor variables, which would inevitably involve a degree of subjective decisions. Alternatively, it is computationally possible to make a composite predictive map of all the viruses in a single analytical exercise. As each virus is added, however, the accuracies for any one will inevitably be compromised, unless all the viruses are in fact limited by precisely the same set of variables. The analytical procedure is as follows: (i) create a single composite predictive map for all viruses together to assess its accuracy and therefore the validity of the selected predictor variables that apply to all the viruses; (ii) create a matrix of Mahalanobis distances based on these variables; and (iii) test for a better than random match between the matrices of Mahalanobis and molecular distances. Note that the Mahalanobis distances are derived from the observed points of presence of each virus (i.e. the training data), not from the predictive maps.

4.2. Data Quantity and Quality

With the super-abundance of information of abiotic factors now available from remotely sensed sources, the problem is to find equally well geo-referenced biotic data at comparable spatial resolutions and geographical ranges. For the tick-borne flaviviruses, at one extreme good point data exist for the observed presence of SSE ($n = 41$), LI ($n = 143$), OHF ($n = 32$) and KFD ($n = 24$)[1] that can be geo-referenced by

[1]Data sources: SSE, map in Juste *et al.* (1997); LI, compiled by Ben McCormick from data provided by P.J. Hudson, and the UK Veterinary Laboratory Agency; P.J. Timoney, pers. comm.; Adams *et al.* (1977); Davidson *et al.* (1991); Hubalek *et al.* (1995); McGuire *et al.* (1998); Walton and Kennedy (1966); OHF, Kharitonova and Leonov (1985); KFD, Work (1958, 1960); Sreenivasan *et al.* (1986); Banerjee (1988); Bhat *et al.* (1978); Dandawate *et al.* (1994).

matching recorded place names (or local grid references) to universal longitude and latitude coordinates using standard atlases. The distribution of WTBE has been mapped in the form of polygons (Immuno, 1997), within which *c.* 900 presence points were selected at random. These data (Figure 3a is Plate 8.3 in the Separate Color Plate Section) are sufficient to generate predictive maps for each virus. At the other extreme, only eight sites of recorded presence are currently available to the authors for Powassan (across North America) (not shown), and up to only three sites for Kadam (in Uganda), Karshi (in Uzbekistan), Langat (in Thailand and Malaya), TSE (in Turkey) and GGE (in Greece and Bulgaria)[2] corresponding to published reports of virus isolation, although more sites could possibly be recovered in the future from local sources (also plotted in Figure 3a).

The situation with regard to TBE in Russia is complicated by the co-circulation of more than one virus type within such a vast area, whose charting is mostly buried in an inaccessible Russian literature. The disease previously known as Russian spring-summer encephalitis (RSSE) is now known to be caused by two genetically distinct virus subtypes, Far Eastern (FETBE) and Siberian (STBE) tick-borne encephalitis (Ecker *et al.*, 1999). The latter has now been isolated over a wide longitudinal range from Estonia and Latvia in the west (26–27°E) (Golovljova *et al.*, 2004; Lundkvist *et al.*, 2001) to Irkutsk near Lake Baikal in Siberia (104°E) in the east (Ecker *et al.*, 1999). It is therefore currently impossible to distinguish the distribution of STBE from that of FETBE, which has also been isolated from Estonia and Latvia (ibid) and throughout Russia to the extreme far east (e.g. Khabarovsk, 135°E) (Hayasaka *et al.*, 1999) and from several sites on Hokkaido Island, Japan (141°E) (Takeda *et al.*, 1999). In the absence of point data representative of the full geographical range of these viruses, but knowing that they occur more or less wherever the vector tick *I. persulcatus* occurs (Korenberg, 1994), 3000 points were selected at random from the predicted distribution of *I. persulcatus* (Katharine Mansell, unpublished MSc thesis), within the broad geographical limits

[2]Powassan, McLean and Donohue (1959); Artsob (1989); Ebel *et al.* (1999); Kadam, Karabatsos (1985); Karshi, Karabatsos (1985); Khutoretskaya *et al.* (1985); Langat, Bancroft *et al.* (1976); Karabatsos (1985); TSE, Hartley *et al.* (1969); Whitby *et al.* (1993); GGE, Papadopoulos (1980); Pavlov (1968).

for this tick as described by (Korenberg *et al.*, 1969) (Figure 3a). This gives a fuller definition of the eco-climatic limits of TBE in Russia than the specific point data available to us for *I. persulcatus* only in Estonia (Vasilenko *et al.*, 1997), Latvia (A. Bormane, pers. comm.) and western Russia as far east as the Urals (E.I. Korenberg, pers. comm.). This approximation is an example of a solution to the problem of data availability.

The eco-climatic conditions characteristic of absence of all the six flaviviruses (SSE, LI, WTBE, Russian TBE, OHF and KFD) were defined by selecting points at random outside the areas of presence, at a distance between 0.05 and 0.5° longitude/latitude from any presence points.

4.3. Composite Predictive Map for Six Viruses

To date it has proved possible to capture the distributions of six tick-borne flaviviruses (LI, SSE, WTBE, all Russian TBE, OHF and KFD) in a single exercise with 70–100% accuracy for all except OHF; 75% of OHF observed points were assigned to the Russian TBE category, but only within the narrow OHF geographical range which overlaps with Russian TBE (Figure 3b is Plate 8.3 in the Separate Colour Plate Section). The overall kappa index (Congalton, 1991) $= 0.729 \pm 0.012$. As expected, the accuracy of the modelled distribution of each virus is considerably less than when each is modelled individually, with a tendency for false predictions of presence beyond the known range. For example, the model identifies conditions suitable for WTBE far to the west and south of its actual location, and even in a few parts of the far east, without the degree of focality known to characterize this virus and captured successfully by an earlier solo model (Randolph, 2000). Likewise, scattered false predictions of SSE appear outside the Basque highlands of Spain and even in Italy. Nevertheless, the overall fit to the known distribution of these viruses across very large parts of the globe is correct. This confirms that the same 10-predictor satellite variables may be used to distinguish areas of presence of each virus from each of the others, and from areas of absence. These 10 predictor variables are shown in

Table 1 The predictor variables selected by forward step-wise selection that define the distinguished areas of presence of each of six tick-borne flaviviruses (SSE, LI, WTBE, Russian TBE, OHF and KFD) from each of the others, and from areas of absence

Satellite signal	Temporal Fourier variable
Normalized Difference Vegetation Index	Annual variance
Land Surface Temperature (LST)	Minimum
Air Temperature (T'air)	Amplitude of the bi-annual cycle
Middle Infra-red (MIR)	Amplitude of the annual cycle
MIR	Mean
T'air	Amplitude of the annual cycle
T'air	Maximum
T'air	Phase of the annual cycle
MIR	Minimum
LST	Amplitude of the bi-annual cycle

Note: For an explanation of the meaning of these temporal Fourier variables, see Hay *et al.* (this volume, pp. 37–77).

Table 1 in the order in which they were selected, i.e. their relative importance in fitting the model to the observations.

Statistically, therefore, each of these six viruses falls within a distinct eco-climatic space defined by factors that are all temporal Fourier variables (i.e. seasonal characteristics) of thermal (nine factors) and moisture (one factor) conditions. To illustrate this on the page, the two most significant variables are chosen as axes for bi-variate plots that illustrate in two dimensions the differences that the model has in fact identified in 10 dimensions (Figure 4 is Plate 8.4 in the Separate Colour Plate Section). In addition, the handful of points representing the observed locations for GGE, TSE, Karshi, Langat and POW are plotted on the same axes. There is clear separation between many viral types even in bi-variate space, and each virus occupies only a subset of the total environmental space at least potentially available to it (note the wider scatter of absence points).

As expected from their more southerly locations, KFD, Langat and even Karshi occupy eco-climatic spaces very different from the northern viruses. However, not all viruses are clearly separated by the two variables used in Figure 4, and some of them are more separated by other variables. For example, LI and WTBE appear to be close together on this graph, but just as the molecular phylogeny shows that LI virus did not evolve directly from WTBE virus, so the Fourier-processed satellite imagery of temperature and moisture conditions shows a barrier of seasonally distinct climate in France. Biologically, relevant climate factors now explain why WTBE has not reached Britain through France (Randolph *et al.*, 2000), but instead the distinct LI virus entered Britain apparently via Spain and Ireland (Gould *et al.*, 2001; McGuire *et al.*, 1998).

4.4. Congruence between Phylogenetic and Eco-Climatic Trees?

Although the above results are only the first, albeit very important, steps towards deriving Mahalonobis distances between all the tick-borne flaviviruses, and so constructing a complete phenetic ecological tree, we now have sufficient data to warrant a first attempt at comparing the matrices of phylogenetic and eco-climatic distances. Eco-climatic distances were calculated in terms of the Mahalanobis distances separating each pair of viruses, calculated using the 10 predictor variables used to produce the map (Figure 3b). The distances were inserted into a matrix of distances in the same order as the matrix of genetic differences (Table 2). Mahalanobis distances are measures of multi-variate separation that allow for the co-variance of the variables concerned (Rogers, 2000). Thus, two centroids (means of multi-variate distributions) may be the same absolute distance apart in multi-variate space (as indicated by the Euclidean distance between them), but the Mahalanobis distance between them will be different if their co-variances differ (especially if they have different orientations in multi-variate space). The calculation of Mahalanobis distances involves inversion of the co-variance matrix, and this can sometimes be difficult if one of the sample sizes is rather small. Thus, a second eco-climatic distance matrix was calculated using just the Euclidean distances (which do not require matrix inversion) between the centroids defining each virus species. In both types of matrix (genetic and environmental), larger numbers indicate a greater genetic or environmental difference between the pairs of viruses being compared.

Table 2 Matrix of genetic distances (top) between the six tick-borne flaviviruses for which sufficient geo-referenced data currently exist to allow estimates of the Mahalonobis and Euclidean distances (middle and bottom) between the eco-climatic spaces occupied by each virus

Genetic distances						
	LI	OHF	POW	FETBE	WTBE	KFD
LI	0					
OHF	0.35275	0				
POW	0.88279	0.99877	0			
FETBE	0.22538	0.29843	0.86321	0		
WTBE	0.15808	0.30600	0.95008	0.19537	0	
KFD	0.65441	0.57301	0.95661	0.58159	0.62972	0
Mahalanobis environmental distances						
	LI	OHF	POW	FETBE	WTBE	KFD
LI	0					
OHF	30.325	0				
POW	25.71	10.43	0			
FETBE	25.05	3.31	6.76	0		
WTBE	6.88	11.13	7.88	8.14	0	
KFD	788.27	870.68	860.11	875.01	833.48	0
Euclidean environmental distances						
	LI	OHF	POW	FETBE	WTBE	KFD
LI	0					
OHF	363.21	0				
POW	383.34	21.76	0			
FETBE	360.85	9.03	24.82	0		
WTBE	191.92	171.41	191.5	169.3	0	
KFD	42.56	374.94	394.85	373.49	204.84	0

Source: Genetic distances supplied by E.C. Holmes.

The pairs of genetic/environmental distance matrices were compared using the non-parametric Mantel test. This provides a test of the H_0 hypothesis, that there is no association between the elements of the two dissimilarity matrices being compared, by carrying out 10 000 randomizations of one of the matrices and estimating each time the test statistic (g, the standard normal variate); collectively, the randomizations define the frequency distribution of g and the position within this frequency distribution of the actual value obtained is then examined. If it is outside the boundary of 95% of the randomizations, then H_0 is rejected in favour of the alternative hypothesis that there is some association between the values in the two matrices ($p<0.05$). For the Genetic distance/Euclidean distance matrix comparison, the Mantel test gave a value of g of 1.76 ($p<0.05$). For the Genetic distance/Mahalanobis distance comparison $g = 2.246$ ($p<0.025$).

The results indicate a significant association between genetic and environmental distances, suggesting that climate may indeed have played a role in directing the evolution of the flaviviruses. Furthermore, the particular environmental factors that appear to be correlated with the origin of new viruses mostly concern the precise seasonal thermal cycle, perhaps determining crucial features of the vector tick's seasonal population dynamics and host relationships as has been shown for WTBE. This result will direct the search for the processes underlying the evolutionary patterns, leading to new interpretations and understanding of the forces that direct and constrain pathogen evolution. These same critical conditions are also likely to direct any shifting incidence of these existing vector-borne diseases under the forces of natural and anthropogenic environmental change.

ACKNOWLEDGEMENTS

We are very grateful to Eddie Holmes for kindly supplying the phylogenetic matrix of the tick-borne flaviviruses, and Simon Hay for helpful comments on the manuscript. SER is a Senior Research Fellow supported by the UK Natural Environment Research Council.

REFERENCES

Adams, K.M.G., Beasley, S.J. and Blewett, D.A. (1977). The occurrence of antibody to *Babesia* and to the virus of louping-ill in deer in Scotland. *Research in Veterinary Science* **23**, 133–138.

Anderson, R.M., Fraser, C., Ghani, A.C., Donnelly, C.A., Riley, S., Ferguson, N.M., Leung, G.M., Lam, T.H. and Hedley, A.J. (2004). Epidemiology, transmission dynamics and control of SARS: the 2002–2003 epidemic. *Philosophical Transactions of the Royal Society of London B* **359**, 1091–1106.

Artsob, H. (1989). Powassan encephalitis. In: *The Arboviruses: Epidemiology and Ecology* (T.P. Monath, ed.), Vol. 4, pp. 29–49. Boca Raton, Florida: CRC Press.

Bancroft, W.H., Scott, R.M., Snitbhan, R., Weaver, R.E. and Gould, D.J. (1976). Isolation of Langat virus from *Haemaphysalis papuana* Thorell in Thailand. *American Journal of Tropical Medicine and Hygiene* **25**, 500–504.

Banerjee, K. (1988). Kyasanur Forest disease. In: *The Arboviruses: Epidemiology and Ecology* (T.P. Monath, ed.), Vol. 3, pp. 93–116. Boca Raton, Florida: CRC Press.

Bhat, H.R., Sreenivasan, M.A., Goverdhan, M.K., Naik, S.V. and Banerjee, K. (1978). Antibodies to Kyasanur Forest disease virus in bats in the epizootic–epidemic area and neighbourhood. *Indian Journal of Medical Research* **68**, 387–392.

Borcic, B., Kaic, B. and Krajl, V. (1999). Some epidemiological data on TBE and Lyme borreliosis in Croatia. *Zentralblatt für Bakteriologie* **289**, 540–547.

Congalton, R.G. (1991). A review of assessing the accuracy of classifications of remotely sensed data. *Remote Sensing of Environment* **37**, 35–46.

Csango, P.A., Blakstad, E., Kirtz, G.C., Pedersen, J.E. and Czettel, B. (2004). Tick-borne encephalitis in southern Norway. *Emerging Infectious Diseases* **10**, 533–534.

Dandawate, C.N., Desai, G.B., Achar, T.R. and Banerjee, K. (1994). Field evaluation of formalin inactivated Kyasanur Forest disease virus tissue culture vaccine in three districts of Karnataka state. *Indian Journal of Medical Research* **99**, 152–158.

Daniel, M. and Kolar, J. (1990). Using satellite data to forecast the occurrence of the common tick *Ixodes ricinus* (L.). *Journal of Hygiene, Epidemiology, Microbiology and Immunology* **34**, 243–252.

Davidson, M.M., Williams, H. and Macleod, J.A.J. (1991). Louping ill in man—a forgotten disease. *Journal of Infection* **23**, 241–249.

Davis, C., Beasley, D., Guzman, H., Raj, P., D'Anton, M., Novak, R., Unnasch, T., Tesh, R. and Barrett, A. (2003). Genetic variation and geographically distinct West Nile virus isolates, United States 2001, 2002. *Emerging Infectious Diseases* **9**, 1423–1429.

Ebel, G.D., Foppa, I., Spielman, A. and Telford, S.R.I. (1999). A focus of deer tick virus transmission in the northcentral United States. *Emerging Infectious Diseases* **5**, 570–574.

Ecker, M., Allison, S.L., Meixner, T. and Heinz, F.X. (1999). Sequence analysis and genetic classification of tick-borne encephalitis viruses from Europe and Asia. *Journal of General Virology* **80**, 179–185.

Gao, G.F., Hussain, M.H., Reid, H.W. and Gould, E.A. (1993). Classification of a new member of the TBE flavivirus subgroup by its immunological, pathogenetic and molecular characteristics: identification of subgroup-specific pentapeptides. *Virus Research* **30**, 129–144.

Gaunt, M.W., Sall, A.A., de Lamballerie, X., Falconar, A.K., Dzhivanian, T. and Gould, E.A. (2001). Phylogenetic relationships of flaviviruses correlate with their epidemiology, disease association and biogeography. *Journal of General Virology* **82**, 1867–1876.

Giladi, M., Metzkor-Cotter, E., Martin, D., Siegman-Igra, Y., Korczyn, A., Rosso, R., Berger, S., Campbell, G. and Lanciotti, R. (2001). West Nile Encephalitis in Israel, 1999: the New York connection. *Emerging Infectious Diseases* **7**, 659–661.

Gilbert, L., Jones, L.D., Hudson, P.J., Gould, E.A. and Reid, H.W. (2000). Role of small mammals in the persistence of Louping-ill virus: field survey and tick co-feeding studies. *Medical and Veterinary Entomology* **14**, 277–282.

Golovljova, I., Vene, S., Sjolander, K.B., Vasilenko, V., Plyusnin, A. and Lundkvist, A. (2004). Characterisation of tick-borne encephalitis virus from Estonia. *Journal of Medical Virology* **74**, 580–588.

Gould, E.A., de Lamballerie, X., Zanotto, P.M. and Holmes, E.C. (2001). Evolution, epidemiology, and dispersal of flaviviruses revealed by molecular phylogenies. *Advances in Virus Research* **57**, 71–103.

Gould, E.A., Zanotto, P.M. and Holmes, E.C. (1997). The genetic evolution of flaviviruses. In: *Factors in the Emergence of Arbovirus Diseases* (J.F. Saluzzo and B. Dodet, eds), pp. 51–63. Paris: Elsevier.

Green, P.E. (1978). *Analyzing Multivariate Data*. Hinsdale, Illinois: The Dryden Press.

Gritsun, T.S., Holmes, E.C. and Gould, E.A. (1995). Analysis of flavivirus envelope proteins reveals variable domains that reflect their antigenicity and may determine their pathogenesis. *Virus Research* **35**, 307–321.

Guirakhoo, F., Radda, A.C., Heinz, F.X. and Kunz, C. (1987). Evidence for antigenic stability of tick-borne encephalitis virus by analysis of natural isolates. *Journal of General Virology* **68**, 859–864.

Hartley, W.J., Martin, W.B., Hakioglu, F. and Chifney, S.T.E. (1969). A viral encephalomyelitis of sheep in Turkey. *Pendik Institute Journal* **2**, 89–100.

Hayasaka, D., Suzuki, Y., Kariwa, H., Ivanov, L., Volkov, V., Demenev, V., Mizutani, T., Gojobori, T. and Takashima, I. (1999). Phylogenetic and virulence analysis of tick-borne encephalitis viruses from Japan and far-eastern Russia. *Journal of General Virology* **80**, 3127–3135.

Heinz, F.X. and Kunz, C. (1981). Homogeneity of the structural glycoprotein from European isolates of tick-borne encephalitis viruses: comparison with other viruses. *Journal of General Virology* **57**, 263–275.

Heinz, F.X. and Kunz, C. (1982). Molecular epidemiology of tick-borne encephalitis virus: peptide mapping of large non-structural proteins of

European isolates and comparison with other flaviviruses. *Journal of General Virology* **62**, 271–285.

Hubalek, Z., Pow, I., Reid, H.W. and Hussain, M.H. (1995). Antigenic similarity of central European encephalitis and louping-ill viruses. *Acta Virologica* **39**, 251–256.

Hudson, P.J. (1992). *Grouse in Space and Time*. Fordingbridge, UK: Game Conservancy Trust.

Hudson, P.J., Norman, R., Laurenson, M.K., Newborn, D., Gaunt, M., Jones, L., Reid, H., Gould, E., Bowers, R. and Dobson, A. (1995). Persistence and transmission of tick-borne viruses: *Ixodes ricinus* and louping-ill virus in red grouse populations. *Parasitology* **111**, S49–S58.

Immuno, A. (1997). *Tick-borne Encephalitis (TBE) and its Immunoprophylaxis*. Vienna: Immuno Ag.

Jones, L.D., Gaunt, M., Hails, R.S., Laurenson, K., Hudson, P.J., Reid, H., Henbest, P. and Gould, E.A. (1997). Transmission of louping-ill virus between infected and uninfected ticks co-feeding on mountain hares. *Medical and Veterinary Entomology* **11**, 172–176.

Juste, R.A., Barral, M., Fernandez de Luco, D., Gonzalez, L., Saez de Ocariz, C., Gomez-Parada, M. and Garcia-Perez, A.L. (1997). Tick-borne encephalitis in Spain. In: *4th International Potsdam Symposium on Tick-Borne Diseases: Tick-borne Encephalitis and Lyme Borreliosis* (O. Kahl and J. Suss, eds), pp. 70–76. Lengerich: Pabst Science Publishers.

Karabatsos, N., ed. (1985). *International Catalogue of Arboviruses Including certain other Viruses of Vertebrates*, 4th edn. San Antonio, Texas: American Society of Tropical Medicine and Hygiene.

Kharitonova, N.N. and Leonov, Y.A. (1985). *Omsk Hemorrhagic Fever, Ecology of the Agent and Epizootiology (translation from the original Russian book of 1978)*. New Delhi: Amerind Publishing Co.

Khutoretskaya, N.V., Aristova, V.A., Rogovaya, S.G., Lvov, D.K., Karimov, S.K., Skvortsova, T.M. and Kondrashina, N.G. (1985). Experimental study of the reproduction of Karshi virus (Togoviridae, Flavivirus) in some species of mosquitoes and ticks. *Acta Virologica* **29**, 231–236.

Korenberg, E.I. (1994). Comparative ecology and epidemiology of Lyme disease and tick-borne encephalitis in the former Soviet Union. *Parasitology Today* **10**, 157–160.

Korenberg, E.I. and Kovaleskii, Y.V. (1994). A model for relationships among the tick-borne encephalitis virus, its main vectors, and hosts. *Advances in Disease Vector Research* **10**, 65–92.

Korenberg, E.I., Zhukov, V.I., Shatkuskas, A.V. and Bushueva, L.K. (1969). Distribution of the taiga tick (*Ixodes persulcatus*) in the USSR. *Zoologicheskii Zhurnal* **48**, 1003–1114.

Labuda, M., Austyn, J.M., Zuffova, E., Kozuch, O., Fuchsberger, N., Lysy, J. and Nuttall, P.A. (1996). Importance of localized skin infection in tick-borne encephalitis virus transmission. *Virology* **219**, 357–366.

Labuda, M., Jones, L.D., Williams, T., Danielová, V. and Nuttall, P.A. (1993a). Efficient transmission of tick-borne encephalitis virus between cofeeding ticks. *Journal of Medical Entomology* **30**, 295–299.

Labuda, M., Nuttall, P.A., Kozuch, O., Eleckova, E., Williams, T., Zuffova, E. and Sabo, A. (1993b). Non-viraemic transmission of tick-borne encephalitis virus: a mechanism for arbovirus survival in nature. *Experientia* **49**, 802–805.

Labuda, M. and Randolph, S.E. (1999). Survival of tick-borne encephalitis virus: cellular basis and environmental determinants. *Zentralblatt für Bakteriologie* **288**, 513–524.

Lanciotti, R.S., Roehrig, J., Deubel, V., Smith, J., Parker, M., Steele, K., Crise, B., Volpe, K., Crabtree, M., Scherret, J., Hall, R., Mackenzie, J., Cropp, C., Panigraphy, B., Ostlund, E., Schmitt, B., Malkinson, M., Banet, C., Weissman, J., Komar, N., Savage, H.M., Stone, W., McNamara, T. and Gubler, D.J. (1999). Origin of the West Nile virus responsible for an outbreak of encephalitis in the northeastern United States. *Science* **286**, 2333–2337.

Laurenson, M.K., Norman, R., Gilbert, L., Reid, H.W. and Hudson, P.J. (2003). Identifying disease reservoirs in complex systems: mountain hares as reservoirs of ticks and louping-ill virus, pathogens of red grouse. *Journal of Animal Ecology* **72**, 177–185.

Lundkvist, A., Vene, S., Golovljova, I., Mavtchoutka, V., Forsgren, M., Kalnina, V. and Plyusnin, A. (2001). Characterization of tick-borne encephalitis virus from Latvia: evidence for co-circulation of three distinct sub-types. *Journal of Medical Virology* **65**, 730–735.

McGuire, K., Holmes, E.C., Gao, G.F., Reid, H.W. and Gould, E.A. (1998). Tracing the origins of louping-ill virus by molecular phylogenetic analysis. *Journal of General Virology* **79**, 981–988.

McLean, D.M. and Donohue, W.L. (1959). Powassan virus: isolation of virus from a fatal case of encephalitis. *Canadian Medical Association Journal* **80**, 708–712.

Moya, A., Holmes, E.C. and González-Candelas, F. (2004). The population genetics and evolutionary epidemiology of RNA viruses. *Nature Reviews Microbiology* **2**, 279–288.

Norman, R., Bowers, R.G., Begon, M. and Hudson, P.J. (1999). Persistence of tick-borne virus in the presence of multiple host species: tick reservoirs and parasite mediated competition. *Journal of Theoretical Biology* **200**, 111–118.

Nuttall, P.A. (1998). Displaced tick-parasite interactions at the host interface. *Parasitology* **116**, S65–S72.

Nuttall, P.A. and Labuda, M. (1994). Tick-borne encephalitides. In: *Ecological Dynamics of Tick-borne Zoonoses* (D.E. Sonenshine and T.N. Mather, eds), pp. 351–391. New York: Oxford University Press.

Ormaasen, V., Brantsaeter, A.B. and Moen, E.W. (2001). [Tick-borne encephalitis in Norway.] (in Norwegian). *Tidsskr Nor Laegeforen* **121**, 807–809.

Papadopoulos, O. (1980). Arbovirus problems in Greece. In: *Arboviruses in the Mediterranean Countries, 6th FEMS Symposium* (J. Vesenjak-Hirjan, J.S. Porterfield, and E. Arslanagic, eds), *Zentralblatt für Bakteriologie, Mikrobiologie und Hygiene* **1**, 117–121.

Pavlov, P. (1968). Studies of tick-borne encephalitis of sheep and their natural foci in Bulgaria. *Zentralblatt für Bakteriologie, Parasitenkiende, Infektionskrankh und Hygeine* **206**, 360–367.

Randolph, S.E. (1998). Ticks are not insects: consequences of contrasting vector biology for transmission potential. *Parasitology Today* **14**, 186–192.

Randolph, S.E. (2000). Ticks and tick-borne disease systems in space and from space. *Advances in Parasitology* **47**, 217–243.

Randolph, S.E. (2004a). Evolution of tick-borne disease systems. In: *Society of General Microbiology Symposium* (S.H. Gillespie, G.L. Smith and A. Osbourn, eds), Vol. 63, pp. 19–42. Cambridge: Cambridge University Press.

Randolph, S.E. (2004b). Tick ecology: processes and patterns behind the epidemiological risk posed by ixodid ticks as vectors. *Parasitology* **129**, S37–S66.

Randolph, S.E., Gern, L. and Nuttall, P.A. (1996). Co-feeding ticks: epidemiological significance for tick-borne pathogen transmission. *Parasitology Today* **12**, 472–479.

Randolph, S.E., Green, R.M., Peacey, M.F. and Rogers, D.J. (2000). Seasonal synchrony: the key to tick-borne encephalitis foci identified by satellite data. *Parasitology* **121**, 15–23.

Randolph, S.E., Miklisová, D., Lysy, J., Rogers, D.J. and Labuda, M. (1999). Incidence from coincidence: patterns of tick infestations on rodents facilitate transmission of tick-borne encephalitis virus. *Parasitology* **118**, 177–186.

Rogers, D.J. (2000). Satellites, space, time and the African trypanosomes. *Advances in Parasitology* **147**, 129–171.

Rogers, D.J., Hay, S.I. and Packer, M.J. (1996). Predicting the distribution of tsetse flies in West Africa using temporal Fourier processed meteorological satellite data. *Annals of Tropical Medicine and Parasitology* **90**, 225–241.

Skarpass, T., Sundoy, A., Bruu, A.L., Vene, S., Pedersen, J., Eng, P.G. and Csángó, P.A. (2002). [Tick-borne encephalitis in Norway] (in Norwegian). *Tidsskr Nor Laegeforen* **122**, 30–32.

Sreenivasan, M.A., Bhat, H.R. and Rajagopalan, P.K. (1986). The epizootics of Kyasanur Forest disease in wild monkeys during 1964 to 1973. *Transactions of the Royal Society of Tropical Medicine and Hygiene* **80**, 810–814.

Stuen, S. (2003). *Anaplasma phagocytophilum* (formerly *Ehrlichia phagocytophila*) infection in sheep and wild ruminants in Norway. PhD thesis. Sandness, Norway: Norwegian School of Veterinary Science.

Takeda, T., Ito, T., Osada, M., Takahashi, K. and Takashima, I. (1999). Isolation of tick-borne encephalitis virus from wild rodents and a seroepizootiologic survey in Hokkaido, Japan. *American Journal of Tropical Medicine and Hygiene* **60**, 287–291.

Ulvund, M.J., Vik, T. and Krogsrud, J. (1983). Louping-ill (tick-borne encephalitis) hos sau i Norge [in Norwegian]. *Norsk Veterinaertidsskrift* **95**, 639–641.

Vasilenko, V., Golovljova, I. and Jogiste, A. (1997). TBE—epidemiology in Estonia. In: *4th International Potsdam Symposium on Tick-borne Diseases: tick-borne encephalitis and Lyme borreliosis* (J. Süss and O. Kahl, eds), pp. 91–99. Lengerich: Pabst Science Publishers.

Walton, G.A. and Kennedy, R.C. (1966). Tick-borne encephalitis virus in southern Ireland. *British Veterinary Journal* **122**, 427–434.

Whitby, J.E., Whitby, S.N., Jennings, A.D., Stephenson, J.R. and Barrett, A.D.T. (1993). Nucleotide-sequence of the envelope protein of a Turkish isolate of tick-borne encephalitis (TBE) virus is distinct from other viruses of the TBE virus complex. *Journal of General Virology* **74**, 921–924.

Wikel, S.K. (1996). Host immunity to ticks. *Annual Review of Entomology* **41**, 1–22.

Work, T.H. (1958). Russian Spring-Summer virus in India. *Progress in Medical Virology* **1**, 248–277.

Work, T.H. (1960). Current perspectives in field and laboratory investigations on the natural history of the tick-borne encephalitis complex viruses. In: *Symposium of the Czech Academy of Sciences: Biology of Viruses of the Tick-borne Encephalitis Complex* (H. Libikova, ed.), pp. 361–379. Smolenice, Czechoslovakia: Academic Press.

Zanotto, P.M., Gao, G.F., Gritsun, T., Marin, M.S., Jiang, W.R., Venugopal, K., Reid, H.W. and Gould, E.A. (1995). An arbovirus cline across the northern hemisphere. *Virology* **210**, 152–159.

Zanotto, P.M., Gould, E.A., Gao, G.F. and Harvey, P.H. (1996). Population dynamics of flaviviruses revealed by molecular phylogenies. *Proceedings of the National Academy of Science of the United States of America* **93**, 548–553.

Global Transport Networks and Infectious Disease Spread

A.J. Tatem[1], D.J. Rogers[1] and S.I. Hay[1,2]

[1]*TALA Research Group, Tinbergen Building, Department of Zoology, University of Oxford, South Parks Road, Oxford OX1 3PS, UK*
[2]*Malaria Public Health & Epidemiology Group, Centre for Geographic Medicine, KEMRI, PO Box 43640, 00100 Nairobi GPO, Kenya*

Abstract 294
1. Introduction 294
2. Global Transport Networks and Pandemics 295
2.1. Plague 295
2.2. Cholera 297
2.3. Influenza 298
2.4. HIV/AIDS 300
2.5. Severe Acute Respiratory Syndrome 301
2.6. Bioterrorism 302
2.7. Predicting, Modelling and Controlling Future Pandemics . . 304
3. Global Transport Networks and Disease Vector Invasions 306
3.1. *Aedes aegypti* 306
3.2. *Anopheles gambiae* 307
3.3. *Aedes japonicus* 307
3.4. *Aedes albopictus* 308
3.5. Predicting Future Disease Vector Invasions 308
4. Global Transport Networks and Vector-Borne Diseases 319
4.1. Yellow Fever 319
4.2. Dengue 322
4.3. West Nile Virus 322
4.4. Malaria 323
4.5. Predicting Future Vector-Borne Disease Movement 326
5. Conclusions 332
Acknowledgements 332
References 333

ADVANCES IN PARASITOLOGY VOL 62
ISSN: 0065-308X $35.00
DOI: 10.1016/S0065-308X(05)62009-X

ABSTRACT

Air, sea and land transport networks continue to expand in reach, speed of travel and volume of passengers and goods carried. Pathogens and their vectors can now move further, faster and in greater numbers than ever before. Three important consequences of global transport network expansion are infectious disease pandemics, vector invasion events and vector-borne pathogen importation. This review briefly examines some of the important historical examples of these disease and vector movements, such as the global influenza pandemics, the devastating *Anopheles gambiae* invasion of Brazil and the recent increases in imported *Plasmodium falciparum* malaria cases. We then outline potential approaches for future studies of disease movement, focussing on vector invasion and vector-borne disease importation. Such approaches allow us to explore the potential implications of international air travel, shipping routes and other methods of transport on global pathogen and vector traffic.

1. INTRODUCTION

For most of human history, regional and continental populations have been relatively isolated from each other. Only comparatively recently has there been extensive contact between peoples, flora and fauna from both old and new worlds (Diamond, 1998). The movement of disease has proved a major force in shaping world history, as wars, crusades, diasporas and migrations have carried infections to susceptible populations. Until World War II, more war victims died of microbes introduced by the enemy, than of battle wounds (Karlen, 1995). More often than not, the victors in past wars were not those armies with the best weapons and generals, but those bearing the deadliest pathogens (Zinsser, 1943; Diamond, 1998).

Initially, new infectious diseases could spread only as fast and far as people could walk. Then as fast and far as horses could gallop and ships could sail. With the advent of truly global travel, the last five centuries have seen more new diseases than ever before become potential pandemics (Karlen, 1995). The current reach, volume and speed of travel are unprecedented, so that human mobility has increased in high-income countries by over 1000-fold since 1800 (Wilson, 1995, 2003). Aviation, in particular, has expanded rapidly as the world economy has grown, though worries about its potential for spreading disease began with the advent of commercial aviation (Massey, 1933). Passenger numbers have grown at nearly 9% per annum since 1960 and are expected to increase at more than 5% per annum for at least the next 10 years, with airfreight traffic showing similar changes (Upham *et al.*, 2003). Similarly, globalization of the world economy has also resulted in a shipping traffic increase of over 27% since 1993 (Zachcial and Heideloff, 2003).

The efficiency, speed and reach of modern transport networks puts people at risk from the emergence of new strains of familiar diseases, or from completely new diseases (Guimera *et al.*, 2005). Additionally, the global growth of economic activity, tourism and human migration is leading to ever more cases of the movement of both disease vectors and the diseases they carry.

This contribution reviews examples of past movement of three categories of disease through global transport networks: (i) pandemics; (ii) disease vector invasions; and (iii) vector-borne diseases. For each, we review examples of past events and their current status, while for sections (ii) and (iii), we put forward novel approaches for the modelling and prediction of future events.

2. GLOBAL TRANSPORT NETWORKS AND PANDEMICS

The past 500 years have provided numerous examples of how the establishment and expansion of worldwide transport networks has facilitated global pandemics of communicable diseases. Here we briefly review a selection of the world's major pandemics, their potential future threat and examine the development of approaches for their prediction and control.

2.1. Plague

More than 200 million people are thought to have been killed by bubonic plague in three major pandemics between the 14th and 17th centuries (Duplaix, 1988; Eckert, 2000). The plague bacillus, *Yersinia pestis*, was transmitted from infected rodents, often rats, by flea bites (Lounibos, 2002). Historical data show occasional major outbreaks of plague separated by long, plague-free periods (Keeling and Gilligan, 2000).

Many medical historians believe that the "Black Death" that killed a third of Europe's central and northern populations in the mid 14th century was brought about by bubonic plague. This view, however, has recently been questioned (Scott and Duncan, 2004). The Black Death arrived in Sicily in 1347 and within two years had swept northwards to Scandinavia, at a time when rats appear to have been absent from much of Europe (Twigg, 1984, 2003). Even if introduced at ports, rat populations could not feasibly have colonized such large areas so quickly. New examination of contemporary accounts, and local parish records, reveal patterns of death that are consistent with a directly transmitted infection among family members and neighbours, with symptoms consistent with some sort of viral haemorrhagia. The real 'vector' appears to have been man: occasional introductions to certain nearby towns and villages coincided with the arrival of travellers on foot or horseback (Scott and Duncan, 2004). People at the time were aware of the effectiveness of isolation and quarantine within affected households, which would not have prevented infection spread by rats and fleas. It seems that only the limited transport and movement patterns of the time prevented even greater impacts on the European human population.

Whether caused by plague or some other pathogen, the Black Death found an entirely susceptible population, devoid of resistance. Its first sweep across Europe afflicted two-thirds of Europe's population before subsiding, though it remained endemic. Population regrowth and renewed travel then reignited the disease, causing further epidemics in 1361, 1371 and 1382, afflicting half, one-tenth and one-twentieth of the population respectively (Zinsser, 1943).

The most recent confirmed plague pandemic occurred early last century, with outbreaks principally in port cities, reflecting the dominant mode of international travel at the time. Initially originating in China, plague-infected rats were transported around the world in sailing vessels, sparking major epidemics at international ports such as Sydney,

Bombay, San Francisco and Rio de Janeiro (Duplaix, 1988). The limited spread inland from port cities remains a mystery, but the control of rats and modernization of housing, food storage and sanitation, have been suggested as preventing large-scale inland rat migration (Zinsser, 1943). Spread did occur from San Francisco, however, resulting in plague becoming endemic in prairie dog colonies of Western USA, causing occasional human cases even today (Markel, 2005).

Bubonic plague is now principally regarded as a disease of only historical importance. There are, however, more frequent reports of increasing incidence locally (Barretto *et al.*, 1994; Kumar, 1995), importation (Perlman *et al.*, 2003), and of antibiotic-resistant strains (Galimand, 1997). Plague is thus re-emerging as a significant public health concern. Since the last pandemic, plague's geographic range has expanded, posing new threats to previously unaffected regions (Gage and Kosoy, 2005) and resulting in hundreds of cases in at least 14 countries (World Health Organization, 2003). Although recent outbreaks have been quickly controlled limiting case numbers, resultant economic disruptions have been severe. India's 1994 outbreak resulted in just 52 deaths, but over 1 billion USD in economic costs, demonstrating how vulnerable the global economy can be to the threat of infectious diseases (Cash and Narasimhan, 2000).

2.2. Cholera

Cholera is caused by an intestinal infection with the bacterium, *Vibrio cholerae*, leading to severe dehydration, shock and often-rapid death (Sack *et al.*, 2004). The bacterium can survive for long periods in water and is commonly transmitted by contaminated water, or food that has been washed in such water. Accounts of cholera-like diseases go back as far as the times of Hippocrates and Buddha (Reidl and Klose, 2002). Over the past 185 years, *Vibrio cholerae* has escaped seven times from its endemic heartland in West Bengal, India, to result in pandemics (Sack *et al.*, 2004). It first started as an epidemic outbreak in 1817 in India, but soon started spreading, in part due to British ship and troop movements, carrying the infection north and east to China, Japan and Indonesia (Reidl and Klose, 2002). The disease also spread along trade routes to the west as far as southern Russia (Karlen, 1995). Each successive pandemic increased in extent and severity, reflecting the expanding reach of the global transport system and increased movements of people, particularly on religious pilgrimages (Rogers, 1919). The 1830s saw Russian troops, English ships, Irish immigrants and Canadian exploration carry *Vibrio cholerae* to the Baltic, England, Ireland, Canada, USA and Mexico (Curtin, 1995). Its arrival in Mecca in 1831 in time for the Muslim pilgrimage is thought to have sparked over 40 epidemics by 1912 (McNeill, 1976). Statistically, the impact of cholera epidemics was rarely severe, with only a small percentage of populations affected, but it developed a fearsome reputation through seemingly bypassing all attempts at quarantine and because it attacked rich as well as poor (McNeill, 1976). Not until John Snow's work on cholera clusters in London in the 1850s, and later acceptance of 'germ' theories of disease, did a proper understanding and effective control develop (McNeill, 1976).

Cholera continues to affect many parts of the world. The most recent, and current, pandemic has resulted in the disease becoming endemic in much of Africa (Naidoo and Patric, 2002; Weir and Haider, 2004), South America (Seas *et al.*, 2000; Chevallier *et al.*, 2004) and southern Asia (Phukan *et al.*, 2004; Weir and Haider, 2004), with antibiotic resistance on the rise (Sack *et al.*, 2004). Newly found serotypes of *Vibrio cholerae* are likely to be the source of the next cholera pandemic, having already resulted in 30 000 cases in just a few months in Dhaka, Bangladesh (Faruque *et al.*, 2003).

2.3. Influenza

The influenza virus is remarkable for the rapidity with which it can spread, the brevity of immunity it confers and its genetic variability (McNeill, 1976; Ferguson *et al.*, 2003a). Despite an often-low fatality rate, the large number of cases makes influenza pandemics and epidemics a major health problem (Palese, 2004). Around 20% of children and 5% of adults worldwide develop symptomatic influenza each year (Nicholson *et al.*, 2003). During the 20th century, influenza was the principal infectious disease to be influenced by the growing global transport network and to display pandemic behaviour. Three major pandemics occurred in 1918, 1957 and 1968 (Cox and Subbarao, 2000).

In 1918, the confluence of American with both European and African troops in northern France, and the emergence of new virus strains, provided the milieu for an epidemic of unprecedented scope (McNeill, 1976). The 1918 influenza pandemic killed around 40 million people in a year (Oxford, 2004), and resulted in an almost 10-year drop in the calculated average life expectancy of the global population (Palese, 2004). This disease is now used as a worst-case scenario for pandemic preparedness planning (Mills *et al.*, 2004). Three influenza viruses with different haemagglutin surface proteins (H_1, H_2, H_3) were responsible for the pandemics of the 20th century. Although it is theoretically possible to create vaccines against any new influenza virus given enough time and production capacity, the genetic instability of the virus means it can evolve quickly, thereby escaping the effects of newly developed vaccines (McNeill, 1976).

The 1957 pandemic originated in mainland China and traversed the globe within six months. This speed was attributable both to a haemagglutin surface protein shift from previous viruses, leaving the global population susceptible to infection; and to the availability of regular air and sea travel via which secondary epidemic sources were established (Thomas, 1992). Before it became widely epidemic in many countries, a vaccine had been developed and produced in sufficient quantities to reduce the incidence and intensity of influenza in developed countries with available vaccine, yet it still claimed over 70 000 lives in the US and many more elsewhere (Palese, 2004). In 1968, another strain was isolated in Hong Kong, and quickly spread around the world causing thousands of deaths, even though antibodies remaining from the 1957 pandemic were thought to have moderated the severity of infections (Cox and Subbarao, 2000). Epidemics continue to occur regularly (Viboud *et al.*, 2004), as do movements of the virus through the global transportation system, causing outbreaks even on aircraft (Marsden, 2003).

Much remains unknown about influenza and the viruses that cause it; how they arise, how the shift to new strains occurs and whether the viruses migrate between northern and southern hemispheres to become epidemic in cold-weather seasons (Cox and Subbarao, 2000). The recent emergence of the avian influenza virus in south-east Asia has led many to fear that the next global pandemic is imminent (Webby and Webster, 2003; Fouchier *et al.*, 2005; Webster and Hulse, 2005). At the time of writing, the H5N1 virus has infected at least 88 people, killing 51 of them (Webster and Hulse, 2005). Should the virus mutate sufficiently to enable sustained human-to-human transmission (Ungchusak *et al.*, 2005), modern transport (Enserink, 2004) and the size of the currently susceptible global population means it could kill millions more than any previous pandemic if adequate preventative measures are not in place (Osterholm, 2005a, b; Oxford, 2005).

2.4. HIV/AIDS

The spread of the HIV/AIDS pandemic worldwide is a travel story whose episodes can be traced by molecular tools and epidemiology (Lemey *et al.*, 2003; Perrin *et al.*, 2003). The exact origin of the HIV/AIDS virus remains unknown, but sero-archaeological studies have documented human infections with HIV prior to 1970 (Mann, 1989). Phylogenetic analysis has indicated that multiple interspecies transmissions from simians introduced two genetically distinct types of HIV into the human population: HIV-1 from chimpanzees and HIV-2 from sooty mangabeys (Lemey *et al.*, 2003). While HIV-2 is mainly restricted to West Africa and is thought to have originated in Guinea-Bissau (Lemey *et al.*, 2003), HIV-1 has spread globally from the first zoonotic transmission from chimpanzee to people around 70 years ago in central Africa (Jonassen *et al.*, 1997; Russell *et al.*, 2000; Perrin *et al.*, 2003).

Data suggest that the current pandemic started in the mid-1970s and by 1980, through air travel, sea travel and human migration, had spread to between 100 000 and 300 000 people on at least five continents (Mann, 1989). Studies have also shown that certain groups of mobile and sexually active individuals are important in seeding local epidemics, including immigrants, intravenous drug users, tourists, truck drivers, military troops and seamen (Perrin *et al.*, 2003; Salit *et al.*, 2005). This is illustrated by the first documented infections in Europe, caused by a Norwegian seaman infected through heterosexual contact in a West African seaport, returning to infect his wife, who transmitted the infection to her daughter (Jonassen *et al.*, 1997). The role of international travel in the spread of HIV was also highlighted by the case of 'Patient Zero', a Canadian flight attendant who travelled extensively worldwide. Analysis of several of the early AIDS cases showed that the infected individuals were the attendant's direct or indirect sexual contacts and could be traced to several different American cities, thereby demonstrating the role of international travel in spreading the virus (Karlen, 1995).

According to UNAIDS estimates, 39.4 million people were living with HIV at the end of 2004, causing around 3 million deaths a year in sub-Saharan Africa (SSA), the worst-hit region (UNAIDS, 2004). The disease's spread can be linked to a variety of factors. In the United States, the rapid spread of AIDS between 1984 and 1990 can be modelled accurately using air traffic flows between cities (Gould, 1995, 1999). The non-homogenous distribution of the global pandemic both between and within countries reflects levels of social vulnerability and mobility (Bronfman *et al.*, 2002). Infection rates are usually higher in urban areas; rural areas, especially in Africa, are affected by the levels of mobility and migration which enable the virus to shift from urban to rural centres, and tend to mean that infection rates are higher near main roads and in trading centres than in isolated villages (Lagarde *et al.*, 2003).

2.5. Severe Acute Respiratory Syndrome

Severe acute respiratory syndrome (SARS) is a coronavirus that adapted from animal hosts to become readily transmissible between humans (Peiris *et al.*, 2004). Southern China has produced one plague pandemic and two influenza pandemics over the last 150 years (McNeill, 1976), and SARS represents the fourth global pandemic originating from the region. SARS probably first emerged in Guangdong province around November 2002 among those in contact with the live-game trade (Peiris *et al.*, 2004). Although unproven, it seems likely that animal to human interspecies transmission occurred at 'wet markets' where a wide range of live poultry, fish, reptiles and other mammals are sold (Peiris *et al.*, 2004).

On 21 February 2003, a physician from Guangdong spent a single day in a Hong Kong hotel, during which time he transmitted an infection to 16 other guests. These guests seeded outbreaks in Hong Kong, Toronto, Singapore and Vietnam, and within weeks SARS infected over 8000 people in 26 countries across 5 continents (Peiris *et al.*, 2004). The WHO invoked traditional public health measures to contain the outbreak, including heightened vigilance, screening of travellers and isolation and quarantine of affected individuals and their close contacts. At the same time, advanced technologies identified the causative agent and informed prevention and treatment options (Skowronski *et al.*, 2005). The exhaustive effort resulted in cessation of transmission by early July 2003.

While the number of fatalities was small (<800) (Skowronski *et al.*, 2005) in comparison with other pandemics, the speed and extent of the proliferation of SARS highlighted the potential for modern globalized economic activity and an ever-expanding air travel network to spread infectious diseases. It also demonstrated how a new and poorly understood disease, with no vaccine and no effective cure, can adversely affect economic growth, trade, tourism and social stability, especially when its perceived risk is many times higher than its actual risk (Heymann, 2004). The economic impact of SARS has been estimated at between US$30–140 billion (Skowronski *et al.*, 2005), largely as a consequence of reduced travel and investment in Asia. SARS also showed how inadequate surveillance and response capacity in a single country can have an impact upon global public health security (Heymann, 2004). While the pandemic ended in late 2003, the fact that three years later there is no vaccine and that the animal reservoir of the disease remains to be identified conclusively (though civet cats are thought to be the source; Song *et al.*, 2005), means a return of SARS cannot be ruled out (Skowronski *et al.*, 2005).

2.6. Bioterrorism

While bioterrorism has been at the forefront of public health planning since the 11 September 2001 attacks on New York, it has a long history. One of the earliest documented uses of biological weaponry occurred in 1346 when wars and plague were decimating the Middle East. A three-year siege of the Crimean walled-port of Jaffa by the Tatars was finally ended soon after plague-infected corpses were catapulted over the city walls, seeding epidemics which led to its downfall (Karlen, 1995).

Today, the list of agents posing major public health risks if acquired and effectively disseminated in a bioterrorist attack is relatively short (Kortepeter and Parker, 1999). This is counterbalanced by the potential of such agents to challenge our abilities to limit the numbers of casualties. The most threatening of infectious agents include botulism, influenza, plague, tularaemia, viral haemorrhagic fevers (e.g. Ebola, Marburg and Lassa fevers), anthrax and smallpox (Lane *et al.*, 2001; Madjid *et al.*, 2003). In the minds of most military and counterterrorism planners, anthrax and smallpox represent the greatest bioterror threats (Kortepeter and Parker, 1999).

Anthrax is one of the great infectious diseases of antiquity, and some of the plagues described in the Bible may have been outbreaks of this disease in cattle and humans (Cieslak and Eitzen, 1999). Caused by infection with *Bacillus anthracis*, it has little potential for person-to-person transmission, and most endemic cases are contracted cutaneously through contact with infected herbivores. As a weapon, it would most likely be delivered by aerosol as seen in 2001 in the US Postal Service attacks (Dewan *et al.*, 2002; Greene *et al.*, 2002) and, consequently, acquired by inhalation. Without rapid treatment, death may occur in as many as 95% of cases (Cieslak and Eitzen, 1999). Although a licensed vaccine and efficient therapy exists, the short incubation period and rapid progression of the disease means identification and treatment of any exposed populations is likely to present a major challenge.

Smallpox is a viral disease spread by inhalation of air droplets or aerosols and is unique to humans (Henderson, 1999). It has played a major part in shaping human history, repeatedly killing millions in Europe in epidemics throughout the 16th century before immunity arose to reduce its severity (Alibek, 2004). Europeans then inadvertently introduced smallpox to the New World, where it aided in wiping out whole ancient civilizations that had no immunity (McNeill, 1976; Karlen, 1995). Aggressive global vaccination programmes led to the eradication of naturally occurring smallpox in 1977. The consequent halt to routine vaccination means that today the world's population again has little immunity to the disease (Lane *et al.*, 2001). Smallpox's case mortality rate of around 30% and its high transmissibility make it a potentially potent bioweapon and although officially just two samples exist, the existence of other sources, particularly from the 1970s Soviet research programme, cannot be ruled out (Lane *et al.*, 2001). In the face of such a possible threat, funding of research examining the scale of possible casualties and optimal control strategies has been increased massively, and various high-profile modelling studies have been undertaken (Meltzer *et al.*, 2001; Halloran *et al.*, 2002; Kaplan *et al.*, 2002; Bozzette, 2003; Ferguson *et al.*, 2003b; Grais *et al.*, 2003b). The following section looks at the approaches and findings of such studies.

2.7. Predicting, Modelling and Controlling Future Pandemics

Numerous approaches have been developed which attempt to capture the possible future movements of newly emergent communicable diseases through global and local transport networks (Thomas, 1992; Haggett, 2000). The growth and movement of an epidemic or pandemic is governed principally by the number of secondary cases generated by an initial or primary case in an entirely susceptible population (the reproductive number of the disease, R_0), and the average time taken for the secondary cases to be infected by a primary case (Bailey, 1967; Anderson and May, 1991). Splitting modelled populations into susceptible, infected and recovered categories forms the basis of much epidemic and pandemic modelling and has provided qualitative insights into the epidemiology of a wide range of pathogens. To obtain quantitative predictions for risk assessment and policy formulation, however, refinement of such basic concepts is required to account for complex biological and behavioural factors, including the variation of infectiveness and susceptibility within a population, and the mixing, movement and socio-economic structure of the population at risk (Ferguson *et al.*, 2003b).

While the movements of pandemics are notoriously unpredictable (Thomas, 1992), those models that can be calibrated using data from previous epidemic events are perhaps the ones that stand the best chance of being used to predict the spread of communicable diseases in the future, enabling the construction of early warning systems and forming a basis for planning control strategies (Haggett, 2000). Such an approach was demonstrated by Rvachev and Longini (1985) who showed the diffusion of the 1968–1969 influenza pandemic to be predictable through a model based on the air travel network of the time. Incidence data from the pandemic origin, Hong Kong, were used to estimate model parameters, such as contact level between susceptible and infectious individuals, time taken in latent and infectious states and the fraction of people susceptible to the virus. Annual average daily air passenger numbers between 52 cities were then used to derive probabilities of travel between the cities. Finally, the seasonality of influenza was taken into account by applying a scaling factor to northern and southern hemisphere city contact level parameters to mimic the hemispheric swing of influenza epidemics. This model was later updated and refined to provide the basis for predictive models of the spread of influenza, smallpox, SARS and other infectious agents through the global transportation network (Longini *et al.*, 1986; Dye and Gay, 2003; Grais *et al.*, 2003a, b; Vogel, 2003; Hufnagel *et al.*, 2004).

Given the vast range of complicating factors, no model can be expected to predict the spread of an infectious disease pandemic with complete accuracy. Modelling can, however, identify possible efficient interventions from a range of available scenarios, taking into account the range of uncertainties of key epidemiological parameters (Ferguson *et al.*, 2003b). Models can also find use during actual outbreaks, as was shown in the real-time statistical modelling of the 2001 UK foot-and-mouth epidemic (Ferguson *et al.*, 2001; Keeling *et al.*, 2003). Haggett (2000) outlines a number of important points relevant to the future control of epidemics and pandemics: (i) pandemic control will rely less and less on conventional spatial barriers as the global transport network continues to expand, (ii) the speed of modern transport means prompt surveillance and rapid reporting now play a critical role in preventing the spatial spread of a disease, (iii) mathematical models will become central in identifying aberrant behaviour in disease trends and (iv) the high cost of surveillance makes sampling design and the development of cost-effective monitoring and testing approaches vital to effective epidemic early warning systems.

3. GLOBAL TRANSPORT NETWORKS AND DISEASE VECTOR INVASIONS

Aircraft and ships are believed to be directly responsible for rapid expansion in the range of many plants and animals via inadvertent transport (Perrings *et al.*, 2005), including some of the world's principal disease vectors (Lounibos, 2002). Here we briefly review some of the major invasion events facilitated by transport, and demonstrate some approaches to the prediction of future disease vector invasions.

3.1. *Aedes aegypti*

Ae. aegypti is known to be a vector of numerous human pathogens, and is the principal vector of both yellow and dengue fever viruses. Though the mosquito is now established throughout the tropical and sub-tropical regions of the world, it was located solely in West Africa until the 15th century (Lounibos, 2002). At some point, the insect adapted to anthropogenic breeding sites, such as water storage jars in ships (Lounibos, 2002). This ability enabled it to take full advantage of the growing slave trade from West Africa to reach the new world or to invade Portugal and Spain before its proliferation elsewhere on European ships. *Ae. aegypti* consequently became established across the tropical and temperate regions of the Americas, where caused yellow fever epidemics at port cities (Haggett, 2000). Intensive control and eradication schemes in the 1950s and 1960s reduced its extent (Gubler, 2004a). Gradual resurgence following the end of these control campaigns has resulted in the species once again becoming widespread across the Americas, and associated with the emergence of dengue and dengue haemorrhagic fever (Gubler and Clark, 1995). *Ae. aegypti* also invaded tropical Asia, where its dispersal has again been associated with a rise in dengue and dengue haemorrhagic fever incidence (Gubler and Clark, 1995).

3.2. *Anopheles gambiae*

Perhaps the most devastating introduction of a disease vector of recent times was that of *An. gambiae*, the most efficient vector of *Plasmodium falciparum* malaria, from West Africa to Natal in North-Eastern Brazil by either steamship or aircraft in 1930 (Soper and Wilson, 1943; Lounibos, 2002). *An. gambiae* proved extremely well suited to parts of Brazil where the temperature, humidity and precipitation patterns match closely malaria endemic regions of East Africa (Killeen *et al.*, 2002; Killeen, 2003). Its gradual spread over 10 years into 54 000 km^2 of Northeast Brazil led to extensive malaria epidemics, costing 16 000 lives and around 3 billion USD (modern day estimate) in healthcare, drugs and the vector eradication programme (Killeen *et al.*, 2002). These epidemics were based solely on the greater vectorial capacity of *An. gambiae* relative to local mosquito species, as malaria was already endemic in Brazil. Similarly, high mortality rates were seen on Mauritius when *An. gambiae* was introduced accidentally in 1866, sparking major malaria epidemics (Lounibos, 2002). Control efforts ended the epidemics, but low-level and localized rural transmission continues.

3.3. *Aedes japonicus*

Ae. japonicus was first recorded in the states of New York and New Jersey, USA in 1998 (Peyton *et al.*, 1999). Native to Japan, Southern China and Korea where the larvae develop in natural and artificial containers, its likely mode of introduction was through tyre shipments. Since the initial introduction, the mosquito has spread to Connecticut (Mustermann and Andreadis, 1999), Maryland, Pennsylvania and Ohio (Lounibos, 2002). The discovery of wild-caught *Ae. japonicus* infected with West Nile virus (WNV) (Turell *et al.*, 2001) and the mosquito's competence as a vector of Japanese encephalitis suggest that it could become of public health importance in

North America. *Ae. japonicus* was also recently detected in Orne Departement in Northern France and in New Zealand, demonstrating its potential to expand into Europe and Australasia (Laird *et al.*, 1994; Schaffner *et al.*, 2003).

3.4. *Aedes albopictus*

The Asian tiger mosquito, *Ae. albopictus*, is a competent vector of 22 arboviruses, including West Nile, dengue and yellow fever viruses (Gubler, 2003; Gratz, 2004). Although *Ae. aegypti* is the principal vector of dengue fever, recent outbreaks of the disease in the absence of *Ae. aegypti* have implicated *Ae. albopictus* as a competent vector (Gratz, 2004; Effler *et al.*, 2005). The range expansion of *Ae. albopictus* from its Old World distribution (Figure 1) over the past 75 years is perhaps the best documented of any vector invasion. The mosquito spread from its range initially to the Pacific Islands (Gratz, 2004) in the 1930s and then, within the last 20 years, to other countries in both the Old and New Worlds (Moore and Mitchell, 1997; Gratz, 2004). This is thought to have been through ship-borne transportation of eggs and larvae in tyres (Reiter and Sprenger, 1987; Reiter, 1998). The spread of *Ae. albopictus* throughout the Eastern USA from its discovery in Harris County, Texas, in 1985 has been monitored extensively (Moore and Mitchell, 1997; Moore, 1999). Figure 2 shows its extent in 2000.

3.5. Predicting Future Disease Vector Invasions

Developing approaches to highlight routes of the greatest risk of invasion by disease vectors within the global transport network is an important prerequisite to planning effective control and disinsection efforts. Although distance may no longer represent a significant barrier to vector traffic, climate at the point of entry still presents a fundamental constraint to establishment, since poikilothermic arthropods are very sensitive to the weather. Here we review an approach, first outlined in Tatem *et al.* (2006a), based on air and sea traffic volume as

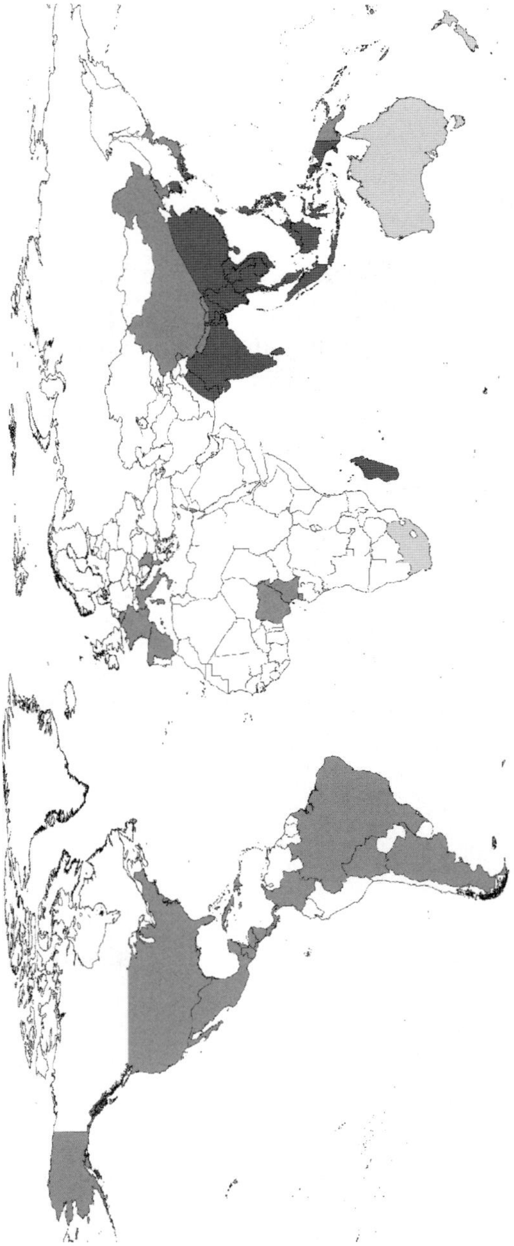

Figure 1 The Old World distribution of *Ae. albopictus* (dark grey); countries reporting established breeding populations of *Ae. albopictus* in the last 30 years (middle grey); countries reporting *Ae. albopictus* interception at ports (light grey). Data sources: Center for International Earth Science Information Network (CIESIN) (http://www.ciesin.org/docs/001-613/map15.gif), supplemented with information from literature sources (Gratz, 2004; Gubler, 2003; Lounibos, 2002; Medlock *et al.*, 2005; Moore, 1999; Moore and Mitchell, 1997).

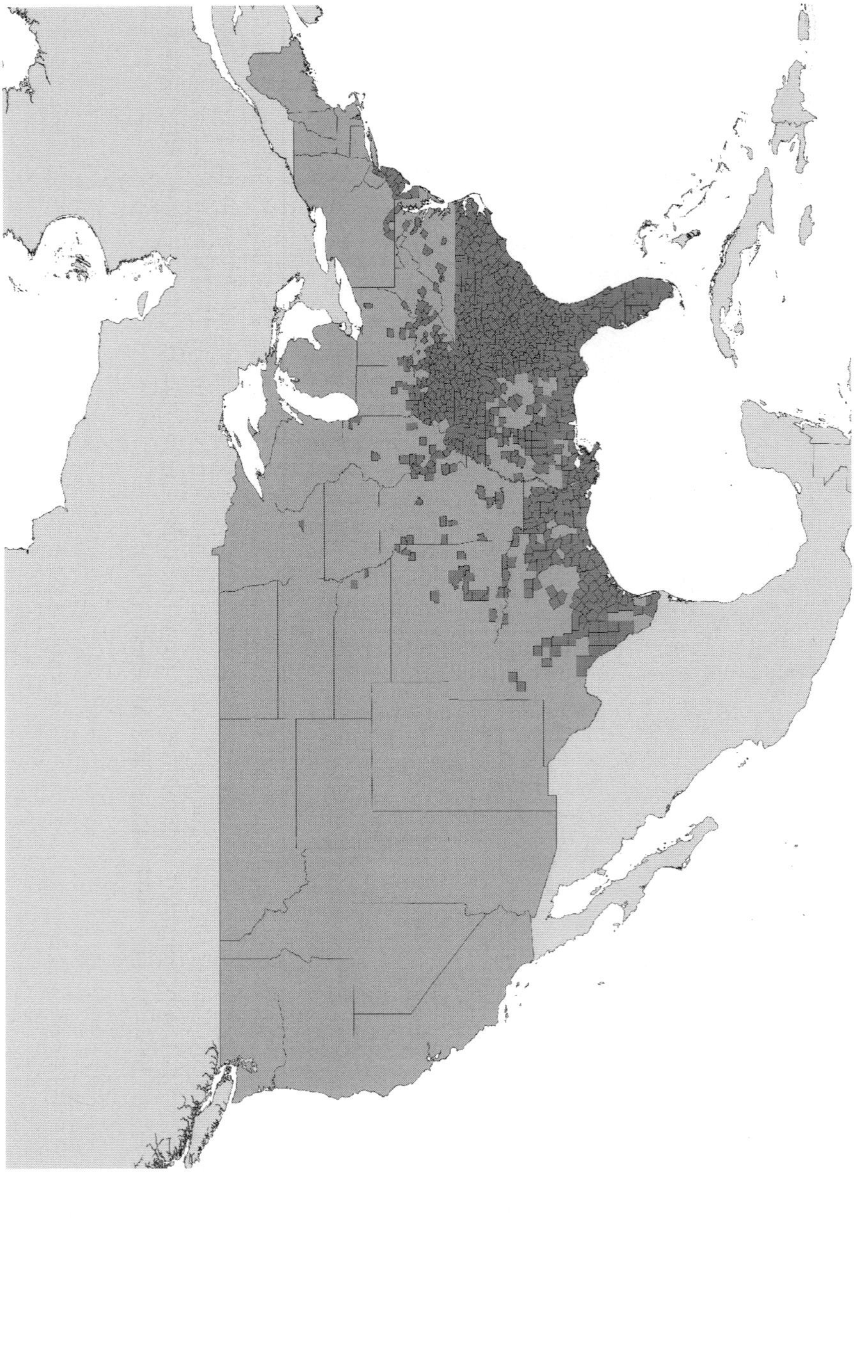

Figure 2 Counties of the United States of America reporting the presence of *Ae. albopictus* in 2000. (Adapted from US Centers for Disease Control and Prevention (CDC); URL: http://www.cdc.gov/ncidod/dvbid/arbor/albopic_97_sm.htm.)

well as climatic data, and tested against the movements of *Ae. albopictus* to examine how well these factors predicted its spread.

3.5.1. Data Sources

Major international seaport names, locations and estimated number of ships per annum between each were obtained from Drake and Lodge (2004). The data consisted of the estimated number of ship visits in 2000 to the 243 most frequently visited ports. Flight data on total passenger numbers moving between major airports in the year 2000 (assuming 100% aircraft capacity) using statistics supplied by all scheduled airlines were obtained from OAG Worldwide Ltd. The database contained data on the world's top 100 airports by traffic, plus the principal airport of 178 other countries if otherwise not included in the top 100. Data on a total of 7129 routes between 278 international airports in the year 2000 were therefore available. Gridded meteorological data (New *et al.*, 2002) at 10′ spatial resolution 1961–1990 were obtained and averaged to produce a synoptic year. The mean, maximum and minimum temperature, rainfall and humidity measurements were extracted to produce nine climatology surfaces.

3.5.2. Climatic Dendrograms

Dendrograms are visual representations of the results of hierarchical clustering (described below) that are commonly used in the fields of evolution and genetics, and occasionally in an environmental context (Sugihara *et al.*, 2003; Rogers and Robinson, 2004). Here, dendrograms are used to provide a climatic representation of the global air and sea transportation networks, a method first outlined in Tatem *et al.* (2006a). The dendrograms provide a way of restructuring the world as a disease vector may see it, given unlimited transport opportunities, with certain spatially distant sea/airports linked closely because of their climatic similarities, while spatially proximate ports may appear poorly linked if their climates differ significantly.

To create global sea and air transport network dendrograms, the locations of the 243 ports and 278 airports were superimposed onto the nine climatology surfaces and each 10′ spatial resolution grid square covering the seaport/airport location identified. To ensure that a representative climate was included, where possible, up to eight land pixels surrounding each seaport/airport grid square were also identified, forming a 3×3 grid square window centred on the airport/port. This requirement was not met for airports located coastally or on small islands or for the majority of seaports. In these cases, reduced numbers of land pixels were extracted. Any seaports/airports located on islands too small to be represented by the climatology surfaces were eliminated from the analysis (reducing the sample size to 241 seaports and 259 airports). The selected grid square data from the nine climatology surfaces thus formed the climate 'signatures' used in the analyses. These signatures represented a quantitative description of the climatic regime at each seaport and airport, in terms of temperature, rainfall and humidity.

Both the lack of sufficient co-variance information in the majority of seaport/airport signatures and the location of many seaports/airports on small islands dictated that only 'Euclidean' distances could be used as a measure of climatic similarities. Euclidean distance is defined as the shortest straight-line distance between two points: in this case it is the distance between the centroids of the port climate signatures in nine-dimensional space (ERDAS, 2003). Euclidean distances between the centroid of each seaport/airport signature and the centroids of every other seaport/airport signature were calculated to derive separate seaport and airport climate dissimilarity matrices. A simple test using other distance measures (e.g. divergence, Jefferies-Matusita; ERDAS, 2003) showed no obvious changes in the dendrogram architecture from the Euclidean one used here.

The climate dissimilarity matrices were subject to hierarchical clustering using an agglomerative algorithm (Webb, 2003). Hierarchical clustering procedures are the most commonly used method of summarising data structure (Webb, 2003). The clustering process produces a hierarchical tree, which is a nested set of partitions represented by a tree diagram, or dendrogram. The clustering results here were translated into dendrograms based on centroid linkage.

Figures 3a and b show the climatic dendrograms for the major seaports and airports, respectively.

The seaport and airport locations were overlaid on the (historical) *Ae. albopictus* distribution map (Figure 1) and were classified as either inside or outside the distribution. Those seaports/airports within the distribution were located on the relevant dendrogram (Figure 3). The dendrogram branch which encompassed at least 90% of the seaports/airports was designated as defining the limits of the 'climatic envelope' of *Ae. albopictus*, i.e. the range of climatic conditions within which it can survive. This allowed for the fact that *Ae. albopictus* has both temperate (diapausing) and tropical (non-diapausing) races with distinct environmental requirements and different original geographical distributions (Hawley *et al.*, 1987). Thus, the 90% cut-off on the seaports dendrogram (Figure 3a) encompassed a single branch, but contained two major sub-branches, with the remaining 10% of ports displaying quite distinct environments (Mormugao, New-Mangalore and Kuching). Ninety percent of airports within the pre-expansion distribution of *Ae. albopictus* can be encompassed in a single branch of the airport dendrogram (Figure 3b), but temperate and tropical races are again distinguishable within this branch. Those seaports/airports not within its historical distribution, but linked by a dendrogram branch within the climatic envelope were therefore identified as being similar enough climatically for there to be a risk of establishment.

3.5.3. Risk Routes

Given that ship/aircraft volume on a transport route, as well as climatic similarity between origin and destination port, is important in determining invasion risk (Lounibos, 2002; Drake and Lodge, 2004; Normile, 2004), the transport and Euclidean climatic distance matrices were used to obtain a relative measure of importation and establishment risk to those seaports/airports identified as being at-risk within the dendrogram. Each matrix was rescaled independently to a range of 0–1 and the results for the climatic matrix then inverted so that values near to 1 represented similar climates and values close to 0 represented dissimilar ones. Values

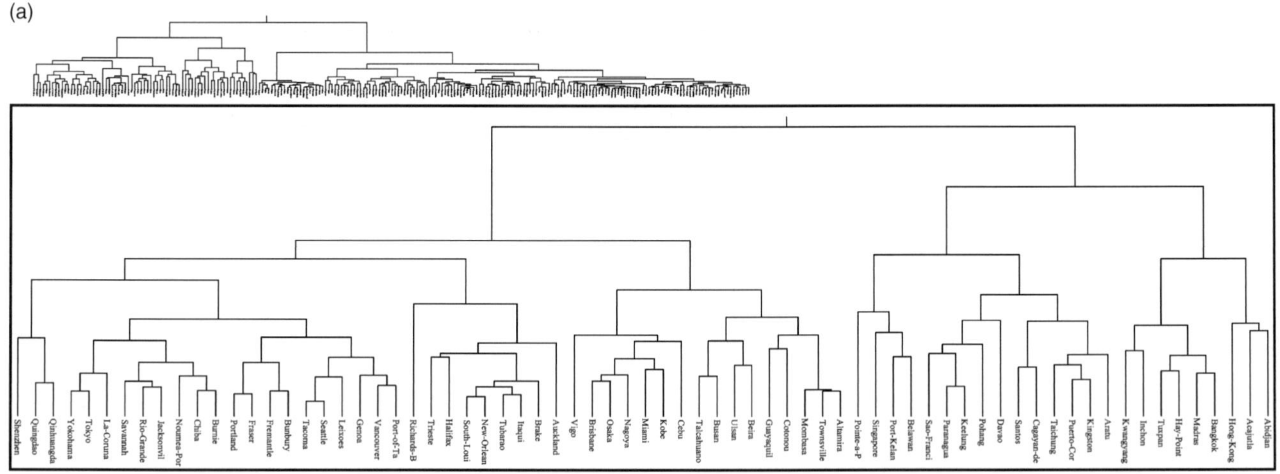

Figure 3 (a) Climatic similarity dendrograms for the major seaports of the World and (b) climatic similarity dendrograms for the major airports of the World. In both figures the inset close-up shows the branches of significance to the dispersal of *Ae. Albopictus*.

(b)

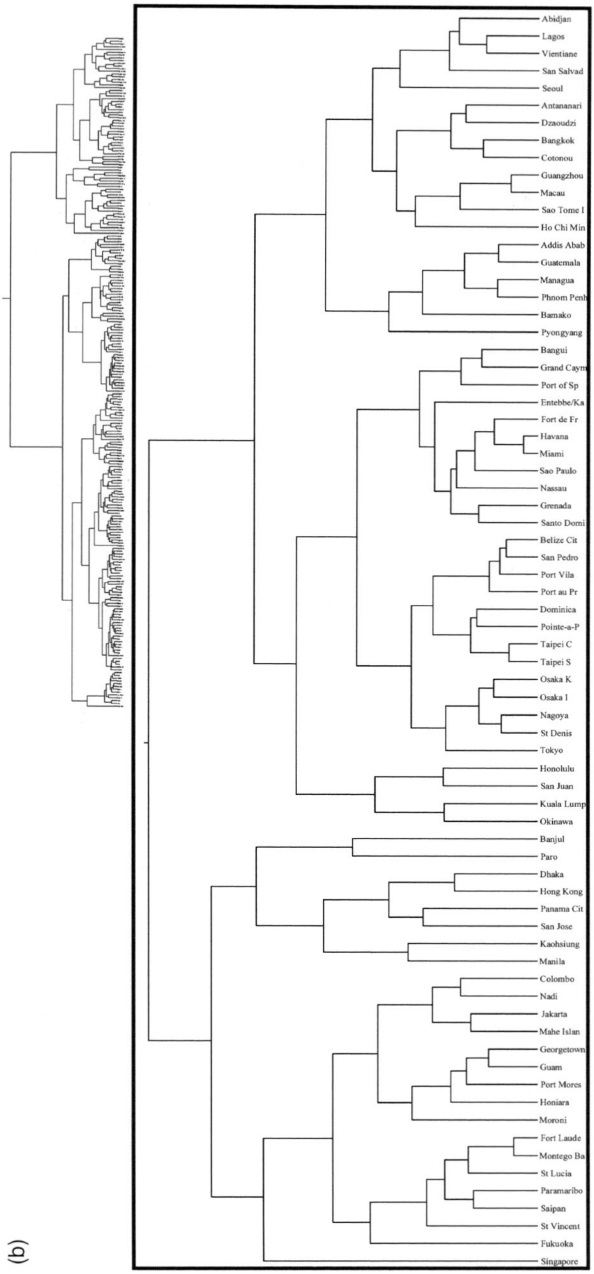

Figure 3 (continued)

in the two rescaled matrices for any pair of ports—one with and one without the invading species in question—were then multiplied together to estimate the relative risk of invasion and establishment. As an example, the rescaled climatic distance between Chiba, Japan and Richards Bay, South Africa was 0.9703, reflecting similar climatic regimes, while the rescaled traffic volume on the route was just 0.0012, a low-traffic volume in global terms. When these are multiplied, we arrive at a disease vector invasion risk of 0.00116, which is low in comparison to the risk value of 0.35 for the Chiba to New Orleans, USA, route and may partly explain why *Ae. albopictus* invaded New Orleans successfully, but not Richards Bay (to date).

Table 1 details the top 20 shipping routes (from over 20 000 possible routes) and Table 2 shows the air travel routes (from over 6000 possible routes) identified as having the highest relative risk of *Ae. albopictus* invasion. The values relative to route 1 and the recorded details of this mosquito's spread are included in both tables. There is correspondence between the predicted risk routes in Table 1 and the global invasions or interception of *Ae. albopictus*. Three of the top 10 shipping routes run from Japan to the south-east United States, where some of the earliest breeding populations of *Ae. albopictus* were found and identified as originating from Japan (Lounibos, 2002). Genoa, the destination of five more routes from Japan in the top 20, was one of the earliest European cities to report *Ae. albopictus* establishment in 1990 (Romi *et al.*, 1999); the vector has since become the most important local nuisance mosquito (Gratz, 2004). Of the remaining routes in Table 1, interceptions of *Ae. albopictus* in tyre shipments from Japan to Australasia have been made in both Brisbane (Kay *et al.*, 1990) and Auckland (Laird, 1990; Laird *et al.*, 1994), but the species appears not to have become established there, probably due to strict inspection and fumigation policies (Lounibos, 2002). No documented evidence exists of invasion by *Ae. albopictus* of Fraser and Vancouver (Canada). On the basis of Fraser's and Vancouver's climatic similarities to, and the seaborne traffic levels from, Japan, these ports should be considered at high risk, especially as *Ae. albopictus* has already been intercepted at nearby Seattle (CDC, 1986).

The relative importance of sea traffic volume and local climate in the establishment of *Ae. albopictus* was estimated by examining the 69

Table 1 *Ae. albopictus* top twenty shipping risk routes

Rank	From		To		*Ae. albopictus* intercepted?	*Ae. albopictus* established?	Risk relative to route 1
1	Chiba	Japan	New Orleans	USA	Y	Y	1.00
2	Chiba	Japan	Genoa	Italy	Y	Y	0.99
3	Chiba	Japan	Fraser	Canada	N	?	0.92
4	Chiba	Japan	Brisbane	Australia	Y	?	0.83
5	Chiba	Japan	Auckland	NZ	Y	?	0.67
6	Chiba	Japan	South Louisiana	USA	Y	Y	0.65
7	Yokohama	Japan	Fraser	Canada	N	?	0.64
8	Kobe	Japan	Fraser	Canada	N	?	0.59
9	Chiba	Japan	Miami	USA	Y	Y	0.59
10	Yokohama	Japan	Genoa	Italy	Y	Y	0.59
11	Chiba	Japan	Vancouver	Canada	N	?	0.57
12	Kobe	Japan	Genoa	Italy	Y	Y	0.53
13	Osaka	Japan	Fraser	Canada	N	?	0.51
14	Yokohama	Japan	Brisbane	Australia	Y	?	0.51
15	Chiba	Japan	Freemantle	Australia	Y	?	0.50
16	Tokyo	Japan	Genoa	Italy	Y	Y	0.46
17	Osaka	Japan	Genoa	Italy	Y	Y	0.45
18	Tokyo	Japan	Fraser	Canada	N	?	0.44
19	Kobe	Japan	Brisbane	Australia	Y	?	0.44
20	Kobe	Japan	New Orleans	USA	Y	Y	0.44

ports within the invasion risk group. Twenty-one ports are within the original range of *Ae. albopictus*, but of the other 47, to-date just over half are in regions reported to be colonized by this species or where breeding populations have been found. Within this group of 47 ports, the average climatic distances of the 24 invaded and 23 non-invaded ports were identical, but sea traffic volumes were 2.43 times greater in the former than the latter (41.84 ship visits per annum per route for invaded ports and 17.24 for non-invaded ones), with the difference found to be significant ($t = 2.343$, $p = 0.024$). Sea traffic volumes, therefore, appear to make an important contribution to relative invasion risk. This result corroborates recent theories regarding direct linkage between biological invasion success and propagule pressure (Vila and Pujadas, 2001; Levine and D'Antonio, 2003; Normile, 2004; Lockwood *et al.*, 2005).

Although air travel was not implicated in the spread of *Ae. albopictus*, 8 of the 20 highest risk air routes link the original range of *Ae. albopictus* with Honolulu in Hawaii, where this species has

Table 2 *Ae. albopictus* top twenty air travel risk routes

Rank	From		To		*Ae. albopictus* intercepted?	*Ae. albopictus* established?	Risk relative to route 1
1	Tokyo N	Japan	Honolulu	Hawaii (USA)	Y	Y	1.00
2	Osaka K	Japan	Honolulu	Hawaii (USA)	Y	Y	0.48
3	Nagoya	Japan	Honolulu	Hawaii (USA)	Y	Y	0.30
4	Tokyo N	Japan	Seattle	USA	Y	?	0.29
5	Tokyo N	Japan	Brisbane	Australia	Y	?	0.14
6	Fukuoka	Japan	Honolulu	Hawaii (USA)	Y	Y	0.13
7	Seoul	South Korea	Honolulu	Hawaii (USA)	Y	Y	0.12
8	Tokyo H	Japan	Honolulu	Hawaii (USA)	Y	Y	0.10
9	Taipei C	Taiwan	Seattle	USA	Y	?	0.10
10	Tokyo N	Japan	Portland	USA	Y	?	0.09
11	Nagoya	Japan	Portland	USA	Y	?	0.09
12	Guam	Guam	Honolulu	Hawaii (USA)	Y	Y	0.09
13	Kuala Lumpur	Malaysia	Brisbane	Australia	Y	?	0.08
14	Tokyo N	Japan	Noumea	New Caledonia			0.08
15	Taipei C	Taiwan	Brisbane	Australia	Y	?	0.08
16	Taipei C	Taiwan	Honolulu	Hawaii (USA)	Y	Y	0.07
17	Osaka K	Japan	Seattle	USA	Y	?	0.07
18	Port Moresby	P.N.G.	Brisbane	Australia	Y	?	0.06
19	St Denis	Reunion	Dzaoudzi	Mayotte			0.06
20	Bangkok	Thailand	Brisbane	Australia	Y	?	0.05

established (Mitchell, 1995) (Table 2). The large air traffic volume running from Tokyo to Hawaii identifies this route as providing over double the risk of invasion as the others in Table 2. Details on Hawaiian ports were not available in the sea traffic database. In every one of the top 20 high-risk air routes, *Ae. albopictus* has been either intercepted or found breeding at the destinations listed.

3.5.4. Climatic Distance Images

In addition to determining at which seaport or airport an accidentally introduced disease vector is likely to become established, it is important to identify the areas to which vectors may subsequently spread. An initial assessment of this can be undertaken by examining the climatic distance of the surrounding area from the disease vector's origin. This was tested for *Ae. albopictus*, again using gridded climatologies (New *et al.*, 2002). Given that the origin of the invasive *Ae. albopictus* was found to be Japan (Moore, 1999; Gratz, 2004), and the above analysis found Chiba to be the likely origin port, the climatic signature of Chiba was used to calculate a global climatic distance image at 10′ spatial resolution. In this analysis, the 'Mahalanobis' distance rather than the Euclidean distance of every 10′ pixel from Chiba's signature was calculated. The Mahalanobis distance differs from Euclidean distance as it accounts for the correlations of the dataset by examining the similarity between each signature's covariance matrix (ERDAS, 2003). The results are shown in Figure 4a, with a close-up of the USA in Figure 4b. Comparison with maps of known *Ae. albopictus* invasion (Figures 1 and 2; Lounibos, 2002) shows good correspondence, demonstrating the broad applicability of the approach in estimating future disease vector movement. The dark regions (indicating climatic similarity to Chiba) across Europe and in South Africa, Australia and New Zealand, indicate that given the opportunity provided by transport networks, *Ae. albopictus* may spread yet further. Future work on invasion risk should focus on the incorporation of data on land transport routes, human population distribution and how these relate spatially to endemic disease regions.

4. GLOBAL TRANSPORT NETWORKS AND VECTOR-BORNE DISEASES

This section reviews briefly the movement and emergence of vector-borne diseases through global human transport, and outlines approaches for highlighting areas at risk from future events.

4.1. Yellow Fever

Outlined in detail in Rogers *et al.*, (this volume, pp. 181–220), yellow fever represents a major public health problem in many regions of tropical Africa and South America, and used to be a disease that was regularly exported elsewhere (Haggett, 2000). The enforcement by the World Health Organization of yellow fever vaccination requirements for travel to endemic regions, and consequent minimal movement of the disease, demonstrates the effectiveness of global control measures. Despite advances in preventative measures and the existence of a

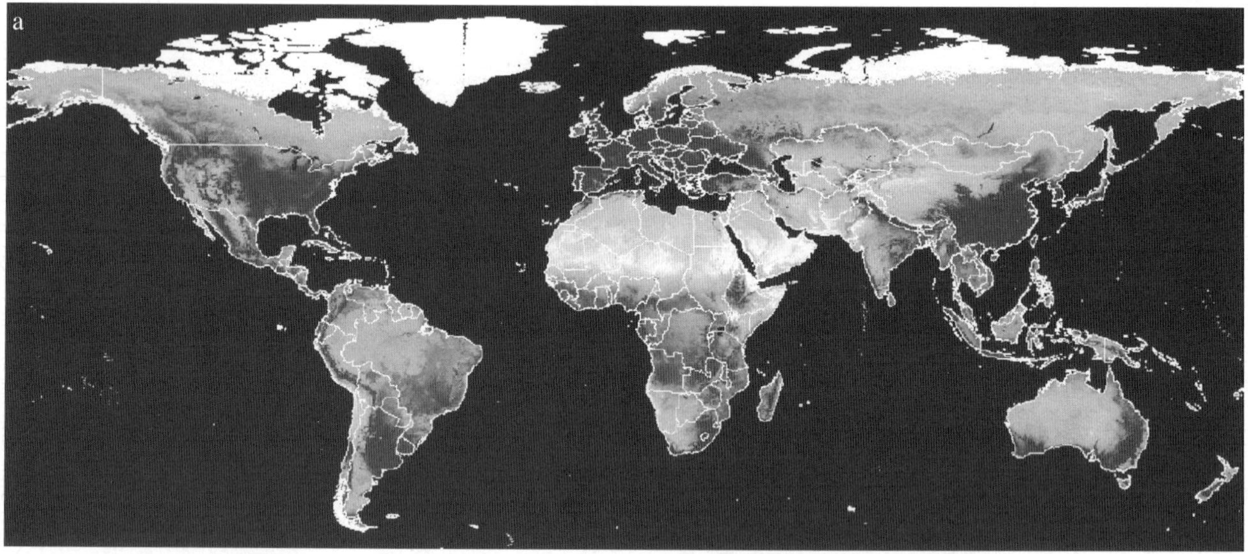

Figure 4 Mahalanobis climatic distance from Chiba Port, Japan, for (a) the world and (b) the United States of America. Darker shades represent areas with climates more similar to that of Chiba.

Figure 4 (continued)

reliable vaccine, yellow fever outbreaks and epidemics have increased in recent years, particularly in Africa (Mutebi and Barrett, 2002). Many of these epidemics have been caused by the large-scale movements of susceptible individuals into high yellow fever risk zones and by the neglect of control procedures (Mutebi and Barrett, 2002).

4.2. Dengue

Outlined in detail in Rogers *et al.*, (this volume, pp. 181–220), dengue fever and dengue haemorrhagic fever are probably the fastest spreading diseases of the modern day (Gubler and Clark, 1995; Gubler, 2002). The past 30 years has seen a marked global emergence and re-emergence of epidemic dengue, with more frequent and larger epidemics associated with more severe disease (Mackenzie *et al.*, 2004). Fuelled by the expansion of the range of its principal vector, *Ae. aegypti* (Lounibos, 2002), the expansion of the range of what may be an urban–rural bridge vector, *Ae. albopictus* (Lounibos, 2002; Gratz, 2004), large-scale global urbanization (Gubler, 2004b; Tatem and Hay, 2004) and increasing movement by air of people incubating its infections (Upham *et al.*, 2003), dengue is fast becoming the most important arboviral disease of humans (Mackenzie *et al.*, 2004). The global air transport network is not only aiding the spread of both dengue vectors and virus serotypes within tropical regions suitable for transmission, but also facilitating a substantial increase of imported cases elsewhere (Jelinek, 2000; Frank *et al.*, 2004; Effler *et al.*, 2005; Shu *et al.*, 2005).

4.3. West Nile Virus

WNV is a mosquito-borne flavivirus native to Africa, Europe and Western Asia (Hayes, 2001). It circulates mainly among birds, but can infect many species of mammals, amphibians and reptiles (Dauphin *et al.*, 2004; Kilpatrick *et al.*, 2005). The appearance of WNV in New York in 1999 and its subsequent spread westwards across the United States (Petersen and Hayes, 2004) and into Central America represents the best-documented movement of a disease in recent times

(Granwehr *et al.*, 2004). Spread by many different species of mosquitoes and the movement of birds, it is less influenced by human transportation networks than other diseases discussed in this review, but its arrival in the US in 1999 is thought to be associated with some form of commerce or human travel. With changing climate, coupled with increased human movement and spread of exotic mosquitoes via global transport networks, the risks of further WNV outbreaks are a distinct possibility (Granwehr *et al.*, 2004; Medlock *et al.*, 2005).

4.4. Malaria

General socio-economic improvement, combined with wetland drainage and other water resource development, improved housing, better animal husbandry and wider availability of quinine resulted in the decline of indigenous malaria in high-income countries during the 19th and 20th centuries (Hay *et al.*, 2004). In contrast, increasing tourism and human migration has meant that malaria continues to be imported into high-income countries, which have been classified as free of malaria transmission. Numbers vary, but it is estimated that as many as 40 000 cases of malaria a year occur in Europe alone (Toovey and Jamieson, 2003). This influx of infected travellers poses two major hazards: (i) a returning traveller who has acquired malaria may be diagnosed incorrectly, as malaria may be unfamiliar to a physician in a country without indigenous malaria, resulting in high *P. falciparum* case-fatality rates (Muentener *et al.*, 1999; Ryan *et al.*, 2002; Spira, 2003). (ii) Favourable climatic conditions may make autochthonous transmission possible through local vectors (Isaäcson, 1989; Layton *et al.*, 1995; Gratz *et al.*, 2000).

4.4.1. Imported Malaria Trends

The last 25 years have seen both the numbers and types of malaria imported to high-income countries change significantly. Figures 5a and b highlight the trends in imported cases to the UK and USA, respectively. Figure 5a shows that the UK consistently had 1500–2500 cases of malaria every year reported since 1977. However the proportion of *P. falciparum* malaria has risen substantially (Malaria Reference Laboratory, 2004). Figure 5b demonstrates that while the USA has fewer imported malaria cases annually than the UK, the trend towards increasing numbers of *P. falciparum* cases continues. The rise in *P. falciparum* cases in the USA is mirrored by a substantial rise in cases acquired in Africa (Figure 5c). The precise causes of the decline in imported *Plasmodium vivax* cases and increase in those of *P. falciparum* in the UK and USA are difficult to determine given the myriad of factors involved in transmission, including changes over time of levels and types of antimalarial use, malaria prevalence, travel activities and reporting efficiency. What is clear, however, is that over the past few decades, travel from the UK and USA to SSA, where *P. falciparum* malaria is highly endemic, has risen steadily. Analysis of UK Civil Aviation Authority (UKCAA) statistics shows that the number of passengers travelling on flight routes between SSA and the UK rose by around a million between 1997 and 2003 (Figure 5d), with the largest rises on routes from West Africa, where the highest levels of *P. falciparum* malaria endemicity are found (see Guerra *et al.*, this volume, pp. 157–179).

4.4.2. Drug Resistance

The increasing level of malaria movement also has the effect of enhancing the possibility and speed of drug-resistant malaria spread. Since the first reports of chloroquine resistance 50 years ago, drug-resistant malaria has posed a major and increasing problem in malaria control (Hastings and Mackinnon, 1998; Anderson and Roper, 2005; Hastings and Watkins, 2005). Chloroquine resistance is now worldwide, while resistance to newer drugs is appearing in many regions, especially in South-east Asia, where multidrug resistance is a major public health problem (Wongsrichanalai *et al.*, 2002; Roper *et al.*, 2004). The growth in global travel and human migration is assumed to have caused this spread, with the speed of spread of resistance mirroring its expansion. Though chloroquine remained effective for some 20 years before signs of resistance emerged, it is unlikely that a similar period of grace can be realistically expected for

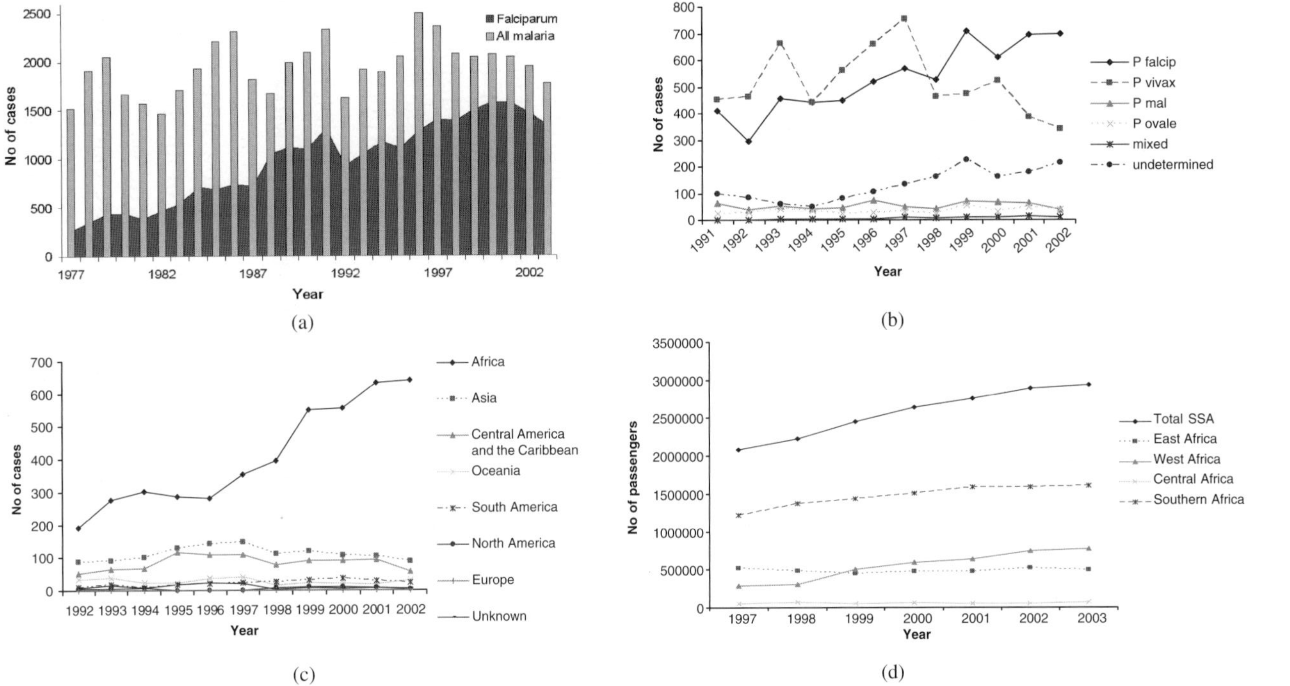

Figure 5 (a) Graph showing the number of UK imported malaria cases 1977–2002 (Data source: UK Health Protection Agency (UKHPA)); (b) Graph showing the number and type of USA imported malaria cases 1991–2002 (Data source: Shah *et al.*, 2003); (c) Graph showing the acquisition region of USA imported malaria cases 1992–2002 (Data source: Shah *et al.*, 2003); and (d) Graph showing the number of passengers travelling on air routes between the UK and SSA, broken down by SSA region 1997–2003 (Data source: UK Civil Aviation Authority (UKCAA)).

recently deployed treatments given the current levels of global travel (Hartl, 2004).

4.4.3. Airport Malaria

Airport malaria is acquired through the bite of an infected tropical *Anopheline* mosquito taken on persons whose geographical history firmly excludes exposure to the vector in its natural habitat (Isaäcson, 1989). Between 1969 and 1999, 87 suspected cases were recorded, with almost all being *P. falciparum* and occurring in the proximity of Paris, Brussels and London airports (Giacomini and Brumpt, 1989; Isaäcson, 1989; Danis *et al.*, 1996; Giacomini, 1998; Gratz *et al.*, 2000), indicative of their air links and sufficient, if only seasonal, climatic similarity with SSA (Tatem *et al.*, 2006b). Figure 6 summarises the locations and time of year of these probable airport malaria cases. *An. gambiae* s.l. mosquitoes have been shown to survive long haul flights (Misao and Ishihara, 1945; Russell, 1987, 1989). For example, in a three-week period in 1994, it was estimated that 2000–5000 *Anopheline* mosquitoes were imported into France by 250–300 aircraft arriving from malaria-endemic regions of Africa, at the rate of 8–20 *Anopheline* mosquitoes per flight (Gratz *et al.*, 2000). The ever expanding global transport network, increased travel to malaria-endemic countries and a decline in aircraft disinsection (Gratz *et al.*, 2000), mean the numbers of malaria-infected *Anopheles* arrivals will only increase, resulting in more cases of airport malaria.

4.5. Predicting Future Vector-Borne Disease Movement

The establishment of a vector-borne disease in a new area from an endemic region can be caused either by movement of an infected host and availability of competent vectors in the new area, or the invasion, if only temporarily, of an infected vector. Here, we extend the methodology outlined in Section 3.5 and review the approaches outlined in Tatem *et al.* (2006b) to examine the latter in terms of a basic exploratory analysis of the monthly risks of *P. falciparum* malaria-infected *An. gambiae* importation by plane from Africa, causing consequent

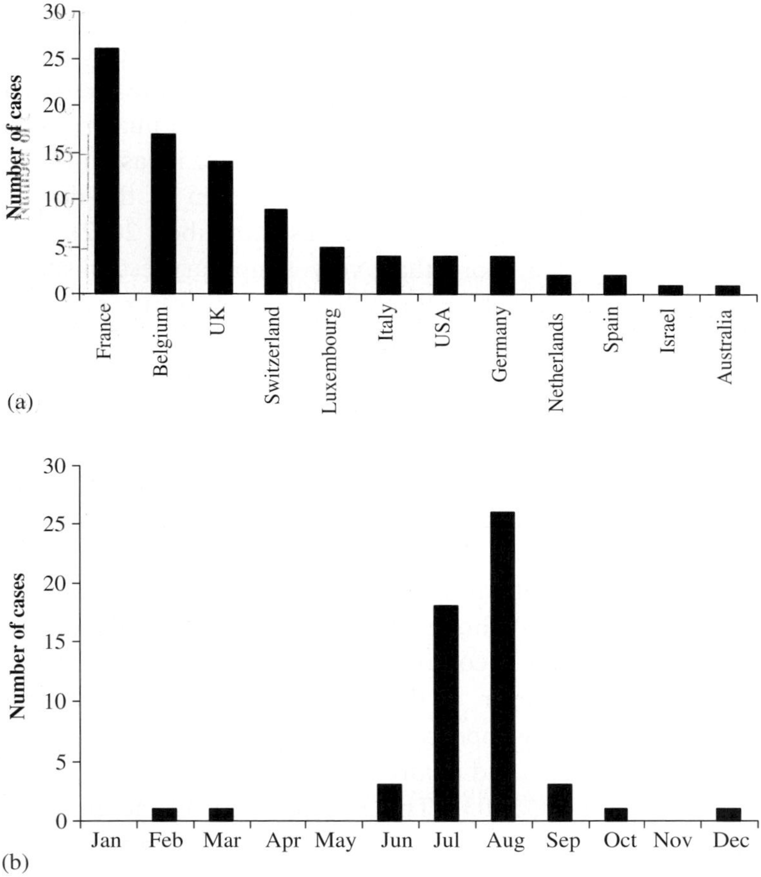

Figure 6 (a) Countries in which confirmed or probable cases of airport malaria have been reported. (b) Month in which suspected European airport malaria cases occurred (where date is provided). (Data taken from Alos *et al.*, 1985; Csillag, 1996; Danis *et al.*, 1996; Giacomini, 1998; Giacomini and Brumpt, 1989; Giacomini and Mathieu, 1996; Giacomini *et al.*, 1995; Gratz *et al.*, 2000; Guillet *et al.*, 1998; Hemmer, 1999; Holvoet *et al.*, 1982; Isaäcson, 1989; Isaäcson and Frean, 2001; Jafari *et al.*, 2002; Karch *et al.*, 2001; Kruger *et al.*, 2001; Lusina *et al.*, 2000; Majori *et al.*, 1990; Mangili and Gendreau, 2005; Mantel *et al.*, 1995; Mouchet, 2000; Praetorius *et al.*, 1999; Shpilberg *et al.*, 1988; Signorelli and Messineo, 1990; Smith and Carter, 1984; Thang *et al.*, 2002; Toovey and Jamieson, 2003; Van den Ende *et al.*, 1996; Whitfield *et al.*, 1984.)

autochthonous transmission. While globally, the movement of *P. vivax* malaria-infected mosquitoes may occur more frequently and pose more of an invasion risk to many regions, it is the movement of *P. falciparum*-infected *Anopheles* that has resulted in numerous airport malaria cases. Additionally, the possible effects of invasion of malaria endemic regions by *An. gambiae* have been seen in the devastating malaria epidemics in Brazil and Mauritius (Lounibos, 2002). The likelihood, however, of a more than very temporary establishment of *P. falciparum* or *An. gambiae* through air travel to Europe and many other locations remains low. Unsuitable year-round climate, *An. gambiae*'s intolerance of urban areas (Hay *et al.*, 2005) and competition from local mosquitoes that are inefficient vectors of *P. falciparum* all provide barriers to establishment.

Malaria caused by *P. vivax* and vectored predominantly by *Anopheles atroparvus* was a common cause of death in the marshes or wetlands of England during the 19th and 20th centuries (Kuhn *et al.*, 2003), with the last autochthonous cases reported in 1953 (Crockett and Simpson, 1953). Although malaria has been eradicated in the UK, re-introduction is theoretically possible (Hay *et al.*, 2000). This risk is negligible, however, as the majority of the former habitats of *An. atroparvus* have disappeared and returning travellers are rapidly diagnosed and treated, and in any case rarely live near suitable vector habitats (Kuhn *et al.*, 2003). This is supported by the analyses presented above and the fact that not one of the 52 000 imported malaria cases reported since 1953 has led to a secondary case. The growth of travel to Africa therefore presents a negligible risk of the establishment of malaria in the UK.

The approach described here follows on from that described in Section 3.5 and full details are given in Tatem *et al.* (2006b). Monthly gridded climatologies were acquired (New *et al.*, 2002) and used to create climatic signatures for each airport, for each month of the year. Climate dissimilarity matrices were created for each month, and these were clustered hierarchically to create 12 global airport dendrograms, one for each month of the year. Airport malaria cases represent instances where imported infected mosquitoes have caused autochthonous malaria transmission and, therefore, were used to define climatic suitability thresholds on each dendrogram. Malaria seasonality maps for

Africa (Tanser *et al.*, 2003) were used to identify the timing of the principal malaria transmission season at each SSA airport to identify when exported *Anopheles* were most likely to be malarious. Finally, the situation in 2000 was examined using the air traffic database described in Section 3.5, and future risks of opening new air routes were examined by analysing climatic similarities between airports and synchrony with malaria transmission seasons.

Table 3 shows the 18 routes identified as at risk for *P. falciparum*-infected mosquito importation and consequent autochthonous transmission for 2000. Comparison with Figure 6 shows excellent correspondence with the timing and location of actual suspected cases of infected-mosquito importation and autochthonous transmission. The relative risks suggest that the high traffic volumes on the Abidjan to Paris route in August, when both have sufficiently similar climates and Côte d'Ivoire is experiencing its principal malaria transmission season, makes this the most likely route for airport malaria occurrence.

Figure 7 shows those sets of SSA airports and non-SSA airports similar enough climatically in January, April, July and October for *P. falciparum*-infected mosquito invasion. The results indicate that many airports in other regions of the world are more favourable climatically for *Anopheles* survival, and for many more months of the year than European destinations. The concentration of major airports in the temperate north means that July (summer in this region) on Figure 7 shows the largest number of potentially suitable destinations climatically, mainly linked to West African airports, including those in Europe where most SSA air traffic is directed, causing cases of airport malaria in the summer months particularly in unusually hot and humid periods. The other three months on Figure 7 show considerably fewer at risk airports, though throughout the year, Caribbean and Central American airports are highlighted consistently, indicating that their climates vary in synchrony most closely with the cycle of African malaria transmission seasons.

The current heavy bias of SSA air traffic to European destinations has resulted in around two cases a year of airport malaria in the summer months, when particularly hot and humid conditions can be suitable for temporary *Anopheles* survival, and occur in synchrony with West African transmission seasons. The effects of opening up new air

Table 3 Year 2000 air travel risk routes for possible temporary *P. falciparum*-infected *An. gambiae* invasion and subsequent autochthonous transmission

Rank	From		To		Month	Risk relative to route 1
1	Abidjan	Côte d'Ivoire	Paris Charles de Gaulle	France	August	1.00
2	Accra	Ghana	Amsterdam Schippol	Netherlands	July	0.26
3	Entebbe/Kampala	Uganda	Brussels	Belgium	July	0.26
4	Accra	Ghana	Amsterdam Schippol	Netherlands	September	0.22
5	Abidjan	Côte d'Ivoire	Brussels	Belgium	August	0.21
6	Accra	Ghana	Rome Fiumicino Apt	Italy	September	0.18
7	Abidjan	Côte d'Ivoire	Zurich	Switzerland	July	0.17
8	Accra	Ghana	Rome Fiumicino	Italy	August	0.17
9	Abidjan	Côte d'Ivoire	London Gatwick	United Kingdom	August	0.12
10	Cotonou	Benin	Brussels	Belgium	August	0.06
11	Libreville	Gabon	Rome Fiumicino	Italy	July	0.06
12	Cotonou	Benin	Paris Charles de Gaulle	France	August	0.05
13	Lome	Togo	Brussels	Belgium	August	0.05
14	Accra	Ghana	London Gatwick	United Kingdom	July	0.05
15	Entebbe/Kampala	Uganda	London Gatwick	United Kingdom	July	0.04
16	Libreville	Gabon	Dubai	United Arab Emirates	July	0.03
17	Abidjan	Côte d'Ivoire	Frankfurt	Germany	August	0.01
18	Entebbe/Kampala	Uganda	London Heathrow	United Kingdom	July	0.01

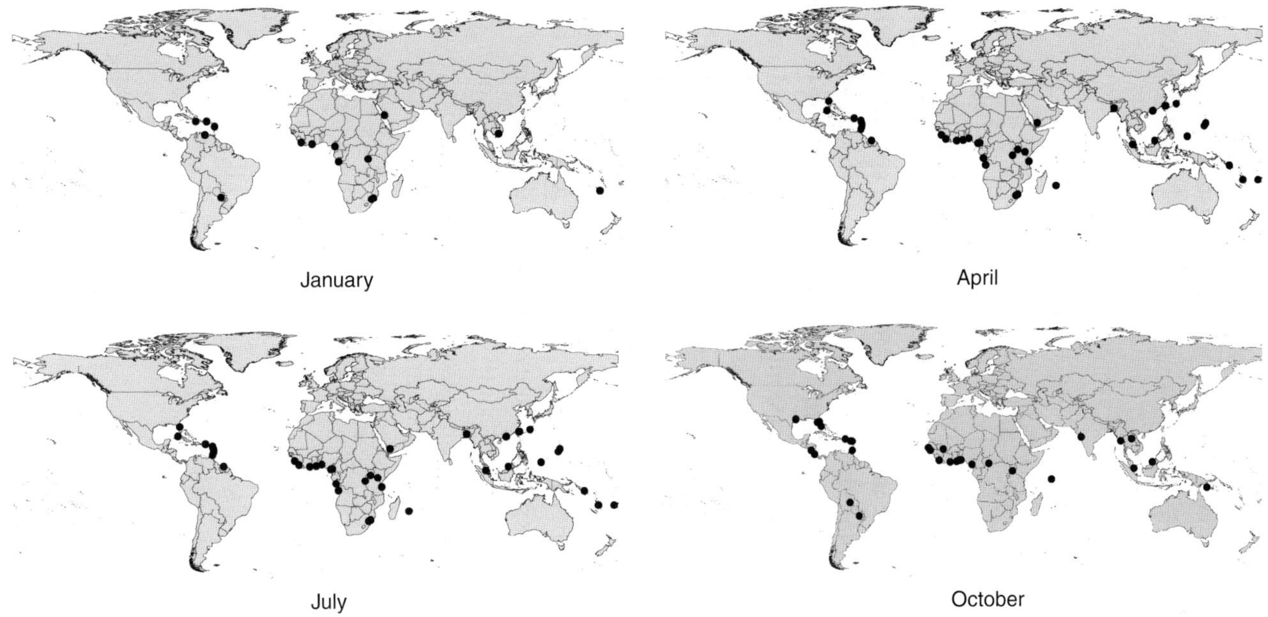

Figure 7 Non-SSA airports that are similar enough climatically to the SSA airports within their primary malaria transmission season for possible *P. falciparum*-infected *Anopheles* invasion to occur.

routes from malaria-endemic African countries to non-European destinations, where conditions are more suitable for *Anopheles* survival and are synchronous with African malaria transmission seasons year-round, could therefore have serious and largely unexpected consequences.

5. CONCLUSIONS

Increases in global travel are occuring simultaneously with many other processes that favour the emergence of disease (Wilson, 1995). Travel is a potent force in disease emergence and spread, whether it is aircraft moving human-incubated pathogens, or insect vectors, great distances in short times, or ships transporting used tyres containing mosquito eggs. The speed and complexity of modern transport make both geographical space and the traditional 'drawbridge' strategy of disease control and quarantine increasingly irrelevant (Haggett, 2000). With no apparent end in sight to the continued growth in global air travel and shipborne trade, we must expect the continued appearance of communicable disease pandemics, disease vector invasions and vector-borne disease movement. Approaches that can model, predict and explain such events can be used to focus surveillance and control efforts efficiently. This review has shown that the risk of movement of infectious diseases and their vectors through the global transportation network can be predicted to provide such information. Future challenges must focus on incorporating information on temporal variations in passenger numbers, stopover risks, intra-species competition, human populations at risk, breeding site availability, possible climate change, disinsection and land transport, as well as quantifying the relative importance of all types of transport for vector and disease movement.

ACKNOWLEDGEMENTS

We are grateful to David Bradley, Dennis Shanks, Robert Snow, Marius Gilbert, Sarah Randolph, Willy Wint, Carlos Guerra, Briony

Boon, Rebecca Freeman-Grais and the series editors for comments on this manuscript. SIH and AJT are funded by a Research Career Development Fellowship (to SIH) from the Wellcome Trust (#069045). We are especially grateful to John Drake for the generous supply of sea traffic data.

REFERENCES

Alibek, K. (2004). Smallpox: a disease and a weapon. *International Journal of Infectious Diseases* **8S2**, S3–S8.

Anderson, R.M. and May, R.M. (1991). *Infectious Diseases of Humans: Dynamics and Control*. Oxford: Oxford University Press.

Anderson, T.J.C. and Roper, C. (2005). The origins and spread of antimalarial drug resistance: lessons for policy makers. *Acta Tropica* **94**, 269–280.

Bailey, N.T.J. (1967). *The Mathematical Approach to Biology and Medicine*. London: Wiley.

Barretto, A., Aragon, M. and Epstein, R. (1994). Bubonic plague outbreak in Mozambique 1994. *Lancet* **345**, 983–984.

Bozzette, S. (2003). A model for smallpox vaccination policy. *New England Journal of Medicine* **348**, 416–425.

Bronfman, M.N., Leyva, R., Negroni, M.J. and Rueda, C.M. (2002). Mobile populations and HIV/AIDS in Central America and Mexico: research for action. *AIDS* **16**, S42–S49.

Cash, R.A. and Narasimhan, V. (2000). Impediments to global surveillance of infectious diseases: consequences of open reporting in a global economy. *Bulletin of the World Health Organization* **78**, 1358–1367.

CDC. (1986). *Aedes albopictus* infestation—United States. *Epidemiologic Notes and Reports Update* **35**, 649–651.

Chevallier, E., Grand, A. and Azais, J.-M. (2004). Spatial and temporal distribution of cholera in Ecuador between 1991 and 1996. *European Journal of Public Health* **14**, 274–279.

Cieslak, T.J. and Eitzen, E.M. (1999). Clinical and epidemiologic principles of anthrax. *Emerging Infectious Diseases* **5**, 552–555.

Cox, N.J. and Subbarao, K. (2000). Global epidemiology of influenza: past and present. *Annual Reviews of Medicine* **51**, 407–421.

Crockett, G.S. and Simpson, K. (1953). Malaria in neighbouring Londoners. *British Medical Journal* **21**, 1141.

Curtin, P.D. (1995). *Death by Migration: Europe's Encounter with the Tropical World in the Nineteenth Century*. Cambridge: Cambridge University Press.

Danis, M., Mouchet, J. and Giacomini, T. (1996). Autochthonous and introduced malaria in Europe. *Medecine et Maladies Infectieuses* **26**, 393–396.

Dauphin, G., Zientara, S., Zeller, H. and Murgue, B. (2004). West Nile: worldwide current situation in animals and humans. *Comparative Immunology Microbiology and Infectious Diseases* **27**, 343–355.

Dewan, P.K., Fry, A.M., Laserson, K., Tierney, B.C., Quinn, C.P., Hayslett, J.A., Broyles, L.N., Shane, A., Winthrop, K.L., Walks, I., Siegel, L., Hales, T., Semenova, V.A., Romero-Steiner, S., Elie, C., Khabbaz, R., Khan, A.S., Hajjeh, R.A. and Schuchat, A. (2002). Inhalational anthrax outbreak among postal workers, Washington, DC, 2001. *Emerging Infectious Diseases* **8**, 1066–1072.

Diamond, J. (1998). *Guns, Germs and Steel: A Short History of Everybody for the Last 13000 Years*. London: Vintage.

Drake, J.M. and Lodge, D.M. (2004). Global hot spots of biological invasions: evaluating options for ballast-water management. *Proceedings of the Royal Society of London B* **271**, 575–580.

Duplaix, N. (1988). Fleas. The lethal leapers. *National Geographic* **173**, 114–136.

Dye, C. and Gay, N. (2003). Modeling the SARS epidemic. *Science* **300**, 1884–1885.

Eckert, E. (2000). The retreat of plague from central Europe, 1640–1720: a geomedical approach. *Bulletin of Historical Medicine* **74**, 1–28.

Effler, P.V., Pang, L., Kitsutani, P., Vorndam, V., Nakata, M., Ayers, T., Elm, J., Tom, T., Reiter, P., Rigau-Perez, J.G., Hayes, J.M., Mills, K., Napier, M., Clark, G.C. and Gubler, D.J. (2005). Dengue fever, Hawaii, 2001–2002. *Emerging Infectious Diseases* **11**, 742–749.

Enserink, M. (2004). Looking the pandemic in the eye. *Science* **306**, 392–394.

ERDAS. (2003). *ERDAS Field Guide*, 7th edn. Atlanta: ERDAS Inc.

Faruque, S.M., Chowdhury, N. and Kamruzzaman, M. (2003). Reemergence of epidemic *Vibrio cholerae* Q139, Bangladesh. *Emerging Infectious Diseases* **9**, 1116–1122.

Ferguson, N.M., Donnelly, C.A. and Anderson, R.M. (2001). The foot and mouth epidemic in Great Britain: pattern of spread and impact of interventions. *Science* **292**, 1150–1160.

Ferguson, N.M., Galvani, A.P. and Bush, R.M. (2003a). Ecological and immunological determinants of influenza evolution. *Nature* **422**, 428–433.

Ferguson, N.M., Keeling, M.J., Edmunds, W.J., Gani, R., Grenfell, B.T., Anderson, R.M. and Leach, S. (2003b). Planning for smallpox outbreaks. *Nature* **425**, 681–685.

Fouchier, R., Kuiken, T., Rimmelzwaan, G. and Osterhaus, A. (2005). Global task force for influenza. *Nature* **435**, 419–420.

Frank, C., Schoneberg, I., Krause, G., Claus, H., Ammon, A. and Stark, K. (2004). Increase in imported dengue, Germany, 2001–2002. *Emerging Infectious Diseases* **10**, 903–906.

Gage, K.L. and Kosoy, M.Y. (2005). Natural history of plague: perspectives from more than a century of research. *Annual Review of Entomology* **50**, 505–528.

Galimand, M. (1997). Multidrug resistance in *Yersinia pestis* mediated by transferable plasmid. *New England Journal of Medicine* **337**, 677–680.

Giacomini, T. (1998). Malaria in airports and their neighbourhoods. *Revue du Praticien* **48**, 264–267.

Giacomini, T. and Brumpt, L.C. (1989). Passive dissemination of *Anopheles* by means of transport: its role in the transmission of malaria (historical review). *Revue Histoire Pharmacie* **36**, 164–172.

Gould, P.R. (1995). The geography of AIDS: study of human movement and the expansion of an epidemic. *La Recherche* **280**, 36–37.

Gould, P.R. (1999). *Becoming a Geographer*. Syracuse: Syracuse University Press.

Grais, R.F., Ellis, J.H. and Glass, G.E. (2003a). Assessing the impact of airline travel on the geographic spread of pandemic influenza. *European Journal of Epidemiology* **18**, 1065–1072.

Grais, R.F., Ellis, J.H. and Glass, G.E. (2003b). Forecasting the geographical spread of smallpox cases by air travel. *Epidemiology and Infection* **131**, 849–857.

Granwehr, B.P., Lillibridge, K.M., Higgs, S., Mason, P.W., Aronson, J.F., Campbell, G.A. and Barrett, A.D.T. (2004). West Nile virus: where are we now? *Lancet Infectious Diseases* **4**, 547–556.

Gratz, N.G. (2004). Critical review of the vector status of *Aedes albopictus*. *Medical and Veterinary Entomology* **18**, 215–227.

Gratz, N.G., Steffen, R. and Cocksedge, W. (2000). Why aircraft disinsection? *Bulletin of the World Health Organization* **78**, 995–1004.

Greene, C.M., Reefhuis, J., Tan, C., Fiore, A.E., Goldstein, S., Beach, M.J., Redd, S.C., Valiante, D., Burr, G., Buehler, J., Pinner, R.W., Bresnitz, E. and Bell, B.P. (2002). Epidemiologic investigations of bioterrorism-related anthrax, New Jersey, 2001. *Emerging Infectious Diseases* **8**, 1048–1055.

Gubler, D.J. (2002). The global emergence/resurgence of arboviral diseases as public health problems. *Archives of Medical Research* **33**, 330–342.

Gubler, D.J. (2003). *Aedes albopictus* in Africa. *Lancet* **3**, 751–752.

Gubler, D.J. (2004a). The changing epidemiology of yellow fever and dengue, 1900 to 2003: full circle? *Comparative Immunology Microbiology and Infectious Diseases* **27**, 319–330.

Gubler, D.J. (2004b). Cities spawn epidemic dengue viruses. *Nature Medicine* **10**, 129–130.

Gubler, D.J. and Clark, G.C. (1995). Dengue/dengue hemorrhagic fever: the emergence of a global health problem. *Emerging Infectious Diseases* **1**, 55–57.

Guimera, R., Mossa, S., Turtschi, A. and Amaral, L.A.N. (2005). The worldwide air transportation network: anomalous centrality, community structure, and cities' global roles. *Proceedings of the National Academy of Sciences of the United States of America* **102**, 7794–7799.

Haggett, P. (2000). *The Geographical Structure of Epidemics*. Oxford: Clarendon.

Halloran, M.E., Longini, I.M., Nizam, A. and Yang, Y. (2002). Containing bioterrorist smallpox. *Science* **298**, 1428–1432.

Hartl, D.L. (2004). The origin of malaria: mixed messages from genetic diversity. *Nature Reviews Microbiology* **2**, 15–22.

Hastings, I.M. and Mackinnon, M.J. (1998). The emergence of drug-resistant malaria. *Parasitology* **117**, 411–417.

Hastings, I.M. and Watkins, W.M. (2005). Intensity of malaria transmission and the evolution of drug resistance. *Acta Tropica* **94**, 218–229.

Hawley, W.A., Reiter, P., Copeland, S., Pumpuni, C.B. and Craig, G.B. (1987). *Aedes albopictus* in North America: probable introduction in used tires from northern Asia. *Science* **236**, 1114–1115.

Hay, S.I., Guerra, C.A., Tatem, A.J., Atkinson, P.M. and Snow, R.W. (2005). Urbanization, malaria transmission and disease burden in Africa. *Nature Reviews Microbiology* **3**, 81–90.

Hay, S.I., Guerra, C.A., Tatem, A.J., Noor, A.M. and Snow, R.W. (2004). Prioritising global efforts for malaria control using regional variation in populations at risk. *Lancet Infectious Diseases* **4**, 327–336.

Hay, S.I., Omumbo, J.A., Craig, M.H. and Snow, R.W. (2000). Earth observation, geographic information systems and *plasmodium falciparum* malaria in sub-Saharan Africa. *Advances in Parasitology* **47**, 173–215.

Hayes, C.G. (2001). West Nile virus: Uganda, 1937, to New York City, 1999. *Annals of the New York Academy of Sciences* **951**, 25–37.

Henderson, D.A. (1999). Smallpox: clinical and epidemiologic features. *Emerging Infectious Diseases* **5**, 537–539.

Heymann, D.L. (2004). The international response to the outbreak of SARS in 2003. *Philosophical Transactions of the Royal Society of London B* **359**, 1127–1129.

Hufnagel, L., Brockmann, D. and Geisel, T. (2004). Forecast and control of epidemics in a globalized world. *Proceedings of the National Academy of Sciences of the United States of America* **101**, 15124–15129.

Isaäcson, M. (1989). Airport malaria: a review. *Bulletin of the World Health Organization* **67**, 737–743.

Jelinek, T. (2000). Dengue fever in international travellers. *Clinical Infectious Diseases* **31**, 144–147.

Jonassen, T.O., Stene-Johansen, K. and Berg, E.S. (1997). Sequence analysis of the HIV-1 group O from Norwegian parents infected in the 1960s. *Virology* **231**, 43–47.

Kaplan, E.H., Craft, D.L. and Wein, L.M. (2002). Emergency response to smallpox attack: the case for mass vaccination. *Proceedings of the National Academy of Sciences of the United States of America* **99**, 4346–4351.

Karlen, A. (1995). *Plague's Progress*. London: Indigo.

Kay, B.H., Ives, W.A., Whelan, P.I., Barker-Hudson, P., Fanning, I.D. and Marks, E.N. (1990). Is *Aedes albopictus* in Australia? *Medical Journal of Australia* **153**, 368.

Keeling, M.J. and Gilligan, C.A. (2000). Metapopulation dynamics of bubonic plague. *Nature* **407**, 903–906.

Keeling, M.J., Woolhouse, M.E., May, R.M., Davies, G. and Grenfell, B.T. (2003). Modelling vaccination strategies against foot-and-mouth disease. *Nature* **421**, 136–142.

Killeen, G.F. (2003). Following in Soper's footsteps: northeast Brazil 63 years after eradication of *Anopheles gambiae*. *Lancet Infectious Diseases* **3**, 663–666.

Killeen, G.F., Fillinger, U., Kiche, I., Gouagna, L.C. and Knols, B.G.J. (2002). Eradication of *Anopheles gambiae* from Brazil: lessons for malaria control in Africa? *The Lancet Infectious Diseases* **2**, 618–627.

Kilpatrick, A.M., Kramer, L.D., Campbell, S.R., Alleyne, E.O., Dobson, A.P. and Daszak, P. (2005). West Nile virus risk assessment and the bridge vector paradigm. *Emerging Infectious Diseases* **11**, 425–429.

Kortepeter, M.G. and Parker, G.W. (1999). Potential biological weapons threats. *Emerging Infectious Diseases* **5**, 523–527.

Kuhn, K.G., Campbell-Lendrum, D.H., Armstrong, B. and Davies, C.R. (2003). Malaria in Britain: past, present and future. *Proceedings of the National Academy of Sciences* **100**, 9997–10001.

Kumar, S. (1995). Bubonic plague in Surat. *Lancet* **345**, 714.

Lagarde, E., van der Loeff, M.S., Enel, C., Holmgren, B., Dray-Spira, R., Pison, G., Piau, J.P., Delaunay, V., M'Boup, S., Ndoye, I., Coeuret-Pellicer, M., Whittle, H. and Aaby, P. (2003). Mobility and the spread of human immunodeficiency virus into rural areas of West Africa. *International Journal of Epidemiology* **32**, 744–752.

Laird, M. (1990). New Zealand's northern mosquito survey, 1988–89. *Journal of the American Mosquito Control Association* **6**, 287–299.

Laird, M., Calder, L., Thornton, R.C., Syme, R., Holder, P.W. and Mogi, M. (1994). Japanese *Aedes albopictus* among four mosquito species reaching New Zealand in used tires. *Journal of the American Mosquito Control Association* **10**, 14–23.

Lane, H.C., Montagne, J. and Fauci, A.S. (2001). Bioterrorism: a clear and present danger. *Nature Medicine* **7**, 1271–1273.

Layton, M., Parise, M.E., Campbell, C.C., Advani, R., Sexton, J.D., Bosler, E.M. and Zucker, J.R. (1995). Mosquito-transmitted malaria in New York City, 1993. *Lancet* **346**, 729–731.

Lemey, P., Pybus, O.G., Wang, B., Saksena, N.K., Salemi, M. and Vandamme, A.-M. (2003). Tracing the origin and history of the HIV-2 epidemic. *Proceedings of the National Academy of Sciences of the United States of America* **100**, 6588–6592.

Levine, J.M. and D'Antonio, C.M. (2003). Forecasting biological invasions with increasing international trade. *Conservation Biology* **17**, 322–326.

Lockwood, J.L., Cassey, P. and Blackburn, T. (2005). The role of propagule pressure in explaining species invasions. *Trends in Ecology and Evolution* **20**, 223–228.

Longini, I.M., Fine, P.E.M. and Thacker, S.B. (1986). Predicting the global spread of new infectious agents. *American Journal of Epidemiology* **123**, 383–391.

Lounibos, L.P. (2002). Invasions by insect vectors of human disease. *Annual Review of Entomology* **47**, 233–266.

Mackenzie, J.S., Gubler, D.J. and Petersen, L.R. (2004). Emerging flaviviruses: the spread and resurgence of Japanese encephalitis, West Nile and dengue viruses. *Nature Medicine* **10**, 98–109.

Madjid, M., Lillibridge, K.M., Mirhaji, P. and Casscells, W. (2003). Influenza as a bioweapon. *Journal of the Royal Society of Medicine* **96**, 345–346.

Malaria Reference Laboratory. (2004). Malaria imported into the UK, 2003: implications for those advising travellers. *CDR Weekly* **14**, 18–23.

Mann, J.M. (1989). AIDS: a worldwide pandemic. In: *"Current Topics in AIDS"* (M.S. Gottlieb, D.J. Jeffries, D. Mildvan, A.J. Pinching and T.C. Quinn, eds). Chichester, UK: Wiley.

Markel, H. (2005). *When Germs Travel: Six Major Epidemics that have Invaded America since 1900 and the Fears they have Unleashed*. New York: Pantheon.

Marsden, A.G. (2003). Influenza outbreak related to air travel. *Medical Journal of Australia* **179**, 172–173.

Massey, A. (1933). *Epidemiology in Relation to Air Travel*. London: H.K. Lewis and Co. Limited.

McNeill, W.H. (1976). *Plagues and People*. New York: Anchor Press.

Medlock, J.M., Snow, K.R. and Leach, S. (2005). Potential transmission of West Nile virus in the British Isles: an ecological review of candidate mosquito bridge vectors. *Medical and Veterinary Entomology* **19**, 2–21.

Meltzer, M.L., Damon, I., LeDuc, J.W. and Millar, J.D. (2001). Modeling potential responses to smallpox as a bioterrorist weapon. *Emerging Infectious Diseases* **7**, 959–969.

Mills, C.E., Robins, J.M. and Lipsitch, M. (2004). Transmissibility of 1918 pandemic influenza. *Nature* **432**, 904–906.

Misao, T. and Ishihara, M. (1945). [An experiment on the transportation of vector mosquitoes by aircraft] (in Japanese). *Rinsho to Kenkyu* **22**, 44–46.

Mitchell, C.J. (1995). Geographic spread of *Aedes albopictus* and potential for involvement in arbovirus cycles in the Mediterranean basin. *Journal of Vector Ecology* **20**, 44–58.

Moore, C.G. (1999). *Aedes albopictus* in the United States: current status and prospects for further spread. *Journal of the American Mosquito Control Association* **15**, 221–227.

Moore, C.G. and Mitchell, C.J. (1997). *Aedes albopictus* in the United States: ten-year presence and public health implications. *Emerging Infectious Diseases* **3**, 329–334.

Muentener, P., Schlagenhauf, P. and Steffen, R. (1999). Imported malaria (1985–95): trends and perspectives. *Bulletin of the World Health Organization* **77**, 560–566.

Mustermann, L.E. and Andreadis, T.G. (1999). *Aedes japonicus* in Connecticut. *Vector Ecology News* **30**, 7–8.

Mutebi, J.P. and Barrett, A.D.T. (2002). The epidemiology of yellow fever in Africa. *Microbes and Infection* **4**, 1459–1468.

Naidoo, A. and Patric, K. (2002). Cholera: a continuous epidemic in Africa. *The Journal of The Royal Society for the Promotion of Health* **122**, 89–94.

New, M., Lister, D., Hulme, M. and Makin, I. (2002). A high-resolution data set of surface climate over global land areas. *Climate Research* **21**, 1–25.

Nicholson, K.G., Wood, J.M. and Zambon, M. (2003). Influenza. *Lancet* **362**, 1733–1745.

Normile, D. (2004). Expanding trade with China creates ecological backlash. *Science* **306**, 968–969.

Osterholm, M.T. (2005a). Preparing for the next pandemic. *New England Journal of Medicine* **352**, 1839–1842.

Osterholm, M.T. (2005b). A weapon the world needs. *Nature* **435**, 417–418.

Oxford, J. (2004). The great influenza: the epic story of the 1918 pandemic. *Nature* **429**, 345–346.

Oxford, J.S. (2005). Preparing for the first influenza pandemic of the 21st century. *Lancet Infectious Diseases* **5**, 129–131.

Palese, P. (2004). Influenza: old and new threats. *Nature Medicine* **10**, 582–587.

Peiris, J.S.M., Guan, Y. and Yuen, K.Y. (2004). Severe acute respiratory syndrome. *Nature Medicine* **10**, S88–S97.

Perlman, D.C., Primas, R., Raucher, B., Lis, R., Weinberg, B., Davilman, A., Yampierre, C., Protic, J., Weiss, D., Ackelsberg, J., Lee, L., Layton, M., Beatrice, S.T., Smith, P.F., Ettestad, P.J., Reynolds, P.J., Sewell, C.M., Enscore, R.E., Kosoy, M.Y., Kubota, K., Lowell, J.L., Chu, M., Kool, J., Gage, K.L., Chow, C.C. and Smelser, C.B. (2003). Imported

plague—New York City, 2002. *CDC Morbidity and Mortality Weekly Report* **52**, 725–728.
Perrin, L., Kaiser, L. and Yerly, S. (2003). Travel and the spread of HIV-1 genetic variants. *Lancet Infectious Diseases* **3**, 22–27.
Perrings, C., Dehnen-Schmutz, K., Touza, J. and Williamson, M. (2005). How to manage biological invasions under globalization. *Trends in Ecology and Evolution* **20**, 212–215.
Petersen, L.R. and Hayes, E.B. (2004). Westward Ho? The spread of West Nile virus. *New England Journal of Medicine* **351**, 2257–2259.
Peyton, E.L., Campbell, S.R., Candeletti, T.M., Romanowski, M. and Crans, W.J. (1999). *Aedes (Finlaya) japonicus japonicus* (Theobald), a new introduction into the United States. *Journal of the American Mosquito Control Association* **15**, 238–241.
Phukan, A.C., Borah, P.K., Biswas, D. and Mahanta, J. (2004). A cholera epidemic in a rural area of northeast India. *Transactions of the Royal Society of Tropical Medicine and Hygiene* **98**, 563–566.
Reidl, J. and Klose, K.E. (2002). *Vibrio cholerae* and cholera: out of the water and into the host. *Federation of European Microbiological Societies Microbiology Reviews* **26**, 125–139.
Reiter, P. (1998). *Aedes albopictus* and the world trade in used tires 1988–95: the shape of things to come? *Journal of the American Mosquito Control Association* **14**, 83–94.
Reiter, P. and Sprenger, D. (1987). The used tire trade: a mechanism for the worldwide dispersal of container breeding mosquitoes. *Journal of the American Mosquito Control Association* **3**, 494–501.
Rogers, D.J. and Robinson, T.P. (2004). Tsetse distribution. In: *The Trypanosomiases* (I. Maudlin, P.H. Holmes and M.A. Miles, eds). Wallingford: CABI Publishing.
Rogers, L. (1919). *Fevers in the Tropics*. Oxford: Oxford University Press.
Romi, R., Di Luca, M. and Majori, G. (1999). Current status of *Aedes albopictus* and *Aedes atropalpus* in Italy. *Journal of the American Mosquito Control Association* **15**, 425–427.
Roper, C., Pearce, R., Nair, S., Sharp, B., Nosten, F. and Anderson, T. (2004). Intercontinental spread of pyrimethamine-resistant malaria. *Science* **305**, 1124.
Russell, K.L., Carcamo, C., Watts, D.M., Sanchez, J., Gotuzzo, E., Euler, A., Blanco, J.C., Galeano, A., Alava, A., Mullins, J.I., Holmes, K.K. and Carr, J.K. (2000). Emerging genetic diversity of HIV-1 in South America. *AIDS* **14**, 1785–1791.
Russell, R.C. (1987). Survival of insects in the wheelbays of Boeing 747B aircraft on flights between tropical and temperate airports. *Bulletin of the World Health Organization* **65**, 659–662.
Russell, R.C. (1989). Transport of insects of public health importance on international aircraft. *Travel Medicine International* **7**, 26–31.

Rvachev, L.A. and Longini, I.M. (1985). A mathematical model for the global spread of influenza. *Mathematical Biosciences* **75**, 3–22.

Ryan, E.T., Wilson, M.E. and Kain, K.C. (2002). Current concepts—illness after international travel. *New England Journal of Medicine* **347**, 505–516.

Sack, D.A., Sack, R.B., Nair, G.B. and Siddique, A.K. (2004). Cholera. *Lancet* **363**, 223–233.

Salit, I.E., Sano, M., Boggild, A.K. and Kain, K.C. (2005). Travel patterns and risk behavious of HIV-positive people travelling internationally. *Canadian Medical Association Journal* **172**, 884–888.

Schaffner, F., Chouin, S. and Guilloteau, J. (2003). First record of *Ochlerotatus japonicus* in metropolitan France. *Journal of the American Mosquito Control Association* **19**, 1–5.

Scott, S. and Duncan, C. (2004). *Return of the Black Death: The World's Greatest Serial Killer*. Chichester: Wiley.

Seas, C., Miranda, J., Gil, A.I., Leon-Barua, R., Patz, J., Huq, A., Colwell, R.R. and Sack, R.B. (2000). New insights on the emergence of cholera in Latin America during 1991: the Peruvian experience. *American Journal of Tropical Medicine and Hygiene* **62**, 513–517.

Shu, P.-Y., Chien, L.-J., Chang, S.-F., Su, C.L., Kuo, Y.-C., Liao, T.-L., Ho, M.-S., Lin, T.-H. and Huang, J.-H. (2005). Fever screening at airports and imported dengue. *Emerging Infectious Diseases* **11**, 460–462.

Skowronski, D.A., Astell, C., Brunham, R.C., Low, D.E., Petric, M., Roper, R.L., Talbot, P.J., Tam, T. and Babiuk, L. (2005). Severe acute respiratory syndrome (SARS): a year in review. *Annual Review of Medicine* **56**, 357–381.

Song, H.D., Tu, C.C., Zhang, G.W., Wang, S.Y., Zheng, K., Lei, L.C., Chen, Q.X., Gao, Y.W., Zhou, H.Q., Xiang, H., Zheng, H.J., Chern, S.W.W., Cheng, F., Pan, C.M., Xuan, H., Chen, S.J., Luo, H.M., Zhou, D.H., Liu, Y.F., He, J.F., Qin, P.Z., Li, L.H., Ren, Y.Q., Liang, W.J., Yu, Y.D., Anderson, L., Wang, M., Xu, R.H., Wu, X.W., Zheng, H.Y., Chen, J.D., Liang, G.D., Gao, Y., Liao, M., Fang, L., Jiang, L.Y., Li, H., Chen, F., Di, B., He, L.J., Lin, J.Y., Tong, S.X., Kong, X.G., Du, L., Hao, P., Tang, H., Bernini, A., Yu,, X.J., Spiga, O., Guo, Z.M., Pan, H.Y., He, W.Z., Manuguerra, J.C., Fontanet, A., Danchin, A., Niccolai, N., Li, Y.X., Wu, C.I. and Zhao, G.P. (2005). Cross-host evolution of severe acute respiratory syndrome coronavirus in palm civet and human. *Proceedings of the National Academy of Sciences of the United States of America* **102**, 2430–2435.

Soper, F.L. and Wilson, D.B. (1943). *Anopheles gambiae in Brazil: 1930 to 1940*. New York: Rockefeller Foundation.

Spira, A.M. (2003). Assessment of travellers who return home ill. *Lancet* **361**, 1459–1469.

Sugihara, G., Bersier, L.-F., Southwood, T.R.E., Pimm, S.L. and May, R.M. (2003). Predicted correspondence between species abundances and

dendrograms of niche similarities. *Proceedings of the National Academy of Sciences* **100**, 5246–5251.
Tanser, F.C., Sharp, B. and le Sueur, D. (2003). Potential effect of climate change on malaria transmission in Africa. *Lancet* **362**, 1792–1798.
Tatem, A.J. and Hay, S.I. (2004). Measuring urbanization pattern and extent for malaria research: a review of remote sensing approaches. *Journal of Urban Health* **81**, 363–376.
Tatem, A.J., Hay, S.I. and Rogers, D.J. (2006a). Global traffic and disease vector dispersal. *Proceedings of the National Academy of Sciences of the United States of America* in press.
Tatem, A.J., Rogers, D.J. and Hay, S.I. (2006b). Out of Africa: estimating the risks of malaria movement by air travel. *Emerging Infectious Diseases* submitted.
Thomas, R. (1992). *Geomedical Systems: Intervention and Control*. London: Routledge.
Toovey, S. and Jamieson, A. (2003). Rolling back malaria: how well is Europe doing? *Travel Medicine and Infectious Diseases* **1**, 167–175.
Turell, M.J., O'Guinn, M.L., Dohm, D.J. and Jones, J.W. (2001). Vector competence of North American mosquitoes (Diptera: Culicidae) for West Nile virus. *Journal of Medical Entomology* **38**, 130–134.
Twigg, G. (1984). *The Black Death: A Biological Appraisal*. London: Batsford Academic.
Twigg, G. (2003). The black death: a problem of population-wide infection. *Local Population Studies* **71**, 40–52.
UNAIDS. (2004). *Report on the global AIDS epidemic*. Switzerland: UNAIDS Geneva.
Ungchusak, K., Auewarakul, P., Dowell, S.F., Kitphati, R., Auwanit, W., Puthavathana, P., Uiprasertkul, M., Boonnak, K., Pittayawonganon, C., Cox, N.J., Zaki, S.R., Thawatsupha, P., Chittaganpitch, M., Khontong, R., Simmerman, J.M. and Chunsutthiwat, S. (2005). Probable person-to-person transmission of avian influenza A (H5N1). *New England Journal of Medicine* **352**, 333–340.
Upham, P., Thomas, C., Gillingwater, D. and Raper, D. (2003). Environmental capacity and airport operations: current issues and future prospects. *Journal of Air Transport Management* **9**, 145–151.
Viboud, C., Boelle, P.-Y., Pakdaman, K., Carrat, F., Valleron, A.-J. and Flahault, A. (2004). Influenza epidemics in the United States, France and Australia, 1972–1997. *Emerging Infectious Diseases* **10**, 32–39.
Vila, M. and Pujadas, J. (2001). Land-use and socio-economic correlates of plant invasions in European and North African countries. *Biological Conservation* **100**, 397–401.
Vogel, G. (2003). SARS outbreak—modelers struggle to grasp epidemic's potential scope. *Science* **300**, 558–559.
Webb, A. (2003). *Statistical Pattern Recognition*. Chichester: Wiley.

Webby, R.J. and Webster, R.G. (2003). Are we ready for pandemic influenza? *Science* **302**, 1519–1522.
Webster, R.G. and Hulse, D. (2005). Controlling avian flu at the source. *Nature* **435**, 415–416.
Weir, E. and Haider, S. (2004). Cholera outbreaks continue. *Canadian Medical Association Journal* **170**, 1092–1093.
Wilson, M.E. (1995). Travel and the emergence of infectious diseases. *Emerging Infectious Diseases* **1**, 39–46.
Wilson, M.E. (2003). The traveller and emerging infections: sentinel, courier, transmitter. *Journal of Applied Microbiology* **94**, 1S–11S.
Wongsrichanalai, C., Pickard, A.L., Wernsdorfer, W.H. and Meshnick, S.R. (2002). Epidemiology of drug-resistant malaria. *Lancet Infectious Diseases* **2**, 209–218.
World Health Organization. (2003). Human plague in 2000 and 2001. *Weekly Epidemiological Record* **78**, 129–136.
Zachcial, M. and Heideloff, C. (2003). *ISL Shipping Statistics Yearbook 2003*. Bremen: Institute of Shipping Economics and Logistics.
Zinsser, H. (1943). *Rats, Lice and History*. London: Routledge.

Climate Change and Vector-Borne Diseases

D.J. Rogers[1] and S.E. Randolph[2]

[1]*TALA Research Group, Tinbergen Building, Department of Zoology, University of Oxford, South Parks Road, Oxford OX1 3PS, UK*
[2]*Oxford Tick Research Group, Tinbergen Building, Department of Zoology, University of Oxford, South Parks Road, Oxford OX1 3PS, UK*

Abstract 346
1. The Mathematics and Biology of Changes in Vector-Borne Diseases. 346
2. Defining the Criteria for Claiming Climate Impacts on Vector-Borne Diseases 351
3. Models for Climate Change Impacts on Vector-Borne Diseases. 353
4. Biological and Statistical Approaches to Vector-Borne Disease Futures 355
   4.1. Malaria: The Biological Approach 355
   4.2. Malaria: The Statistical Approach 357
   4.3. Malaria: Further Developments of Biological Models 358
   4.4. Tick-Borne Encephalitis (TBE) in Europe 363
5. Recent Changes in Vector-Borne Diseases: Has Climate Change Already had an Impact?. 366
   5.1. Increased Incidence of TBE: Coincidence or Causality of Climate Change?. 366
   5.2. Increased Incidence of Malaria in the East African Highlands 370
   5.3. Northern Spread of Bluetongue Virus into Europe. 374
6. Conclusions. 376
Acknowledgements 377
References 377

ADVANCES IN PARASITOLOGY VOL 62
ISSN: 0065-308X $35.00
DOI: 10.1016/S0065-308X(05)62010-6

ABSTRACT

In this review we examine formally the conditions under which vector-borne diseases are likely to change, and the directions of those changes, under various scenarios of climate change. We specify the criteria that must be met in order to conclude that climate change is having an effect on vector-borne diseases. We then take several examples from the literature and show how some of them meet these criteria, while others do not. For those that do not, there are alternative explanations that involve much more plausible drivers of the recorded changes in the diseases concerned.

1. THE MATHEMATICS AND BIOLOGY OF CHANGES IN VECTOR-BORNE DISEASES

The debate about the impact of climate change on vector-borne diseases is informed and stimulated by increases (or decreases) in the numbers of disease cases through time. It is therefore essential to adopt a quantitative, modelling approach to understand these changes, and this is most usefully based on the Basic Reproductive Number (R_0) equation for such diseases, defined as follows:

$$R_0 = \frac{mbca^2e^{-\mu T}}{\mu r} \tag{1}$$

where

m = the ratio of vector numbers (V) to host numbers (N),
b = the transmission coefficient from vertebrate to vector,
c = the transmission coefficient from vector to vertebrate,
a = the biting rate of the vectors on the hosts of interest,
μ = the mortality rate of the vectors,
T = the extrinsic incubation period of the infection in the vectors and
r = the rate of recovery of the host from infection.

The R_0 equation defines the number of new cases of the disease that arise at some time in the future from one case of the disease at the present time when introduced into a population of totally susceptible hosts. If the vertebrate hosts experience a significant natural mortality rate and/or a pronounced parasite-induced increase in mortality rate during the course of infections, this (combined) rate should be added to the recovery rate term in the denominator of Eq. (1), since it determines the rate at which the infections disappear from the host population.

It can be seen from Eq. (1) that there is a strong link between the R_0 equation and the Vectorial Capacity (VC) equation familiar to many medical entomologists. The latter generally assumes perfect transmission efficiency between vertebrate and vector and vector back to vertebrate (i.e. $b = c = 1$ in Eq. (1)) in which case the relationship between R_0 and VC is as follows:

$$\mathrm{VC} = R_0 r \tag{2}$$

Thus, the VC equation (as it was intended to do) contains all the information about the rate of transmission of the disease from the vector to the vertebrate population, but none of the information about the vertebrate's response to that infection (in the form of the recovery rate, r).

The R_0 equation is useful throughout epidemiology because many important quantities may be derived from it, such as the equilibrium disease prevalence or the level of vaccination required to suppress or even eradicate any disease. Clearly, if the value of R_0 lies below 1.0, the disease will decline, and eventually will be eradicated. If R_0 is above 1.0, the disease will increase from very low levels to reach an equilibrium prevalence that is determined *inter alia* by the strength of the host's immune response. We emphasize that this critical threshold of an R_0 value of 1.0 is an *absolute* figure—not a *relative* one. To define R_0 for any disease, therefore, we must have accurate estimates of each of the contributory parameters and variables. We should never drop parameters or variables out of the equation simply because we have no estimates of their values, as this effectively sets a default value of 1.0 for them.

It is clear from Eq. (1) that climate will be vital in determining the R_0 values of most or all vector-borne diseases for the simple reason that six of the seven variables and parameters in that equation are related to vector abundance, biology or physiology. Vectors are generally small arthropods that are exquisitely sensitive to weather conditions at many stages of their life cycles. Even those parameters that might be expected to remain constant (such as b and c in Eq. (1)) are known to be affected by temperature in one or more vector species. For example, *Culicoides nubeculosus* is a midge that is normally refractory to the strains of bluetongue virus circulating in Europe at the present time (i.e. for this species $b = 0$ in Eq. (1)). When its larvae or pupae are kept at higher than normal temperatures, however, the emerging adults of this species are capable of being infected with bluetongue, and can transmit the disease (Baylis *et al.*, 2001, 2002). Also, the availability to vectors of microbes as they circulate in the vertebrate host's peripheral blood may vary with temperature and the degree of vasodilation.

If climate change is to have any effect at all on vector-borne diseases, it must do so through the intermediaries of the parameters and variables of the R_0 equation. That is to say that unless climate change affects the value of one or more of the variables or parameters of Eq. (1), there will be no impact of climate change on vector-borne diseases. It is important to distinguish two concepts that are often confused in the contemporary literature. Few vector biologists would deny the importance of climate/weather in determining the natural distribution and intensity of vector-borne diseases, as is obvious from Eq. (1). The present debate concerns how much current *changes* in climate are affecting *changes* in vector-borne diseases. Denial (or questioning) of the second idea does not imply denial (or questioning) of the first.

The question now arises as to the direction of change in R_0 that might arise from climate-induced changes in any of the contributing parameters or variables of the R_0 equation. Should we expect R_0 always to increase with global warming, or to decrease (imagining, of course, that the prevalence of the disease will increase or decrease as a consequence)? One thing is certain; the direction of change will be determined by the outcome of interactions between effects that tend to push in opposite directions. For example, it is likely that increasing temperature will lead to an increase in the vector-biting rate (a), tending to increase R_0, but also to an increase in the mortality rate (μ), tending to decrease R_0. How can we quantify these effects formally?

Let us imagine that climate, for simplicity in the form only of temperature, affects only four parameters and variables of Eq. (1); m, a, μ and T (these quantities will certainly be affected by climate, and the argument presented here can easily be extended to the others). The effect of temperature on R_0 is then the sum of the effects of temperature on each component of R_0, as follows:

$$\frac{dR_0}{dt} = \frac{dm}{dt}\frac{dR_0}{dm} + \frac{da}{dt}\frac{dR_0}{da} + \frac{d\mu}{dt}\frac{dR_0}{d\mu} + \frac{dT}{dt}\frac{dR_0}{dT} \quad (3)$$

where t is temperature and all the other variables/parameters are as defined in Eq. (1). Each term on the right-hand side of Eq. (3) contains two parts, one for the effect of temperature on the variable/parameter of interest, and one for the effect of varying this variable/parameter on R_0. Effectively, therefore, the first part is biology and the second part is mathematics. For example, the first part of the first term (dm/dt) describes how vector/host ratios change with temperature, while the second part (dR_0/dm) describes how R_0 changes in response to this changing ratio. Figure 1 (Figure 1 is Plate 10.1 in the Separate Color Plate Section) shows sketches of the likely effect of temperature on each of the biological parts of Eq. (3). The figure shows that there is the greatest uncertainty about the direction of change in vector/host ratio with temperature increases. It is not clear whether this ratio will increase or decrease with increasing global temperatures (in fact, it is likely to increase in some areas and decrease in others). In all other cases, it is possible to guess the likely direction of the effect (positive or negative), but not its precise shape or level. Figure 1 indicates with a positive or negative sign the direction of the change shown in each panel without, for the moment, quantifying such changes any further.

The impacts of these biological changes on the R_0 equation are defined mathematically as follows:

$$\frac{dR_0}{dm} = \frac{bca^2e^{-\mu T}}{\mu r} = \frac{R_0}{m}$$

$$\frac{dR_0}{da} = \frac{mbc2ae^{-\mu T}}{\mu r} = \frac{2R_0}{a} \quad (4)$$

$$\frac{\mathrm{d}R_0}{\mathrm{d}\mu} = -R_0\left[\frac{1+\mu T}{r}\right]$$

$$\frac{\mathrm{d}R_0}{\mathrm{d}T} = -\mu R_0$$

It should be noted again that some of these effects are positive, while others are negative. For example, R_0 increases with increase in the vector/host ratio (the first result in Eq. (4)), but decreases with increases in the vector mortality rate (the last result in Eq. (4)).

The biological effects in Figure 1 and the mathematical effects in Eq. (4) are brought together in Table 1. The biological effects are shown in the second column of the table and the mathematical effects in the third column. The fourth column, titled 'Net effect', is simply the product of the second and third columns, as required by Eq. (3), and these net effects are summed in the very last cell of the last column, again as required by Eq. (3). We conclude that the net effect of global warming on vector-borne diseases may be positive or negative, depending on the relative contributions of variations in each of the contributory variables/parameters. Table 1 shows that even if we were certain of the impact of global warming on the vector/host ratio (the second row in Table 1), the final summed term might still be

Table 1 Effects of temperature on the biological ingredients of the R_0 equation

	Biology	Mathematics	Net effect
m, Vectors per host	+ ? −	+	+/−?
a, Biting rate	+	+	+
μ, Mortality rate	+	−	−
T, Incubation period	−	−	+
Net effect of global warming = Σ =			+ or −?

Note: The effects of temperature on the biological ingredients of the R_0 equation are shown in the second column (from the panels in Figure 1). The effects of these changes on the contribution of each ingredient to R_0 are shown in the third column (from text Eq. (4)), leading to a prediction of the net effect of each ingredient in the fourth column (the product of columns 2 and 3). The overall effect of changes in all ingredients is the sum of the individual effects, and is shown in the last cell of the last column.

negative or positive, depending on the relative strengths of the effects on biting and mortality rates in the third and fourth rows.

While all of the above are fairly obvious with a little careful thought applied to Eq. (1), this formal treatment shows just how far we have to go before we can predict the *quantitative* impact of climate change on vector-borne diseases. It is not good enough to imagine that climate change affects only one climatic variable (temperature) and always has the same effect (producing an increase) on all vector-borne diseases. As the examples below show, disentangling cause and effect in changing vector-borne disease incidence and prevalence is an art requiring a mixture of sound biology and robust mathematics.

2. DEFINING THE CRITERIA FOR CLAIMING CLIMATE IMPACTS ON VECTOR-BORNE DISEASES

In this section we move from the theoretical treatment of the previous section towards the application of theory to real-world practice. To conclude that there has been an impact of climate change on any disease we need to quantify precisely how varying climate affects each and every parameter and variable of Eq. (1). We should be able to show a 'significant' change in one or more variables in response to a measured change in climate. In practice this has not yet been done. Instead, we have meteorological records that show changing weather patterns recorded at a series of weather stations, and records of changing numbers of disease cases diagnosed in local clinics and hospitals.

Several cautions must be applied in interpreting these datasets. First, the network of stations that have recorded climatic conditions globally has not been constant. In general, there was an increase in the number of stations over the course of the last century. Notable regional decreases did occur in the last quarter of the century, such as in eastern Europe, as an effect of political changes occurring at the time. Second, meteorological stations historically tended to be situated in or near towns and major airports. Expansion of the cities locally, and the associated 'heat island' effects could have caused an increase in recorded instrumental temperatures that apply only to the urban areas, not to the rural ones where much vector-borne disease transmission occurs. Third, the spatial coverage of meteorological stations was variable, adequate in most of the USA and Europe, but patchy elsewhere, especially in the tropics. While the interpolation techniques applied to these meteorological station (i.e. point) data were shown to be robust by the original authors, debates have since centred upon the confidence that can be placed in the gridded, 0.5° resolution climate surfaces produced from the original point data. In some cases, (especially the tropics) the gridded data are aggregated into regional estimates of climate variables before further analysis, while in others, conclusions are drawn on analyses carried out at the individual grid cell (i.e. 0.5°) level.

Disease data from hospitals and clinics are also prone to a variety of hidden errors. Improved diagnostic methods can produce an apparent increase in numbers of cases without any increase in transmission. In many parts of the tropics, where health service provision has not kept pace with human population increases, increasing numbers of cases might simply reflect an increasing population rather than an increase in disease prevalence. There are rarely any denominator data to correct for these effects (for example, total population within the catchment area of each health centre).

With these caveats in mind we seek here to define how we might operationally link climate changes to increases or decreases in vector-borne diseases through time, to say beyond reasonable doubt that climate has 'caused' the observed changes in vector-borne diseases. We thus seek to go beyond the mindless correlation of one dataset with another that is all too prevalent in the climate change literature.

We suggest that climate change might reasonably be judged to be the cause of changing vector-borne disease incidence, prevalence or distribution if three conditions are met simultaneously. The observed climate change should have occurred at the right time, in the right place and in the 'right' direction (according to our current understanding of climate/disease linkages) for it to be the likely cause of the observed changes in disease over time (Randolph, 2004). These are logical criteria that exclude many of the current examples of climate-induced changes in vector-borne diseases, which generally exhibit only one or two of the required characteristics, not all three. Some of these are discussed below.

3. MODELS FOR CLIMATE CHANGE IMPACTS ON VECTOR-BORNE DISEASES

In the absence of any direct evidence for climate impacts on disease, considerable modelling effort has recently been directed towards predicting the potential impacts of climate change, and this modelling has been of one of two major types. The first sort of model is biologically based and seeks to describe some aspect of the process of transmission, and how this will be affected by climate change; these are referred to here as mechanistic, biological or 'process-based' models. The second sort of model attempts to match the current distribution of the disease in question to current climate variables in a statistical framework. Within this framework the impact of future climate changes is explored by interpolation and extrapolation of the results obtained; these are referred to here as statistical or 'pattern matching' models.

Because, for example, the distribution of a disease or its vector is predicted from known relationships of biological to climatic variables, biological or process-based models do not use the recorded distribution of the disease or vector in the model-building process itself. In theory, the recorded distribution could therefore be used to measure model accuracy. In practice, however, this is rarely done for the reason (given above) that the observed distribution has often been affected by human intervention of one sort or another and is therefore an inaccurate measure of the (unknown or unrecorded) original geographic extent of the disease (which is what the biological model claims to provide). This situation is of course less than satisfactory, since a model that cannot be tested against an existing dataset is irrefutable, and therefore effectively worthless.

By the same token that biological models cannot use observed, present-day distributions to test their accuracy, it is sometimes stated that statistical models should also not use present-day distributions as a basis for their pattern-matching approach. A distribution that has been affected by intervention cannot be used to define the full 'environmental envelope' of the disease, which is assumed to be the goal of the statistical exercise. In defence of this approach, however, we believe we can only start with whatever records we have for any disease. Human intervention is likely to make inroads into the distribution of vector-borne diseases only in those places where the diseases are near the natural limits of their distribution in the first place. This suggestion is supported by a recent analysis showing that the impact of human intervention on areas of holo-endemic malaria has been negligible, while the impact on seasonally malarial areas in temperate regions has been considerable (Hay *et al.*, 2004). There is indeed a link between the natural ecology of a disease system and the extent to which its distribution has been diminished by human intervention: the surviving core range indicates where environmental conditions are particularly favourable for transmission. We thus question the rejection of the use of current maps of disease distribution, on the grounds of their unnaturalness, for both biological and statistical approaches to modelling vector-borne disease distributions. It seems strange to reject the only piece of information we have for many diseases (i.e. a map) in our attempts to understand the distribution of the same diseases!

Unquestionably, the biological or process-based modelling route is to be preferred for two major reasons. First, only process-based models can give us real insight into the quantitative relationship between biological and meteorological variables. Second, only such models can be used to explore quantitatively the effects of interventions of one sort or another (e.g. environmental sanitation, insecticide applications, drug prophylaxis or vaccine administration) on future levels of disease. Only through use of such models can the cost-effectiveness of various alternative strategies for disease control be investigated. There is, however, a serious downside to such models; they must be complete or else they will mislead, or fail altogether.

For all their biological inadequacies, statistical models hold out more promise in situations where biological knowledge is incomplete. It is generally always possible to build a statistical model of one sort or another for any disease situation. For example, logistic regression and discriminant analytical models have been used to describe the distributions of vectors and the diseases they transmit, and can achieve impressive (statistical) accuracy. Care should be taken when building such models, however, and great attention should be paid to investigating the contribution to the obtained fit of each of the included variables, as explained in detail in Rogers (this volume, pp. 1–35). Statistical packages can be applied more or less mindlessly to describe distributions; but they should not be applied thoughtlessly.

4. BIOLOGICAL AND STATISTICAL APPROACHES TO VECTOR-BORNE DISEASE FUTURES

To illustrate many of the above points we examine features of the current debate of the impact of global change on malaria, malaria in Africa and tick-borne encephalitis in Europe.

4.1. Malaria: The Biological Approach

In a series of papers Martens and colleagues have applied a version of Eq. (1) for predicting the impact of climate change on malaria (Martens, 1997, 1998; Martens *et al.*, 1999).

Equation (1) is first re-arranged at the value of R_0 equal to 1.0 to determine the vector/host ratio (m) at this critical threshold for disease persistence:

$$m = \frac{\mu r}{bca^2 e^{-\mu T}} \tag{5}$$

A high value of m means that more vectors per host are required for the disease to persist, and therefore implies inefficient transmission compared with the situation when m is low. Equation (5) can be re-expressed as follows:

$$\frac{1}{m} = \frac{bca^2 e^{-\mu T}}{\mu r} \tag{6}$$

The left-hand side of this equation was originally called the Epidemic Potential (EP), but is now referred to as the Transmission Potential (TP) and is taken as a direct measure of the environmental favourability of conditions for disease persistence; the higher the TP (i.e. the lower the m), the more favourable are local conditions for disease persistence. In applying this equation to climate change scenarios (that affect some of the parameters and variables on the right-hand side of Eq. (6), as indicated schematically in Figure 1), Martens and colleagues effectively omitted parameters *r*, *b* and *c* by setting them equal to 1.0 because their precise values were unknown and they were assumed to be unaffected by climate change. In so doing, of course, an original equation that requires absolute measures of all parameters and variables (i.e. the R_0 equation) has been reduced to one that gives an unquantified relative measure of the same critical number. The authors felt they had overcome this problem by calculating the ratio of TP after climate change to TP before climate change, to give a measure of relative change in TP over time. Constant variables that appear both on the top and bottom line of this ratio cancel each other out and thus disappear from the final result. However, ratios of this sort can be very misleading. Consider two places on the globe. In one, the absolute value of R_0 increases from 5 (before climate change) to 10 (after climate change), while in the other, R_0 increases from 0.1 to 0.2. In both cases the critical ratio of R_0 after change to R_0 before change (or some linear function of these quantities) is 2. Malaria was present in the first place, and absent in the second, both before and after climate change; the change itself did not bring about any change in the status (i.e. distribution) of malaria at the two points, although in a third place (with absolute R_0 values of 0.75 before and 1.5 after climate change) it might do so, again with the same critical ratio of 2. The point to stress, therefore, is that the ratio of the paired values of R_0 (or some measure of this quantity) is a totally inadequate guide to the changing status of malaria under climate change scenarios, although maps of the $TP_{\text{after climate change}}/TP_{\text{before climate change}}$ ratio, showing high values in temperate regions are regularly interpreted (and explicitly described, e.g. Epstein, 2000) as predicting malaria spread into temperate regions under conditions of global warming.

Predictions of dire malaria futures in the face of climate changes cannot, by definition, be tested at the present time, but their repetition leads to a view that they are both consensual and correct (McCarthy *et al.*, 2001). "*One can be certain, but one can be wrong*" (John Kerry, during the US presidential campaign in 2004) applies not just to American politicians.

More recent assessments of future impacts using Martens' approach and HadCM3 scenarios predict less dramatic impacts of warming on malaria globally, but continue to use an incompletely parameterized TP model, in this case to identify areas where the climate is suitable for malaria transmission for at least three months of the year (van Lieshout *et al.*, 2004). The paper does not reveal how TP estimates are used to determine whether or not any particular month is suitable for malaria transmission.

4.2. Malaria: The Statistical Approach

With a view to overcoming these methodological problems in predicting malaria futures, Rogers and Randolph (2000) applied the statistical or pattern-matching approach to the same problem, beginning with the recorded distribution map of global malaria (WHO, 1997). A geo-referenced version of this map was sampled by taking 1500 points at random within the recorded distribution and 1500 other points at random outside the distribution, but within 10° of any recorded presence site. This limits absence sites to being within a reasonable geographical (and hence climatic) distance of presence sites. Processed climatic data from the interpolated meteorological station records held by the Climatic Research Unit CRU in the University of East Anglia, UK, were extracted for each presence and absence point. These data, at 0.5° grid resolution, form the global climate norms for the period 1961 to 1990 used by the Intergovernmental Panel on Climate Change (IPCC) [http://ipcc-ddc.cru.uea.ac.uk] and are hereafter referred to as the 'CRU global climate norms dataset'. The mean, maximum and minimum values of the mean (TM), maximum (TX) and minimum (TN) monthly temperatures, rainfall (PR) and saturation vapour pressure (SVP) variables were available in the climate dataset, and variables were selected in a step-wise inclusive fashion to maximize the Mahalanobis distance (a multi-variate measure of distance, adjusting for co-variance of the variables) between any pairs of dissimilar (absence to presence) clusters. Only five variables (minimum TM, minimum *R*, minimum SVP, mean SVP and mean TX) were required to give a satisfactory fit, with 78% of points accurately identified, 14% false positives and 8% false negatives (Figure 2a). (Figures 2a–c are Plates 10.2a–c in the separate Color Plate

Section). The fit of the statistical model to the global malaria was significantly better than the fit of two alternative, biological models to the same map, although this is not particularly surprising because the map was used in the construction of the statistical model, but not the biological one. We repeat, however, that the biological model has not been tested for its accuracy against any independent datasets. The model is assumed to be correct and therefore, it is further assumed, the map resulting from it must also be a correct picture of where malaria was distributed before human intervention.

The statistically based model of global malaria distribution under present-day climate was then re-run with various climate scenarios for the future (Rogers and Randolph, 2000). Even under a relatively extreme scenario of climate change (the HadCM2 'High' scenario: http://www.met-office.gov.uk/sec5/CR_dic/Brochure97/), there was remarkably little change in the predicted global distribution of malaria in the future, compared with the present day (Figures 2b and c). Unsurprisingly, areas predicted to be the most affected are those near the current edges of malaria's global distribution; the southern United States, Turkey, Turkmenistan and Uzbekistan, Brazil and China. They also include some highland areas, for example, in East Africa, where malaria is predicted to appear for the first time, and some presently marginal areas that become too dry in the future and from which malaria is predicted to disappear (e.g. the eastern seaboard of India). In global terms, malaria is predicted to appear for the first time in areas in which about 360 million people live at the present time, and to disappear from areas where about 330 million people live at present (Figure 2c). The net difference (30 million people) is almost certainly not significant given the uncertainties in the modelling. Zero net differences, however, hide the fact that almost 700 million people will be affected, one way or another, by even the modest changes predicted by the statistical model.

4.3. Malaria: Further Developments of Biological Models

Using a different biological model for malaria in Africa, Thomas *et al.* came to conclusions rather more like those of the statistical model just described (Thomas *et al.*, 2004). These authors used the model produced by the Mapping Malaria Risk in Africa (MARA/ARMA) Collaboration [http://www.mara.org.za], which is based on a series of fuzzy class membership rules that each generate a value from zero (conditions unsuitable for malaria transmission) to 1.0 (conditions suitable for transmission) using monthly precipitation, monthly mean temperature and annual frost data from the CRU global climate norms dataset.

As Thomas *et al.* point out, the MARA model does not incorporate any multi-dimensional statistical relationship between the climate variables in generating the fuzzy logic values. In other words, each climate variable is used independently to calculate a fuzzy value, and the minimum fuzzy value (across all variables) for any month determines whether or not malaria transmission is possible in that month. This minimum value is referred to as a Malaria Transmission Climate Suitability Index (MTCSI). The MTCSI is obviously on a continuous scale and the operator needs to decide a threshold value on this scale that distinguishes suitable from unsuitable conditions for the period in question. Four consecutive months of suitability (MTCSI $\geqslant 0.9$) were deemed necessary by Thomas *et al.* for malaria persistence in any area, and it is generally accepted that these conditions define areas of stable malaria transmission and not those with unstable transmission associated with epidemics. Neither the original MARA/ARMA model nor Thomas *et al.*'s version of it has been tested statistically against any map of malaria for the continent, although it has "*performed well against observed data*" (Thomas *et al.*, 2004).

Under future climates described by the HadCM2 medium-high scenario [http://ipcc-ddc.cru.uea.uk] Thomas *et al.*'s version of the MARA/ARMA model predicted reduced climatic suitability for malaria transmission in areas of central/southern Africa centred on Zambia/Zimbabwe/western Mozambique for 2025, 2055 and 2085, the area becoming more extensive during this time, and in parts of the dry savannah regions of West Africa centred on southern Mali for 2055 and 2085. The model predicted increases in climatic suitability for transmission in many parts of the Horn of Africa, especially Ethiopia, Kenya and Somalia and in southern Africa (Botswana and

South Africa) with smaller patches elsewhere between these two regions, and in Angola. For the 2025 prediction of malaria status in Africa (Thomas *et al.*, 2004) the MARA/ARMA model was run specifying a minimum of five consecutive months with minimum fuzzy logic values greater than thresholds of 0.5 or 0.9 (i.e. two models) to classify an area as suitable for stable malaria; this was done in an attempt to "*test the model predictions regarding the highlands*" (Thomas *et al.*, 2004).

Despite the very different approaches of the biological model of Thomas *et al.* and the statistical model of Rogers and Randolph, and the different scenarios of climate change chosen by these authors (HadCM2 'medium-high' and 'high', respectively), both models predict that similar, scattered areas in East Africa will become climatically suitable for malaria transmission by 2025 or 2050, respectively. On the other hand, the biological model predicts a much more extensive disappearance of malaria from central/southern Africa than is predicted by the statistical model (compare Figures 2e and f with Figure 1c in the respective publications). Nevertheless, overall, there seems to be a much greater chance of reconciling the biological and statistical approaches to modelling malaria futures in Africa than was the case earlier. We stress again the importance of considering carefully what is being modelled in such exercises. A change in malaria suitability is itself not a good measure of whether or not malaria will newly appear in an area, or disappear from part of its historical distribution. Suitability can increase in an area already suitable for the disease, or decrease in an already unsuitable area, with no change in the malaria status (presence or absence) in either area. The change must be categorical, from predicted absence to predicted presence, or *vice versa*, for the distribution map of malaria to change in the future. Of course, within the distributional limits of the disease, its severity might also change, for example, if the duration of seasonal transmission is prolonged.

A different approach again to predicting malaria futures in Africa was taken by Tanser *et al.*, who developed a threshold-based model predicting whether or not any particular month is suitable for *Plasmodium falciparum* malaria transmission in Africa (Tanser *et al.*, 2003). The model predicted monthly malaria suitability in areas with centred

3-month moving average mean temperatures of $\geqslant$19.5 °C plus the local yearly standard deviation of mean monthly temperature, minimum yearly temperatures of $\geqslant$5 °C and retrospective 3 monthly average rainfall of $\geqslant$60 mm, with at least one of the 3-monthly averages during the year of $\geqslant$80 mm (that acts as a 'catalyst', thought by the authors to be essential for malaria persistence). Single months predicted unsuitable for transmission but bounded on either side by suitable months were automatically assigned transmission status on the assumption that the parasite reservoir would persist through the short period of climatic unsuitability. The various threshold values were determined by inspection of climatic and malaria data from 15 sites across Africa (encompassing the full range from holoendemic malaria to malaria-free conditions), further informed by the literature on malaria development in both vectors and vertebrate hosts, and finally refined with area-specific expert knowledge and historical published and unpublished maps. Historical climatic data for Africa, derived from records taken between 1920 and 1980 and spatially interpolated to 0.05° resolution were used in model development (Hutchinson *et al.*, 1995). All of the threshold criteria had to be met for an area to be classed as malarial in any particular month and the model was run to predict present and future durations of the malarial seasons in Africa (Tanser *et al.*, 2003). Here, therefore, both the geographic extent and the seasonal duration of malaria could be examined. It is possible that there may be more malaria in the future in Africa only in those regions already affected by the disease; this increase would not be detected in the other statistical or biological models described in this section. Tanser *et al.* used their model to predict the current malaria status (i.e. the duration of the transmission season) throughout the continent and validated these predictions against a database of 3791 (3199 positive and 592 negative) parasite ratio surveys carried out between 1929 and 1994. These surveys were completed within a single month; the model was deemed accurate if it predicted, plus or minus one month, malaria transmission for each survey month. Based on these criteria, the model's spatial sensitivity (the ability of the model to predict areas of transmission to within a month) was 63% (95% CI 61–65%), temporal sensitivity (the ability to predict malaria occurrence in any month) was 90% (95% CI

89–91%) and specificity within 1-month temporal accuracy was 96% (95% CI 91–100%). It predicted a current total of 3.1 billion person-months of exposure to Africa every year, among a total exposed population of 445 million people. Using a variety of HadCM3 future climate scenarios (Johns *et al.*, 2003), Tanser *et al.* predict that even by 2100, there is surprisingly little change in the latitudinal extent of malaria in Africa (an exception being the Limpopo Province of South Africa which becomes prone to occasional epidemics), but a 5–7% (mainly altitudinal) increase within this area and an overall continental increase of 16–28% in person-months of exposure across all three scenarios investigated. Much of this increase, therefore, occurs because of an increase in the duration of the transmission season within areas that are already malarial; changes are as much due to changing (both decreasing and increasing) rainfall as they are to increasing temperature (Tanser *et al.*, 2003). Predicted increases by the year 2100 in malaria in the highlands of East Africa (Ethiopia and Kenya), into parts of South Africa and in Angola parallel the predicted changes in climate suitability of Thomas *et al.* (and to some extent the statistical model of Rogers and Randolph), as do predicted decreases in the length of the transmission season in West Africa (central Mali), but there is no equivalent predicted decrease or disappearance of malaria in Zambia/Zimbabwe/Mozambique (Thomas *et al.*, 2004); instead this region is predicted to have more extensive areas of transmission suitability of 4–6 months' duration (Tanser *et al.*, 2003).

While the Tanser *et al.* model appears to be the only biological model that has been independently validated, criticism has been directed at this study because of the small number of reference sites used to derive the thresholds, the ways in which model accuracy was gauged, the inappropriateness of parasitological surveys to estimate current transmission of malaria (parasites are not cleared from bloodstreams the instant transmission ceases each year) and finally the use of 'person-months' exposure as a measure of transmission (an increase from a low base level is likely to create a greater increase in clinical malaria than is an equivalent increase from a high base level in areas where population immunity is already high) (Reiter *et al.*, 2004). The commentary accompanying the Tanser *et al.* paper (Hales and Woodward, 2003) is also criticized because of its assumptions about stability and seasonality of malaria transmission (Reiter *et al.*, 2004).

Of course, we have no proof that the predictions of malaria futures by either the statistical or biological models are correct. What these exercises show, with alarming clarity, is the great ignorance that we still have of one of the major killer diseases of children in the developing world. It seems somewhat misguided to argue the future of a disease that we are so significantly failing to control at the present time (Guerra *et al.*, this volume, pp. 157–179).

4.4. Tick-Borne Encephalitis (TBE) in Europe

It is not only the tropics that are affected by severe vector-borne diseases. Tick-borne encephalitis (TBE), caused by two subtypes of flavivirus (TBEV) transmitted by the ticks *Ixodes ricinus* and *I. persulcatus*, is the most significant vector-borne disease in Europe and Eurasia. Central nervous system pathology causes a case morbidity rate of 10–30%, and a case mortality rate typically of 1–2%, but as high as 24% in the Far East (Immuno, 1997). TBE is a typical zoonosis, with enzootic cycles maintained in natural rodent-tick cycles (Labuda and Randolph, 1999); humans may be infected if accidentally bitten by an infected tick, or, less commonly, by drinking untreated milk from infected sheep or goats (which cannot themselves pass the virus to ticks (Labuda *et al.*, 1997)). Increased risk to humans may therefore arise in three ways: improved conditions for natural transmission cycles resulting in higher densities of infected ticks; changed human behaviour resulting in greater exposure to ticks and increased consumption of raw milk. As both of the first two factors are climate dependent, TBEV is commonly included in the list of vector-borne pathogens anticipated to become more of a threat to humans in a predicted warmer world (Lindgren, 1998; Martens *et al.*, 1999).

In Europe, TBEV is highly focal in its distribution, limited to a well-defined region in central Europe, where the distribution is very patchy, and a quasi-separate region covering the Baltic States and the south-eastern rim of Scandinavia, where the distribution is more continuous. This pattern has been predicted with 85% accuracy from satellite-derived data on environmental conditions (Randolph, 2000), and its biological basis is now clearly understood. TBEV occurs in only a narrow subset of suitable tick habitats because its cycles are maintained by transmission of non-systemic infections between infected nymphs and uninfected larval ticks co-feeding on small rodent hosts, principally mice of the genus *Apodemus* (Labuda *et al.*, 1993). The force of transmission is sufficiently high only when there is a high degree of coincident feeding of larval and nymphal ticks on individual hosts (Randolph *et al.*, 1999), which is only possible when the seasonal activity periods of these two tick stages occur in synchrony (Randolph *et al.*, 2000). This is determined by rather precise climatic conditions, specifically a particular seasonal land-surface temperature profile. A pattern of higher than average rate of cooling in the autumn relative to midsummer peak temperatures is statistically significantly associated with larval–nymphal synchrony and with the presence of TBE (Randolph, 2000; Randolph *et al.*, 2000), although the precise biological basis for this phenomenon is not yet fully understood. At the same time, moisture availability to ticks on the ground, as measured by a remotely sensed vegetation index, must be sufficient to ensure good tick survival.

Like many vector-borne pathogen cycles that depend on the interaction of so many biotic agents with each other and with their abiotic environment, enzootic cycles of TBEV have an inherent fragility. Their continuing survival or expansion cannot be predicted from simple univariate correlations, for example, with temperature. As the biological models needed to capture the complexity of such systems are not yet in place, we applied the same two-step statistical approach as described for malaria above to predict the effects of forecast climate change (Randolph and Rogers, 2000). TBE distribution was matched first to present climatic conditions and then to future climate scenarios, using the same sources of information on present and future climate as for the malaria study (Rogers and Randolph, 2000).

The extent, although not the focality, of the present distribution was predicted with 86% accuracy by five climatic variables, minimum TN, minimum TX, mean TX, maximum TM and maximum SVP from the CRU global climate norms dataset, selected by discriminant analysis as the most important variables. These variables are all consistent with those selected from satellite imagery for predictive mapping of TBEV

(Randolph, 2000). Climate surfaces, however, are very much cruder than satellite imagery (offering a narrower range of climatic variables and much coarser spatial resolution), and they falsely predict TBEV presence through southern Poland and in southwestern Sweden, although new foci have indeed been identified in this latter region since 2000.

Future rises in temperature and decreases in moisture in the summer, as predicted under the HadCM2 medium-high scenario, appear set to drive the distribution of TBEV into higher latitude and higher altitude regions progressively through the 2020s, 2050s and 2080s. The Alps, however, are always too high to become a region of risk. In the 2020s, France, Switzerland, Slovenia, Hungary and much of Austria may be cleared of TBEV and the range of this virus (though not necessarily its vector) may contract to inland regions of the Baltic States. By the 2050s, TBEV may move into areas at present free of infection, notably the mountains on the Slovak/Polish border and further northwest in Scandinavia, but central Europe may be virtually cleared of TBEV. The final toe-hold in the 2080s may be confined to a small part of Scandinavia, including new foci in southern Finland.

Although the predictions can only be as good as the climate scenarios upon which they are based, this analysis disproves the common perception that a warmer world will necessarily be a world under greater threat from vector-borne diseases. In the case of TBEV that change appears to be to our advantage. Despite the untestability of these predictions, they are at least consistent with our understanding of the inherent fragility of this particular system: climate change seems likely to disrupt the fine balance between precise seasonal temperature profiles, summer moisture conditions and the tick's seasonal population dynamics. Furthermore, the prediction that the distribution of TBEV may expand north and west of Stockholm is consistent with the conclusion that increased temperatures have already extended the northern and western limits of *I. ricinus* in Sweden (Talleklint and Jaenson, 1998; Lindgren *et al.*, 2000). There have also been recent observations of ticks and cases of TBE above the previously defined altitudinal limit of 700 m in the Czech Republic (Daniel *et al.*, 2003).

Yet, in apparent contradiction to our predictions that the overall distribution of TBE will contract at some time between 1990 and the

2020s, TBE has in fact increased in incidence by an order of magnitude in some parts of Europe over the past one or two decades. It is of course perfectly possible for incidence in persisting foci within a contracted range to increase. There are also many reasons to believe that this increase has been caused by factors other than climate, as the next section discusses.

5. RECENT CHANGES IN VECTOR-BORNE DISEASES: HAS CLIMATE CHANGE ALREADY HAD AN IMPACT?

Predictions about the future of vector-borne diseases are inherently untestable but, given that there are already significant signs of anthropogenic effects on global climates, we may reasonably ask if there is any evidence to date that vector-borne diseases have changed their incidence or prevalence in the last few decades and according to the three criteria specified in Section 2. In the following examples, we shall see signs of significant changes in vector-borne diseases, but these do not always meet the specified requirements because one or more of the criteria are not met. Instead there are often other, non-climatic reasons for the changes observed.

5.1. Increased Incidence of TBE: Coincidence or Causality of Climate Change?

The changing incidence of TBE in Europe offers an excellent case study to distinguish between coincidence and causality of climate change by testing for our strict criteria (the right sorts of climate change, at the right time and in the right places), because the pattern of change has been temporally and spatially highly heterogeneous. At the northern extreme, in Sweden, the annual case numbers of TBE showed a 3-fold step increase between 1983 and 1986, followed by a stable higher incidence from 1986 and then a further doubling from 2000 (Figure 3). Along the southern boundary of Europe, in Slovenia, Croatia and Hungary, recorded case numbers of TBE changed very little from the mid-1970s, until there was a 60–70% decrease in Croatia and Hungary from 1997 (Figure 3). In eastern Europe, TBE cases

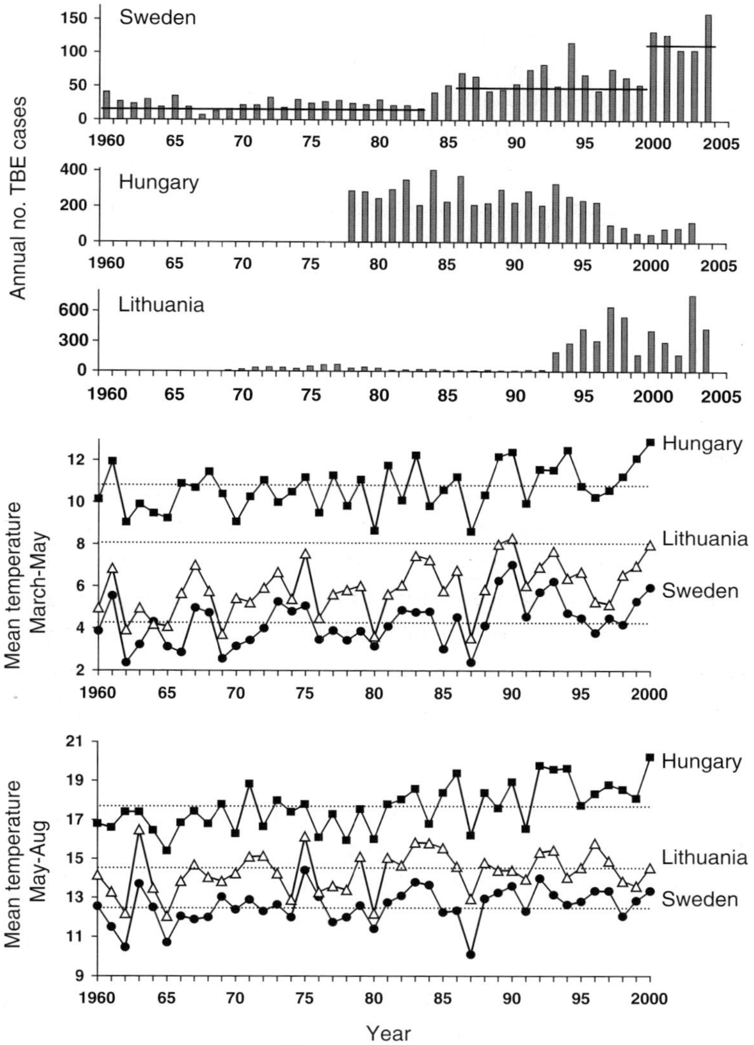

Figure 3 Top: changes in the annual numbers of cases of tick-borne encephalitis in Sweden, Hungary and Lithuania, 1960–2004. The step increases in Sweden from 1983 to 1986 and again in 2000 are highlighted by horizontal lines showing mean levels in each period. Bottom: changes in mean spring (upper) and summer (lower) temperatures (taken from the interpolated climate surfaces prepared by the CRU, University of East Anglia) for 0.5° grid squares centred on Zala county Hungary, Siauliu in Lithuania and Stockholm in Sweden. Dotted horizontal lines show the 1960–2000 mean levels for each site.

increased dramatically, but not until 1992 or 1993: the Czech Republic and Slovakia suffered 2-fold increases, Estonia a 5-fold increase, and Poland, Lithuania and Latvia increases of an order of magnitude each (Figure 3). In Latvia, however, this increase has been largely reversed since 1999, with some evidence that this is due to vaccination (Sumilo *et al.*, 2006). Furthermore, the degree of TBE increase varied markedly in different regions within each country (Zabicka, 1994; Danielová and Benes, 1997; Vaisviliene *et al.*, 2002; Sumilo *et al.*, 2006).

Based on our understanding of the biology and ecology of the TBE system, we can identify the following factors as most likely to be the 'right sort' of climate change to account for these epidemiological patterns: warmer temperatures during the spring may cause earlier onset of both tick host-seeking activity and human work or leisure activity in tick-infested forests; warmer springs and summers may accelerate tick development rates and therefore the virus transmission cycle; but drier summers may cause higher tick mortality. Of the three possible sources of information on climate change (satellite imagery, ground records from meteorological stations and climate surfaces constructed by interpolation between ground stations), each has its own strengths and weaknesses. Following the IPCC, here we use the latter, available as mean monthly records for the last century at a spatial resolution of 0.5° longitude/latitude, compiled by the CRU at the University of East Anglia in the UK (New *et al.*, 2000). This dataset is hereafter referred to as the 'CRU global climate time series'.

A commonly observed pattern in long-term temperature records (IPCC, 2001) is a gradual increase in mean temperatures over the past 40 years, but a more marked increase in spring from 1989 onwards and in winter from 1988, possibly due to increased solar radiation at the Earth's surface caused by less polluted air (Wild *et al.*, 2005). This pattern is indeed seen at all sites so far examined in the Czech Republic, Lithuania, Poland, Hungary and Sweden (Randolph, 2004), irrespective of the variable changes in TBE incidence (Figure 3). Higher absolute temperatures in Hungary where TBE decreased, however, might have caused greater tick mortality over the summer (Figure 3), even though they were accompanied by lesser decreases in spring and summer rainfall over the same period than in NE Poland and Lithuania where TBE increased.

In addition, more detailed analysis of ground meteorological station records in Latvia has revealed a sudden and sustained increase in mean maximum temperatures only for the third dekad (days 21–30) of April starting in 1993 (Sumilo *et al.*, 2006). Although as yet unexplained, it could be epidemiologically significant, because this is precisely the time of year when rising spring temperatures cross the threshold necessary for larval tick activity. From 1993 onwards, conditions have been consistently favourable for the synchronous onset of nymphal and larval activity and therefore TBEV transmission. Again, this pattern is common to all sites examined in Latvia, however, and therefore cannot on its own explain regional differences in TBE epidemiology.

In Sweden, the temporal mismatch between TBE epidemiology and recent climate change undermines claims that warmer spring and winter conditions caused increased TBE incidence in Stockholm county (Lindgren and Gustafson, 2001). The increases in these seasonal temperatures from 1989 occurred *after* the step increase in TBE cases from 1984. The weak statistical correlation between TBE case numbers and spring temperatures, due entirely to two extreme points, cannot therefore signify causality, and indeed after 1984 there were more cases of TBE in many years despite similar temperature conditions (Randolph, 2001). However, it is intuitively likely that low temperatures could be the limiting factor in TBEV transmission at the northern extreme of its range. Evidence that climate change has driven the increase in TBE incidence here is not yet convincing.

Nevertheless, a seasonal shift in reported cases of TBE in the Czech Republic and Poland, with proportionally more cases earlier in the summer and later in the autumn (Zabicka, 1994; Danielová and Benes, 1997), suggests that TBEV transmission dynamics have changed somewhat. This could be due as much to reduced host seeking by ticks in drier summers (Randolph *et al.*, 1999) as to warmer spring and autumn temperatures.

Climate is just one of the many factors, biological (abiotic and biotic) and non-biological (socio-economic and public health), that influence tick-borne disease dynamics. Ultimately, all cause variable rates of contact between ticks, transmission hosts and humans, resulting in greater densities of infected ticks and greater incidental spill-over from enzootic cycles to humans. Although changing climate factors may have played a role in the up-surge of TBE, none has yet been identified that can satisfactorily explain both the spatial variation and the major temporal discontinuities in the incidence of this disease. Increased abundance of wildlife hosts for ticks, especially deer that feed adult reproductive ticks, and sociological changes that followed the collapse of communism are currently the subject of detailed investigation at regional levels on a pan-European scale to seek tight correlations with tick-borne disease patterns.

5.2. Increased Incidence of Malaria in the East African Highlands

Given the model-based predictions of malaria spread in a future, warmer world, it is perhaps not surprising that several workers have claimed evidence for this already, in the form of increased malaria, especially in the highlands of East Africa. In doing so, they combine an assumption (increased highland temperatures) with a fact (increased malaria) and conclude that the one is the 'cause' of the other. As pointed out earlier, no one disputes that malaria and other vector-borne diseases are affected by temperature. What is in dispute is how many of the three key criteria for judging an impact of global change on disease have been met in these cases. Has the temperature really increased? Has malaria incidence or prevalence (rather than simply the number of malaria cases) increased? Have the increases in temperature been in the 'right' direction', in the 'right' places and at the 'right' times to explain the increases in malaria? Hay and colleagues examined the long-term climate records for East Africa, both for the last century as a whole, and also for the crucial last decades of the last century, and could find very few sites where recorded temperatures (from the CRU global climate time series) have shown any significant change of any sort, while in many places malaria has increased substantially (Hay *et al.*, 2002a). The augmented Dickey–Fuller test was used to conclude that temperature has not changed significantly. This test is drawn from the field of econometrics and is not necessarily favoured by all time-series analysts (Chatfield, 2004), but the point to stress here is that those who claimed that there *is* an effect of global warming on malaria had carried out no statistical tests of any sort at all! Hay *et al.*'s analysis revealed quite clearly that even when average temperatures (and other climate variables) have changed by a certain amount, these changes are not spatially uniform. According to the CRU global climate time series, some areas of East Africa actually became cooler during parts of the last century, while others warmed at faster-than-average rates (Hay *et al.*, 2002). Whether this is an artefact of a rather sparse set of meteorological stations in East Africa, or a genuine phenomenon remains unclear. We shall see in the case of bluetongue below that Europe (much better provided with meteorological stations) also showed signs of spatial patchiness in the degree and even direction of global warming in the last 20–30 years. We conclude that 'global warming' does not mean 'warming globally' and urge great caution in the bringing together of sets of meteorological and epidemiological data to look for impacts of climate change.

Hay *et al.*'s analysis was initially criticized on two main grounds: a claimed inappropriate use of a 0.5° grid climate dataset to estimate climate changes at the clinic sites in East Africa and the assumption that only statistically significant changes in climate will give rise to significant changes in malaria (Patz *et al.*, 2002). Patz *et al.* suggested that increased climate variability without any increases in climate mean values may give rise to malaria outbreaks, although they gave no examples of this. In response to these criticisms, Hay and colleagues first stated that the malaria case data recorded at each point (the location of a clinic) came from an area around that point approximately equivalent in size to one of the 0.5° grid squares (Hay *et al.*, 2002b) and later showed that independent climate records from a meteorological station at Kericho also showed no long-term trends in either mean climate values (Shanks *et al.*, 2002) or climate variability (Rogers *et al.*, 2002). Later analyses using a different statistical approach showed no increase in climate variability either globally or locally, at the East African sites from where the malaria data came (Vinnikov and Robock, 2002; SIH, personal communication).

Further light was thrown on the potential contribution of climate change to changes in malaria in Africa during the last century by

Small *et al.* who applied the MARA/ARMA model (Craig *et al.*, 1999) to the African data from the CRU global climate time series to derive the MTCSI on a monthly basis for each of the 0.5° grid squares from 1911 to 1995 (Small *et al.*, 2003). Excluding areas that were never suitable (MTCSI always $\leqslant$0.1) or perennially suitable (MTSCI always $\geqslant$0.9) left a total of 4603 grid cells with potentially variable climatic suitability. Each of these was subjected to the same sort of time-series analysis as used by Hay *et al.* (2002a). In this analysis, 83.2% of the tested grid squares showed no significant change over time in their MTSCI values (78.8% classified as stationary with no trend and 4.4% as random walk with no drift). Further, 11.8% of the cells showed deterministic trends, nearly equally split between positive (5.7%) and negative (6.1%) and the remaining 5.1% of cells showed significant stochastic trends. Strong positive deterministic trends were found mostly in East Africa, with a cluster of grid cells in southern Mozambique. Strong negative trends were found along the Sahelian border across Africa, with clusters in south-eastern Ethiopia and the southern part of the common border of Côte d'Ivoire and Ghana. Grid cells with stochastic trends were scattered throughout the moist and semi-moist areas of the continent, with a concentration of high values in the southern Congo basin (Small *et al.*, 2003). The authors' conclusion (p. 15344) is simple and clear: "*Climate warming, expressed as a systematic temperature increase over the 85 year period, does not appear to be responsible for an increase in malaria suitability over any region in Africa.*" They also point out that where predicted suitability has changed over the last century, this usually has more to do with changing precipitation patterns than with changing temperature, a point also hinted at by both Thomas *et al.* (2004) and Tanser *et al.* (2003).

The issue of climate variability in Africa was later revisited by Zhou *et al.* (2004) who first showed that there were effectively no significant changes in meteorological station averages for three climate variables at seven sites above an altitude of 2000 m in East Africa (i.e. 21 observations) between 1978–1988 ($n = 11$) and 1989–1998 ($n = 10$); 19/21 t-tests for three climate variables were not significant; no allowance was made for multiple t-tests which would diminish further the small number of 'significant' differences. There were, however, significant increases in the variances of the maximum monthly temperatures between the two time periods at five sites (Alaba, Kericho, Eldoret, Nandi and Kabale), in the minimum monthly temperature at Kericho and in rainfall at Eldoret, and a significant decrease in the variance of minimum temperature at Ziway. Zhou *et al.* then fitted a seasonal autoregressive model to the monthly malaria inpatient (wrongly claimed to be outpatient) data (Step 1) and the residuals of this fit were used as the dependent variables in a step-wise linear regression model using the climate variables (Step 2). For reasons that are not explained in the original paper, those variables at lags found to be significant during Step 2 were included in the final model with equal weights (of 1.0) and were summed across all significant lag periods (Zhou *et al.*, 2004). This practice, of course, tends to overemphasise the importance of each of the climate variables at higher lags. The proportions of the variance of the raw data extracted sequentially by Steps 1 and 2 were taken by the authors to indicate the relative roles of seasonal autoregression components (essentially the intrinsic determinants of malaria case numbers) and climatic variability (essentially the extrinsic determinants of malaria case numbers) at each site (Zhou *et al.*, 2004). However, this latter component in the model included only measures of climate (T_{min}, T_{max}, R_{ain} and their cross-products) not of climate variability, so the authors do not seem to have produced a model that tests their hypothesis. Approximately equal mean amounts (38.6 and 36.1%) of variance of the monthly malaria figures were attributed to the seasonal autoregressive components and the climatic variables, respectively (ranges 18–63% and 12–63%, respectively), the combination summing to between 65 and 81% of the total across the seven study sites. These authors also highlight the joint importance of both rainfall and temperature; "*the use of either temperature or rainfall alone is not sensitive enough for the detection of anomalies that are associated with malaria epidemics*" (Zhou *et al.*, 2004). Zhou *et al.*'s paper has been criticized along the lines indicated here by Hay *et al.* (2005) who also highlight a number of other difficulties with the paper, including the critical assumption that public health provision has kept pace with population increase in Kenya (almost certainly, it has not).

There remains, of course, the possibility that statistically insignificant changes in either mean climatic variables or variation in these variables may be responsible for increasing malaria (Patz *et al.*, 2002), but, given the wealth of evidence for other causative factors for these increases, until proven otherwise, it seems more parsimonious to rule out climate as the driving variable.

Evidence for increasing malaria in many parts of Africa is overwhelming, but the more likely causes for most of these changes to date include land-cover and land-use changes and, most importantly, drug resistance rather than any effect of climate. The recrudescence of malaria in the tea estates near Kericho, Kenya, in East Africa, where temperature has not changed significantly (Hay *et al.*, 2002a), shows all the signs of a disease that has escaped drug control following the evolution of chloroquine resistance by the malarial parasite. Twelve year windowed Fourier analysis of these changes through time provided strong support for the (re-)emergence of annual (i.e. seasonal) and 3-year cycles in disease cases (Rogers *et al.*, 2002). Neither local temperature nor rainfall nor an *El Niño* proxy showed any signs of similar changes in the amplitude of their corresponding annual and 3-yearly cycles that might have caused these changes. We conclude that malaria waxes and wanes to the beat of two rhythms: an annual one dominated by local, seasonal weather conditions and a *ca.*3-yearly one dominated by herd immunity (Hay *et al.*, 2000). Effective drugs suppress both cycles before they can be expressed. This produces a population which is mainly or entirely dependent on drug effectiveness, and which suffers the consequence of eventual drug failure, during which the rhythms re-establish themselves, as they appear to have done in Kericho.

5.3. Northern Spread of Bluetongue Virus into Europe

The one good example of the impact of climate change to date on an observed emergence of a vector-borne disease concerns the expanded distribution of bluetongue virus (BTV). This account is taken from the review by Purse *et al.*, (2005). This virus infects ruminants, causing a devastating but non-contagious disease in sheep and some deer species, and long-lived sub-clinical infections in cattle that therefore act as the principal reservoir host. It is transmitted by certain species of *Culicoides* biting midges within tropical and sub-tropical parts of the world (35°S–40°N), where the competent species occur. These weak flyers can be transported by the wind for several kilometres within one night. This, together with traditional livestock trade routes, has long offered the potential for BTV to enter Europe from sub-Saharan Africa, Turkey and the Middle East. Yet BTV appeared only occasionally and briefly in southern Europe prior to 1998, with outbreaks confined to Portugal, southwest Spain and some Greek islands within the northern limits of the range of the major Old World vector, *C. imicola*.

Since 1998, three separate incursions into Europe by several different bluetongue serotypes have occurred: from the east, some via Turkey, spreading westwards and northwards into Greece, the Balkans as far as Croatia, and Italy and the central Mediterranean islands; others through Greece westward into Italy; and from the south via Tunisia and/or Algeria into Italy and the Mediterranean islands including the Balearics. In summary, over the past 7 years six strains of BTV from two different directions have achieved an overall northward range extension of some 800 km into 12 new countries.

Purse *et al.* (2005) plausibly discounted most possible biotic, sociological or agricultural factors as likely causes of the precisely observed molecular epidemiology and movement patterns. On the other hand, coincident in time and space with the emergence of BTV in Europe, there were pronounced increases in nighttime and winter temperatures, therefore fewer frost days, and changes in moisture conditions (Purse *et al.*, 2005). In addition to the temperature-sensitive parameters common to insect-borne disease systems (see Section 1 above), the competence of *Culicoides* vectors is also directly enhanced by warm temperatures (Paweska *et al.*, 2002). As Purse *et al.* (2005) point out, it is the interacting effects of multi-variate climate that determine the 'normal' spatial and seasonal distributions of *Culicoides* vectors and their competence to transmit BTV. They provide detailed evidence that, on balance, the biological responses ('independent and sometimes opposing') to the observed changes in climate in this part of Europe between the 1980s and the 1990s could have permitted the following observed novel features since 1998: (i) increased virus persistence during the winter, (ii) the northward expansion of the major BTV vector, *Culicoides imicola*, from its previous limits within N Africa, Portugal,

SW Spain, and some Greek islands and (iii) new transmission by indigenous European *Culicoides* species.

This latter phenomenon is particularly significant. It represents a qualitative change, presumably arising out of small quantitative responses to gradual climate change such as increased midge survival rates and population sizes and possibly increased transmission efficiency, to novel vectors that are common and widespread across the whole of central and northern Europe. If *C. imicola* continues to expand its own range, it will become increasingly easy to hand over the baton of the vector role within greater areas of overlap with northern midge species.

6. CONCLUSIONS

Biologists should take a leaf out of the climatologists' work on what is known of the 'problem of attribution'; this means establishing the correct quantitative relationship between cause and effect. A recent analysis of the heat wave in Europe in 2003 that resulted in thousands of excess human deaths concluded that there was a *ca.* 50% chance that it was due to anthropogenic effects (Stott *et al.*, 2004). This example shows that it will probably never be possible to state with absolute certainty that effect B is the result of cause A. In our view the causal chain between global warming and increases or decreases in malaria and other vector-borne diseases is even longer than the link between greenhouse gas emissions and global climate change, and likely has many more non-linearities of the sort outlined at the start of this review. Not only will it never be possible to say that effect B (e.g. an increase in disease) is the direct result of cause A (e.g. an increase in temperature) and cause A only, it also seems likely that the predicted regional mean changes in climatic variables hide a great deal of local spatial variation, to which any particular disease might respond in quite different ways. Knowing that this is likely to happen should alert epidemiologists to take a great deal more care in interpreting disease data than has been the case to date.

The theoretical treatment given in this review shows that our understanding of the dynamics of vector-borne diseases is considerably greater than our understanding of other infectious diseases, and we need to turn this understanding into a protocol for determining how and in what direction climate change might affect vector-borne diseases. The examples given here show that a good statistical model is better than a bad biological model; that biological models for the same disease do not always agree with each other; that climate may have 'opposite' effects on different diseases; that there is often a multitude of causes for changes in vector-borne diseases and that in a number of cases already reported in the literature, these causes are clearly not climatological.

It is important, therefore, to adopt a landscape-epidemiological approach in which the 'landscape' includes not just the physical environment and its changing climate, but the economic and sociological environments as well. A recent analysis of the history of malaria in Europe shows how changing land-use and livestock practices were more responsible for diminishing malaria's geographic extent in the continent than were any concurrent climatological changes (Kuhn *et al.*, 2003). As for the future of vector-borne diseases, we could do worse than heed Einstein: "*I never think of the future. It comes soon enough*" (Partington, 1996).

ACKNOWLEDGEMENTS

We thank Drs. Simon Hay, Alastair Graham, Bethan Purse and Dave Smith for reading and commenting on the manuscript.

REFERENCES

Baylis, M., Mellor, P.S., Wittmann, E.J. and Rogers, D.J. (2001). Prediction of areas around the Mediterranean at risk of bluetongue by modelling the distribution of its vector using satellite imaging. *Veterinary Record* **149**, 639–643.

Chatfield, C. (2004). *The Analysis of Time Series: An Introduction*, 6th edn. London: Chapman & Hall.

Craig, M.H., Snow, R.W. and le Sueur, D. (1999). A climate-based distribution model of malaria transmission in sub-Saharan Africa. *Parasitology Today* **15**, 105–111.

Daniel, M., Danielová, V., Kriz, B., Jirsa, A. and Nozicka, J. (2003). Shift of the tick *Ixodes ricinus* and tick-borne encephalitis to higher altitudes in central Europe. *European Journal of Clinical Microbiology and Infectious Diseases* **22**, 327–328.

Danielová, V. and Benes, C. (1997). [Increased incidence of tickborne encephalitis in the Czech Republic] (in Czech). *Prakticky Lekar* **77**, 580–583.

Epstein, P.R., (2000). Is global warming harmful to health? *Scientific American* August 2000, 50–57.

Hales, S. and Woodward, A. (2003). Climate change will increase demands on malaria control in Africa. *Lancet* **362**, 1775.

Hay, S.I., Cox, J., Rogers, D.J., Randolph, S.E., Stern, D.I., Shanks, G.D., Myers, M.F. and Snow, R.W. (2002a). Climate change and the resurgence of malaria in the East African highlands. *Nature* **415**, 905–909.

Hay, S.I., Cox, J., Rogers, D.J., Randolph, S.E., Stern, D.I., Shanks, G.D., Myers, M.F. and Snow, R.W. (2002b). Reply to Patz *et al.* 2002. *Nature* **420**, 628.

Hay, S.I., Guerra, C.A., Tatem, A.J., Noor, A.M. and Snow, R.W. (2004). The global distribution and population at risk of malaria: past, present, and future. *Lancet Infectious Diseases* **4**, 327–336.

Hay, S.I., Myers, M.F., Burke, D.S., Vaughn, D.W., Endy, T., Ananda, N., Shanks, D., Snow, R.W. and Rogers, D.J. (2000). The aetiology of interepidemic periods of mosquito-borne disease. *Proceedings of the National Academy of Sciences of the United States of America* **97**, 9335–9339.

Hay, S.I., Shanks, G.D., Stern, D.I., Snow, R.W., Randolph, S.E. and Rogers, D.J. (2005). Climate variability and malaria epidemics in the highlands of East Africa. *Trends in Parasitology* **21**, 52–53.

Hutchinson, M.F., Nix, H.A., McMahan, J.P., Ord, K.D. (1995). *Africa—A Topographic and Climatic Database, CD-ROM (1)* Centre for Resource and Environmental Studies, Australian National University.

Immuno Ag. (1997). *Tick-Borne Encephalitis (TBE) and its Immunoprophylaxis*. Vienna: Immuno Ag. 1997.

IPCC. (2001). *Climate Change 2001: The Scientific Basis*. Cambridge: Cambridge University Press.

Johns, T.C., Gregory, J.M., Ingram, W.J., Johnson, C.E., Jones, A., Lowe, J.A., Mitchell, J.F.B., Roberts, D.L., Sexton, D.M.H., Stevenson, D.S., Tett, S.F.B. and Woodage, M.J. (2003). Anthropogenic climate change for 1860 to 2100 simulated with the HadCM3 model under updated emissions scenarios. *Climate Dynamics* **20**, 583–612.

Kuhn, K.G., Campbell-Lendrum, D.H., Armstrong, B. and Davies, C.R. (2003). Malaria in Britain: past, present, and future. *Proceedings of the National Academy of Sciences of the United States of America* **100**, 9997–10001.

Labuda, M., Kozuch, O. and Lysy, J. (1997). *Tick-borne Encephalitis Virus Natural Foci in Slovakia: Ticks, Rodents and Goats*. Proceedings of 4th

International Potsdam Symposium on tick-borne diseases: tick-borne encephalitis and Lyme borreliosis, Berlin, (ed. J. Suss & O. Kahl), pp. 36–46. Lengerich, Germany: Pabst Science Publishers.

Labuda, M., Nuttall, P.A., Kozuch, O., Eleckova, E., Williams, T., Zuffova, E. and Sabo, A. (1993). Non-viraemic transmission of tick-borne encephalitis virus: a mechanism for arbovirus survival in nature. *Experientia* **49**, 802–805.

Labuda, M. and Randolph, S.E. (1999). Survival of tick-borne encephalitis virus: cellular basis and environmental determinants. *Zentralblatt für Bakteriologie* **288**, 513–524.

Lindgren, E. (1998). Climate change, tick-borne encephalitis and vaccination needs in Sweden—a prediction model. *Ecological Modelling* **110**, 55–63.

Lindgren, E. and Gustafson, R. (2001). Tick-borne encephalitis in Sweden and climate change. *Lancet* **358**, 16–18.

Lindgren, E., Talleklint, L. and Polfeldt, T. (2000). Impact of climatic change on the northern latitude limit and population density of the disease-transmitting European tick *Ixodes ricinus*. *Environmental Health Perspectives* **108**, 119–123.

Martens, P. (1997). *Health Impacts of Climate Change and Ozone Depletion—An Eco-Epidemiological Modelling Approach*. Maastricht, Netherlands: Ph.D. thesis, University of Maastricht.

Martens, P., Kovats, R.S., Nijhof, S., de Vries, P., Livermore, M.T.J., Bradley, D.J., Cox, J. and McMichael, A.J. (1999). Climate change and future populations at risk of malaria. *Global Environmental Change* **9**, S89–S107.

McCarthy, J.J., Canziani, O.F., Leary, N.A., Dokken, D.J. and White, K.S. (2001). *Climate Change 2001: Impacts, Adaptation and Vulnerability—Contribution of Working Group II to the Third Assessment Report of the Intergovernmental Panel on Climate Change*. Cambridge: Cambridge University Press.

New, M., Hulme, M. and Jones, P. (2000). Representing twentieth-century space-time climate variability. Part II: Development of 1901–1996 monthly grids of terrestrial surface climate. *Journal of Climate* **13**, 2217–2238.

Partington, A., ed. (1996). *The Oxford Dictionary of Quotations*, 4th edn. Oxford: Oxford University Press.

Patz, J.A., Hulme, M., Rosenzweig, C., Mitchell, T.D., Goldberg, R.A., Githeko, A.K., Lele, S., McMichael, A.J. and Le Sueur, D. (2002). Regional warming and malaria resurgence. *Nature* **420**, 627–628.

Paweska, J.T., Venter, G.J. and Mellor, P.S. (2002). Vector competence of South African *Culicoides* species for bluetingue virus serotype 1 (BTV-1) with special reference to the effect of temperature on the rate of virus replication in *C. imicola* and *C. bolitinos*. *Medical and Veterinary Entomology* **16**, 10–21.

Purse, B.V., Mellor, P.S., Rogers, D.J., Samuel, A.R., Mertens, P.P.C. and Baylis, M. (2005). Climate change and the recent emergence of bluetongue in Europe. *Nature Reviews Microbiology* **3**, 171–181.

Randolph, S.E. (2000). Ticks and tick-borne disease systems in space and from space. *Advances in Parasitology* **47**, 217–243.

Randolph, S.E. (2001). Tick-borne encephalitis in Europe. *The Lancet* **358**, 1731–1732.

Randolph, S.E. (2004). Evidence that climate change has caused 'emergence' of tick-borne diseases in Europe? *International Journal of Medical Microbiology* **293**, 5–15.

Randolph, S.E., Green, R.M., Peacey, M.F. and Rogers, D.J. (2000). Seasonal synchrony: the key to tick-borne encephalitis foci identified by satellite data. *Parasitology* **121**, 15–23.

Randolph, S.E., Miklisová, D., Lysy, J., Rogers, D.J. and Labuda, M. (1999). Incidence from coincidence: patterns of tick infestations on rodents facilitate transmission of tick-borne encephalitis virus. *Parasitology* **118**, 177–186.

Randolph, S.E. and Rogers, D.J. (2000). Fragile transmission cycles of tick-borne encephalitis virus may be disrupted by predicted climate change. *Proceedings of the Royal Society of London B* **267**, 1741–1744.

Reiter, P., Thomas, C.J., Atkinson, P.M., Hay, S.I., Randolph, S.E., Rogers, D.J., Shanks, G.D., Snow, R.W. and Spielman, A. (2004). Global warming and malaria: a call for accuracy. *The Lancet Infectious Diseases* **4**, 323–324.

Rogers, D.J. and Randolph, S.E. (2000). The global spread of malaria in a future, warmer world. *Science* **289**, 1763–1766.

Rogers, D.J., Randolph, S.E., Snow, R.W. and Hay, S.I. (2002). Satellite imagery in the study and forecast of malaria. *Nature* **415**, 710–715.

Shanks, G.D., Hay, S.I., Stern, D.I., Biomndo, K. and Snow, R.W. (2002). Meteorologic influences on *Plasmodium falciparum* malaria in the highland tea estates of Kericho, western Kenya. *Emerging Infectious Diseases* **8**, 1404–1408.

Small, J., Goetz, S.J. and Hay, S.I. (2003). Climatic suitability for malaria transmission in Africa, 1911–1995. *Proceedings of the National Academy of Sciences of the United States of America* **100**, 15341–15345.

Stott, P.A., Stone, D.A. and Allen, M.R. (2004). Human contribution to the European heatwave of 2003. *Nature* **432**, 610–612.

Sumilo, D., Bormane, A. and Randolph, S.E. (2006). Tick-borne encephalitis in the Baltic States. *International Journal of Medical Microbiology* **296**, S 40 (in press).

Talleklint, L. and Jaenson, T.G.T. (1998). Increasing geographical distribution and density of *Ixodes ricinus* (Acari: Ixodidae) in central and northern Sweden. *Journal of Medical Entomology* **35**, 521–526.

Tanser, F.C., Sharp, B. and le Sueur, D. (2003). Potential effect of climate change on malaria transmission in Africa. *The Lancet* **362**, 1792–1798.

Thomas, C.J., Davies, G. and Dunn, C.E. (2004). Mixed picture for changes in stable malaria distribution with future climate in Africa. *Trends in Parasitology* **20**, 216–220.

Vaisviliene, D., Kilciauskiene, V., Zygutiene, M., Asokliene, L. and Caplinskas, S. (2002). TBE in Lithuania: epidemiological aspects and laboratory diagnosis. *International Journal of Medical Microbiology* **293**, S179–S181.

van Lieshout, M., Kovats, R.S., Livermore, M.T.J. and Martens, P. (2004). Climate change and malaria: analysis of the SRES climate and socio-economic scenarios. *Global Environmental Change* **14**, 87–99.

Vinnikov, K.Y. and Robock, A. (2002). Trends in moments of climatic indices. *Geophysical Research Letters* **29**, art. no. 1027, doi:10.1029/2001GLO14025.

WHO. (1997). World malaria situation in 1994. *Weekly Epidemiological Record* **72**, 285–292.

Wild, M., Gilgen, H., Roesch, A., Ohmura, A., Long, C.N., Dutton, E.G., Forgan, B., Kallis, A., Russak, V. and Tsvetkov, A. (2005). From dimming to brightening: decadel changes in solar radiation at Earth's surface. *Science* **308**, 847–850.

Zabicka, J. (1994). [The tick-borne encephalitis in Poland] (in Polish). *Przegl Epidemiology* **48**, 197–203.

Zhou, G., Minakawa, N., Githeko, A.K. and Yan, G. (2004). Association between climate variability and malaria epidemics in the East African highlands. *Proceedings of the National Academy of Sciences of the United States of America* **101**, 2375–2380.

Index

Accessibility modeling, in global population distribution determination, 129–131
ADDS (Africa Data Dissemination Service), 67
ADEOS (Advanced Earth Observation Satellite) program, 63–64
Administrative boundary data, 68
A. duodenale, in STH infections, 221–224, 232–239, 243, 246–247
Ae. aegypti, 183, 185–197, 207, 209
  global distribution of, 243
  in dengue, 207, 307
  in yellow fever, 210
Ae africanus, in yellow fever, 184–185
Ae. albopictus, 185
  air travel risk routes of, 318, 330
  global distribution of, 243
  in yellow fever, 210
  shipping risk routes of, 317
Ae. japonicus, 307–308
Africa
  east African highlands, increased incidence of malaria in, 370–374
  infectious disease contagion in, 377
  malaria in, 355, 370–377
  Rift Valley Fever (RVF) in, 16, 29
  WNV in, 307, 322–323
  yellow fever and dengue fever in, 186–187, 194–195, 198, 200–209
AIC (Akaike Information Criterion), 23, 199
  in yellow fever and dengue fever, 200–209
AIDS, 300–301
Airport malaria, 326–329
Albendazole, 234, 239, 247, 250
Altitudinal mask, in malaria transmission, 161–167
A. lumbricoides (roundworm), in STH infections, 222
An. atroparvus, 328
Ancillary data, in global population distribution determination, 124, 135–136, 148
An. gambiae, 307, 326–330
Anthrax and bioterrorism, 302–304
Asia, 298, 300
  cholera in, 297–298
  influenza in, 298–305
  WNV in, 307, 322–323
  yellow fever and dengue fever in, 186–187, 194–195, 198, 200–209
ASTER (Advanced Spaceborne Thermal Emission and Reflection) radiometer, in infectious disease distribution mapping, 55–56
AVHRR (Advanced Very High Resolution Radiometer) sensor, 37, 82, 198, *see also* NOAA
  and MODIS, 55–68
  archives, 39

B. anthracis, 303
Biological maps, 3
Bioterrorism, 302
Black Death, due to bubonic plague, 295–297
Bootstrap sampling, 15–17, 30, 32
Brazil, yellow fever in, 184, 207–209

ADVANCES IN PARASITOLOGY VOL 62
ISSN: 0065-308X $35.00
DOI: 10.1016/S0065-308X(05)62013-1

BTV (bluetongue virus), into Europe, northern spread of, 374
Bubonic plague, 295–297
BZAs (benzimidazole anthlemintics), 223

Chad, STH infections in, 228
Chiba, disease vector invasions in, 306
China, 223, 244–245, 296–297, 299, 301, 307, 358
Chloroquine, 324, 374
Cholera, 297–298
CIESIN (Center for International Earth Science Information Network), 187, 309
C. imicola, 375–376
Climate suitability mask, and malaria transmission, 167
Clustering data, for yellow fever and dengue fever, 199
CNSD (conditional negative semidefinite) models, 91
Contingency table, 99–101

Data-based assessment of accuracy, in global prediction of diseases, 99–103
  for categorical variables, 99–105
  for cross-validation, 98, 102–103
DEM (Digital Elevation Model), 47, 66, 160
  in distribution mapping, 26, 31, 101
  in malaria limit mapping, 159, 161, 167–168
Democratic Republic of Congo, 248–249
Dendrograms, 311–314, 328
DHF (Dengue Haemorrhagic Fever), 190–193, 306–307, 322
  and yellow fever, global distribution of, 186–187, *see also separate entry*
  archive and literature searches of, 194–196
  bootstrap results of, 203–205
  control of, 183–197
  dengue fever data, 197–198
  dengue model predictor variables, 204
  distribution and impacts of, 197
  environmental data for, 198
  epidemiology of, 184, 191
  existing maps of, 193, 210
  individual variables in, importance of, 202–205
  materials and methods in, 193–200
  modelling approach to, 199
  outbreaks of, per year per continent, 209
  overall model accuracy of, 201
  populations at risk of, 205–208
  results of global distribution of, 186–187
  risk maps for, 200–201
  symptoms of, 184–190
Diethylcarbamazine (DEC), 238
Discriminant analysis models, 10–12
Diseases, global prediction of, *see also* global disease maps
  data-based assessment of accuracy in, 99–103, *see also separate entry*
  dynamic models of, 85
  global disease maps, validation of, 105–106
  image processing chain in, 85–86
  indirect relations in, 81–82
  model-based accuracy assessment in, 103–105, *see also separate entry*
  modelling framework choice in, 84–85
  scale and spatial resolution issues, 86–96, *see also separate entry*
  scale and uncertainty issues in, 80–108
  sensor characteristics in, 82–84
  uncertainty issues in, 81–107, *see also separate entry*
Disease vector invasions, prediction, 306–319
  climatic dendrograms in, 311–313
  climatic distance images in, 318–319
  data sources in, 309, 311
  risk routes in, 313–319, 330
Distribution models/modelling, 4–12, 14
  accuracy metrics applicable to, 8–9
  environmental envelope expansion in, 17–18
  families of, 4–12

for vectors and vector-borne diseases, 1–32
model selection and multi-model inference, 21–30
predictor variable selection in, 12–14
spatial information incorporation into, 19–21
DVD imagery contents, 46–47

EMR (ElectroMagnetic Radiation) sensors, 81
Environmental data, for yellow fever and dengue fever, 198–199
Environmental envelope expansion, in distribution modelling, 17–18
Europe, tick-borne disease systems in, 264–285

FETBE (Far Eastern Tick-Borne Encephalitis) virus, 271
Flaviviridae virus, in yellow fever and dengue, 182
Fourier data products, in TFA technique, 49
FRESH (Focus Resources on Effective School Health), 235–236

GAC (Global Area Coverage) data, 38
Geohelminths, *see* STH
Geo-registration and projection, in infectious disease distribution mapping, 41–42
Geostationary satellites, 60–61
Gibbs' sampler, 20–21
GIS (Geographic Information Systems), in infectious disease distribution mapping, 36, 127 *see also* GIS/RS
GIS/RS, 237–251
control applications of, in STH infections, 237–242
in global distributions of STH infections, 243–246
in STH infections, 221–248
GLCF (Global Landcover Facility), 68
Global disease maps, validation of, 105–106
Global environmental data, for infectious disease distribution mapping, 41–42
Global epidemiology, ecology and control, of soil-transmitted Helminth infections, 221–252
Global grid, rendering population on, 121–122
Global population distribution determination, 119–151, *see also* global grid; global rural urban mapping project; gridded population of the world
accessibility modeling in, 130–132
adjusting population estimates to target years in, 133–136
analysis of, methods and issues in, 150–151
ancillary data in, limitations of, 136–138
and general health studies, 139–140
boundary matching over time in, 135
data dissemination in, 148–149
data of, 162–166
data selection in, 149
health applications of, 138–144
highly modeled surfaces in, 132–133
ideal spatial resolution in, 145–147
in specific diseases understanding, 140–144
institutional stewardship in, 122–124
methodology of, 133–138
modeled population grids in, 136–137
population databases in, 121, 133, 135
Global rural urban mapping project, 128–130
Global spatial limits, of malaria transmission in 2005, 158–175
Global transport networks
and disease vector and pandemics, 306–319
and infectious disease contagion, 293–332
and vector borne diseases, 319–332
GLOBE (Global Land One-kilometre Base Elevation), 66
GOES (Geostationary Operational Environmental Satellite), 61

GPW (Gridded Population of the World), 47, 127–128, 146, 148, 151, 246
  and related databases, 125–126
Gridded climatologies, 67, 318, 328
Gridded human population data, 67–68
Gridded population of the world, 127–128
GRUMP (Global Rural Urban Mapping Project), 126, 128–130, 146, 161
  allocation mechanism for, 128
Guadeloupe, yellow fever in, 183
Guangdong, SARS in, 301–302

Haemagogus species, in yellow fever, 184
Highly modeled surfaces approach, in global population distribution determination, 132–133
HIV/AIDS pandemic, 300
Hookworms, *see N. americanus*
HRPT (High Resolution Picture Transmission), 38
Hungary, malaria in, 365–368

ICT (immunochromatographic card tests), 238
Ideal spatial resolution, in global population distribution determination, 145–147
Image processing chain, in global prediction of diseases, 85–86
India
  epidemic outbreak in, 297
  yellow fever in, 198, 201, 207
Indonesia
  STH infections in, 236, 244
  yellow fever in, 201, 208
Infectious disease contagion, and global transport networks, 293–332
Infectious disease distribution mapping, 35–70
  from digital to environmental data for, 39–45
  geographic information systems (GIS) in, 36
  geo-registration and projection in, 41–42
  global environmental data for, 35–70
  quality control and data pre-processing in, 43–45
  remote sensing in, 36
Influenza, 298–305
Institutional stewardship, in global population distribution determination, 122–124
Invasions, 306–332
I. persulcatus, 280–281, 363
I. ricinus, 273–275, 363, 365
  and WTBE virus, 265, 272–276, 283
ITHG (International Travel Health Guideline) information, on malaria, 159–173
Ivermectin, 237–238

Japan, 297, 307, 316–320
  STH infections in, 371

Kenya, 104–106
  STH infections in, 235, 248
Kriging, 92–106

LAC (Local Area Coverage), 38
Landsat satellite, 61–62
Lithuania, malaria in, 367–368
L. loa, in STH infections, 238
Logistic models, 4–10
Louping Ill (LI) virus, 268
LQAS (Lot Quality Assurance Sampling) method, 239
LST (Land Surface Temperature) data, 44

Mahalanobis climatic distance, 320
Malaria, 323–326
  biological approach of, 355–366
  biological model for, 358–363
  drug resistance in, 374
  imported malaria trends, 323–324
  in the east African highlands, increased incidence of, 370–374
  statistical approach of, 357–358
Malaria risk, 157–175
  distribution of, 159–161

Malaria transmission in 2005, *see also* malaria risk
  and altitudinal mask, 161–167
  and climate suitability mask, 167
  and population density mask, 167–169
  biological limits of, 161–169
  distribution of, from travel guidelines, 159–161
  global spatial limits of, 157–175
  regional analysis of, 169–173
Malaysia
  infections in, 201
  yellow fever in, 208
Maps, biological, *see* biological maps
MARA (Mapping Malaria Risk in Africa) model, 359
MARA/ARMA model, 141, 359–360, 372
MCMC (Markov Chain Monte Carlo) model, 104
Mexico, yellow fever in, 183, 196, 201
Microwave remote sensing, in global prediction of diseases, 82
MIR (middle-infrared) sensor, 38
Model-based accuracy assessment, in global prediction of diseases, 103–105
  for categorical variables, 99–102, 104–105
  for continuous variables, 102–104
Modeled population grids, ancillary data for, 135–136
Modelling approach, to yellow fever and dengue fever, 199–200
Modelling framework, choice in global prediction of diseases, 84–85
'Modified Gibbs' sampler, 20
MODIS (MODerate resolution Imaging Spectroradiometer), in infectious disease distribution mapping, 55–58, 63–68
Mosquitoes and malaria transmission, 161
MSG (Meteosat Second Generation) satellites, 60
MTCSI (Malaria Transmission Climate Suitability Index), 359
Multi-model inference, and model selection, 21–30
  application of, to vector and disease mapping, 26–30
MVIRI (Meteosat Visible and Infra-Red Imager), 60

N. americanus (hookworms), in STH infections, 223
NASA (National Aeronautics and Space Administration), 82
NDVI (Normalised Difference Vegetation Index), 40
NGA (National Geospatial-intelligence Agency), 66
Nigeria, STH infections in, 247, 249
NOAA-6 satellite, 37
Norway, vector-borne diseases in, 351–355
NPOESS (National Polar-orbiting Operational Environmental Satellite Systems), 64–65
  and data continuity, 64–65

O. volvulus, in STH infections, 238

Pakistan, yellow fever in, 201
Pandemics
  and global transport networks, 295–306
  predicting, modelling and controlling future pandemics, 304–306
PCA (Principal Components Analysis), 45
Peri-domestic mosquitoes, in yellow fever, 184–185
P. falciparum, 326
  and malaria transmission, 360
  versus *P. vivax*, 173
Philadelphia, yellow fever in, 183
Plague, 295–297
Plate-Carrée projection, 41
Population databases, in global population distribution determination, 133–135
Population density mask, and malaria transmission, 167–169

P. vivax, 328
and malaria transmission, 157–175
versus *P. falciparum*, 169

REMO (Rapid Epidemiological Mapping of Onchocerciasis), 237
Rescaling, in infectious disease distribution mapping, 35–37, 66, 68
Rift Valley Fever (RVF), 199, 205
in Africa, 16
rift valley fever model predictor variables, 29
Risk maps, for yellow fever and dengue, 200–201
RNA flaviviruses, as tick-borne species, 269
ROC (Receiver Operating Characteristic) curves, 101, 232
RS (Remote Sensing), in infectious disease distribution mapping, 36–37, *see also* GIS/RS
RSSE (Russian Spring-Summer Encephalitis), 280
Russia, tick-borne disease systems in, 270–271, 280–282

SAR (Synthetic Aperture Radar) imagery, in global prediction of diseases, 82
SARS (Severe Acute Respiratory Syndrome) virus, 301–305
Satellite sensors
for epidemiological applications, technical specifications, 56
satellite sensor data sets, 106–107
SaZA (satellite zenith angle) data, 42–43
Scale and spatial resolution issues, in global prediction of diseases, 86–96
measurement scales, 88–89
spatial variation characterization, 90–92
upscaling and downscaling the model in, 95–96, 108
Scandinavia, Black Death in, 296
School-age children, STH in, 222, 225, 230, 241, 250
Sensor characteristics, in global prediction of diseases, 82–84
SEVIRI (Spinning Enhanced Visible and Infrared Imager) sensor, 60
S. haematobium, in STH infections, 241–242
Sicily, Black Death in, 296
Simulium blackflies, in STH infections, 239
Smallpox, 303–305
S. mansoni, in STH infections, 239–241
in Uganda, 240
Soil-transmitted helminth infections, *see* STH
South Korea, 236, 243
Sparse data sets, 14–18
Spatial information incorporation, into models, 19–21
Spatial variation characterization, in global prediction of diseases, 90–92
SPOT (Satellite Pour l'Observation de la Terre) programme, 56, 62–63
SRTM (Shuttle Radar Topography Mission), 66, 83
SSA (sub-Saharan Africa), STH in, 222–223
STBE (Siberian Tick-Borne Encephalitis) virus, 271
S. aegypti, in dengue fever, 192
STH (Soil-Transmitted Helminth) infections
among schoolchildren in urban and rural communities, 233
and urbanization, 232–234
chronic and intense STH infections, 223
ecological correlations in, 227–231
global control strategies, 234–237
global distributions of, 243–246
global epidemiology, ecology and control of, 221–252
in Asia, 243–246
infective stages, 225, 230
predicted numbers of, 246–250
predictions of infection patterns of, 231–234

transmission dynamics and environment in, 224–227
Sweden, malaria in, 366–369

TALA RS data, geo-referencing information for, 42
Tanzania, STH infections in, 228, 233, 239, 242, 247, 249, 251
TBE (Tick-Borne Encephalitis), *see also* vector-borne diseases; *individual entries*
biotic liberators, abiotic constraints in, 272–276
biotic selectors in, 269–270
climate in the evolution of, 276–285
composite predictive map for six viruses, 281–283
evolutionary time scales of, 264–267
geography and phylogeny of, 270–272
in Europe, 348, 351–352
increased incidence of, 366–370
phenetic eco-climatic trees construction for, 277–279
phylogenetic and eco-climatic trees for, 283–285
phylogenetic tree of, 264
phylogenic patterns, correlation, 269–276
quantity and quality of data on, 279–281
TBE complex viruses, 274–276
and sheep, 275
TBEV, 363–366, 369
Terra, Aqua and Modis satellites, 55–56
TFA (Temporal Fourier Analysis) techniques, 45–55
Fourier data products in, 49
history and application, 45–49
TFA surfaces, inter-comparison of, 49–55
Thailand, STH infections in, 280
TIROS-N satellite, *see* NOAA-6 satellite
Topotypes, in yellow fever, 182
Tromøya, vector-borne diseases in, 263–265, 285
Tsetse flies, 2
T. trichiura (whipworm), in STH infections, 222

Uganda, STH infections in, 228, 230, 239–242, 250–251, 280
UK, 221
Uncertainty issues, in global prediction of diseases, 80, 88
data-based assessment of accuracy in, 99–103, *see also separate entry*
in data and methods, 97–103
USA, 352
yellow fever and dengue fever in, 181–211
yellow fever transmission cycles for, 185

Vaccination, for yellow fever, 200–201
VC (Vectorial Capacity) equation, 347
V. cholerae, 297–298
Vector-borne diseases, 346–377, *see also individual entries*
and climate change, 353–355
biological and statistical approaches to, 355–366
climate change impacts of, models for, 353–355
climate impacts on, 351–353
evolutionary emergence of, 265–267
evolutionary time scale for, 264–267
in Norway, 261, 272, 276
in Tromøya, 265–266
mathematics and biology of changes in, 346–351
recent changes in, 366–376
vector-borne disease movement, prediction, 326–332
Vectors and vector-borne diseases, models for, 1–32
VIIRS (Visible Infrared Imager Radiometer Suite), 65

W. bancrofti, in STH infections, 238
WNV (West Nile Virus), 307, 322–323
WTBE (Western Tick-Borne Encephalitis) virus, 265–285
and *I. ricinus*, 275
with rodents, fragile cycles of, 272–274

Yellow fever, 319–322, *see also individual entries*
  and dengue, global distribution of, 189–193
  archive and literature searches of, 194–196
  bootstrap results of, 203–204
  control of, 183, 197
  data, 196–197
  distribution and impacts of, 186–187
  environmental data for, 198–199
  epidemiology of, 184–186
  existing maps of, 193–194
  individual variables in, importance of, 202–203
  materials and methods in, 193–200
  model predictor variables, 202
  modelling approach to, 199–200
  outbreaks of, per year per continent, 209
  overall model accuracy of, 201
  pathogens in, 183–193
  populations at risk of, 205–208
  results of global distribution of, 243–246
  risk maps for, 200–201
  symptoms of, 184
  vaccination certificate requirements, 189
Y. pestis, plague bacillus, 296

Contents of Volumes in This Series

Volume 41

Drug Resistance in Malaria Parasites of Animals and Man 1
W. Peters
Molecular Pathobiology and Antigenic Variation of *Pneumocystis carinii* 63
Y. Nakamura and M. Wada
Ascariasis in China. 109
P. Weidono, Z. Xianmin and D.W.T. Crompton
The Generation and Expression of Immunity to *Trichinella spiralis* in Laboratory Rodents. 149
R.G. Bell
Population Biology of Parasitic Nematodes: Application of Genetic Markers 219
T.J.C. Anderson, M.S. Blouin and R.M. Brech
Schistosomiasis in Cattle. 285
J. De Bont and J. Vercruysse

Volume 42

The Southern Cone Initiative Against Chagas Disease 1
C.J. Schofield and J.C.P. Dias
Phytomonas and Other Trypanosomatid Parasites of Plants and Fruit. 31
E.P. Camargo
Paragonimiasis and the Genus *Paragonimus*. 113
D. Blair, Z.-B. Xu and T. Agatsuma
Immunology and Biochemistry of *Hymenolepis diminuta* 223
J. Anreassen, E.M. Bennet-Jenkins and C. Bryant
Control Strategies for Human Intestinal Nematode Infections 277
M. Albonico, D.W.T. Crompton and L. Savioli
DNA Vaccines: Technology and Applications as Anti-parasite and Anti-microbial Agents 343
J.B. Alarcon, G.W. Waine and D.P. McManus

ADVANCES IN PARASITOLOGY VOL 62
ISSN: 0065-308X $35.00
DOI: 10.1016/S0065-308X(05)62014-3

Volume 43

Genetic Exchange in the Trypanosomatidae 1
W. Gibson and J. Stevens
The Host–Parasite Relationship in Neosporosis 47
A. Hemphill
Proteases of Protozoan Parasites 105
P.J. Rosenthal
Proteinases and Associated Genes of Parasitic Helminths. 161
J. Tort, P.J. Brindley, D. Knox, K.H. Wolfe and J.P. Dalton
Parasitic Fungi and their Interaction with the Insect Immune System 267
A. Vilcinskas and P. Götz

Volume 44

Cell Biology of *Leishmania* 1
B. Handman
Immunity and Vaccine Development in the Bovine Theilerioses 41
N. Boulter and R. Hall
The Distribution of *Schistosoma bovis* Sonaino, 1876 in Relation to Intermediate Host Mollusc–Parasite Relationships 98
H. Moné, G. Mouahid and S. Morand
The Larvae of Monogenea (Platyhelminthes) 138
I.D. Whittington, L.A. Chisholm and K. Rohde
Sealice on Salmonids: Their Biology and Control 233
A.W. Pike and S.L. Wadsworth

Volume 45

The Biology of some Intraerythrocytic Parasites of Fishes, Amphibia and Reptiles 1
A.J. Davies and M.R.L. Johnston
The Range and Biological Activity of FMRFamide-related Peptides and Classical Neurotransmitters in Nematodes 109
D. Brownlee, L. Holden-Dye and R. Walker
The Immunobiology of Gastrointestinal Nematode Infections in Ruminants. 181
A. Balic, V.M. Bowles and E.N.T. Meeusen

Volume 46

Host–Parasite Interactions in Acanthocephala: a Morphological Approach 1
H. Taraschewski
Eicosanoids in Parasites and Parasitic Infections 181
A. Daugschies and A. Joachim

Volume 47

Remote Sensing and Geographic Information Systems in Epidemiology
An Overview of Remote Sensing and Geodesy for Epidemiology and Public Health Application 1
S.I. Hay
Linking Remote Sensing, Land Cover and Disease 37
P.J. Curran, P.M. Atkinson, G.M. Foody and E.J. Milton
Spatial Statistics and Geographic Information Systems in Epidemiology and Public Health 83
T.P. Robinson
Satellites, Space, Time and the African Trypanosomiases 133
D.J. Rogers
Earth Observation, Geographic Information Systems and *Plasmodium falciparum* Malaria in Sub-Saharan Africa 175
S.I. Hay, J. Omumbo, M. Craig and R.W. Snow
Ticks and Tick-borne Disease Systems in Space and from Space 219
S.E. Randolph
The Potential of Geographical Information Systems (GIS) and Remote Sensing in the Epidemiology and Control of Human Helminth Infections 247
S. Brooker and E. Michael
Advances in Satellite Remote Sensing of Environmental Variables for Epidemiological Applications 293
S.J. Goetz, S.D. Prince and J. Small
Forecasting Diseases Risk for Increased Epidemic Preparedness in Public Health 313
M.F. Myers, D.J. Rogers, J. Cox, A. Flauhalt and S.I. Hay
Education, Outreach and the Future of Remote Sensing in Human Health 335
B.L. Woods, L.R. Beck, B.M. Lobitz and M.R. Bobo

Volume 48

The Molecular Evolution of Trypanosomatidae 1
J.R. STEVENS, H.A. NOYES, C.J. SCHOFIELD AND W. GIBSON
Transovarial Transmission in the Microsporidia 57
A.M. DUNN, R.S. TERRY AND J.E. SMITH
Adhesive Secretions in the Platyhelminthes 101
I.D. WHITTINGTON AND B.W. CRIBB
The Use of Ultrasound in Schistosomiasis 225
C.F.R. HATZ
Ascaris and Ascariasis. 285
D.W.T. CROMPTON

Volume 49

Antigenic Variation in Trypanosomes: Enhanced Phenotypic Variation in a Eukaryotic Parasite 1
H.D. BARRY AND R. MCCULLOCH
The Epidemiology and Control of Human African Trypanosomiasis. 71
J. PÉPIN AND H.A. MÉDA
Apoptosis and Parasitism: from the Parasite to the Host Immune Response 132
G.A. DOSREIS AND M.A. BARCINSKI
Biology of Echinostomes except *Echinostoma* 163
B. FRIED

Volume 50

The Malaria-Infected Red Blood Cell: Structural and Functional Changes 1
B.M. COOKE, N. MOHANDAS AND R.L. COPPEL
Schistosomiasis in the Mekong Region: Epidemiology and Phytogeography 88
S.W. ATTWOOD
Molecular Aspects of Sexual Development and Reproduction in Nematodes and Schistosomes 153
P.R. BOAG, S.E. NEWTON AND R.B. GASSER
Antiparasitic Properties of Medicinal Plants and Other Naturally Occurring Products 200
S. TAGBOTO AND S. TOWNSON

Volume 51

Aspects of Human Parasites in which Surgical Intervention May Be Important 1
D.A. Meyer and B. Fried
Electron-transfer Complexes in *Ascaris* Mitochondria 95
K. Kita and S. Takamiya
Cestode Parasites: Application of *In Vivo* and *In Vitro* Models for Studies of the Host–Parasite Relationship 133
M. Siles-Lucas and A. Hemphill

Volume 52

The Ecology of Fish Parasites with Particular Reference to Helminth Parasites and their Salmonid Fish Hosts in Welsh Rivers: A Review of Some of the Central Questions 1
J.D. Thomas
Biology of the Schistosome Genus *Trichobilharzia* 155
P. Horák, L. Kolářová and C.M. Adema
The Consequences of Reducing Transmission of *Plasmodium falciparum* in Africa 234
R.W. Snow and K. Marsh
Cytokine-Mediated Host Responses during Schistosome Infections: Walking the Fine Line Between Immunological Control and Immunopathology 265
K.F. Hoffmann, T.A. Wynn and D.W. Dunne

Volume 53

Interactions between Tsetse and Trypanosomes with Implications for the Control of Trypanosomiasis 1
S. Aksoy, W.C. Gibson and M.J. Lehane
Enzymes Involved in the Biogenesis of the Nematode Cuticle 85
A.P. Page and A.D. Winter
Diagnosis of Human Filariases (Except Onchocerciasis) 149
M. Walther and R. Muller

Volume 54

The Evolution of Parasitism – a Phylogenetic Perspective
Introduction – Phylogenies, Phylogenetics, Parasites and the Evolution of Parasitism 1
D.T.J. LITTLEWOOD
Cryptic Organelles in Parasitic Protists and Fungi 9
B.A.P. WILLIAMS AND P.J. KEELING
Phylogenetic Insights into the Evolution of Parasitism in Hymenoptera . . . 69
J.B. WHITFIELD
Nematoda: Genes, Genomes and the Evolution of Parasitism 101
M.L. BLAXTER
Life Cycle Evolution in the Digenea: a New Perspective from Phylogeny . . . 197
T.H. CRIBB, R.A. BRAY, P.D. OLSON, D. T. J. LITTLEWOOD
Progress in Malaria Research: the Case for Phylogenetics 255
S.M. RICH AND F.J. AYALA
Phylogenies, the Comparative Method and Parasite Evolutionary Ecology. . . 281
S. MORAND AND R. POULIN
Recent Results in Cophylogeny Mapping 303
M.A. CHARLESTON
Inference of Viral Evolutionary Rates from Molecular Sequences 331
A. DRUMMOND, O.G. PYBUS AND A. RAMBAUT
Detecting Adaptive Molecular Evolution: Additional Tools for the Parasitologist 359
J.O. MCINERNEY, D.T.J. LITTLEWOOD AND C.J. CREEVEY

Volume 55 (Cumulative Subject Index – Vols. 28–52)

Contents of Volumes 28–52. 1
Cumulative Subject Indexes for Volumes 28–52. 15
Contributors to Volumes 28–52 213

Volume 56

Glycoinositolphospholipid from *Trypanosoma cruzi*: Structure, Biosynthesis and Immunobiology 1
J.O. PREVIATO, R. WAIT, C. JONES, G.A. DOSREIS, A.R. TODESCHINI, N. HEISE AND L.M. PREVIATA
Biodiversity and Evolution of the Myxozoa 43
E.U. CANNING AND B. OKAMURA

The Mitochondrial Genomics of Parasitic Nematodes of Socio-Economic Importance: Recent Progress, and Implications for Population Genetics and Systematics 133
M. Hu, N.B. Chilton and R.B. Gasser
The Cytoskeleton and Motility in Apicomplexan Invasion 213
R.E. Fowler, G. Margos and G.H. Mitchell

Volume 57

Canine Leishmaniasis 1
J. Alvar, C. Cañavate, R. Molina, J. Moreno and J. Nieto
Sexual Biology of Schistosomes 89
H. Moné and J. Boissier
Review of the Trematode Genus *Ribeiroia* (Psilostomidae): Ecology, Life History, and Pathogenesis with Special Emphasis on the Amphibian Malformation Problem 191
P.T.J. Johnson, D.R. Sutherland, J.M. Kinsella and K.B. Lunde
The *Trichuris muris* System: a Paradigm of Resistance and Susceptibility to Intestinal Nematode Infection 255
L.J. Cliffe and R.K. Grencis
Scabies: New Future for a Neglected Disease 309
S.F. Walton, D.C. Holt, B.J. Currie and D.J. Kemp

Volume 58

Leishmania spp.: on the Interactions They Establish with Antigen-Presenting Cells of their Mammalian Hosts 1
J.-C. Antoine, E. Prina, N. Courret and T. Lang
Variation in *Giardia*: Implications for Taxonomy and Epidemiology 69
R.C.A. Thompson and P.T. Monis
Recent Advances in the Biology of *Echinostoma* species in the "revolutum" Group 139
B. Fried and T.K. Graczyk
Human Hookworm Infection in the 21st Century 197
S. Brooker, J. Bethony and P.J. Hotez
The Curious Life-Style of the Parasitic Stages of Gnathiid Isopods 289
N.J. Smit and A.J. Davies

Volume 59

Genes and Susceptibility to Leishmaniasis 1
EMANUELA HANDMAN, COLLEEN ELSO AND SIMON FOOTE
Cryptosporidium and Cryptosporidiosis. 77
R.C.A. THOMPSON, M.E. OLSON, G. ZHU, S. ENOMOTO, MITCHELL S.ABRAHAMSEN AND N.S. HIJJAWI
Ichthyophthirius multifiliis Fouquet and Ichthyophthiriosis in Freshwater Teleosts. 161
R.A. MATTHEWS
Biology of the Phylum Nematomorpha 245
B. HANELT, F. THOMAS AND A. SCHMIDT-RHAESA

Volume 60

Sulfur-Containing Amino Acid Metabolism in Parasitic Protozoa. 1
TOMOYOSHI NOZAKI, VAHAB ALI AND MASAHARU TOKORO
The Use and Implications of Ribosomal DNA Sequencing for the Discrimination of Digenean Species. 101
MATTHEW J. NOLAN AND THOMAS H. CRIBB
Advances and Trends in the Molecular Systematics of the Parasitic Platyhelminthes 165
PETER D. OLSON AND VASYL V. TKACH
Wolbachia Bacterial Endosymbionts of Filarial Nematodes 245
MARK J. TAYLOR, CLAUDIO BANDI AND ACHIM HOERAUF
The Biology of Avian *Eimeria* with an Emphasis on their Control by Vaccination 285
MARTIN W. SHIRLEY, ADRIAN L. SMITH AND FIONA M. TOMLEY

Volume 61

Control of Human Parasitic Diseases: Context and Overview. 1
DAVID H. MOLYNEUX
Malaria Chemotherapy 47
PETER WINSTANLEY AND STEPHEN WARD
Insecticide-Treated Nets 77
JENNY HILL, JO LINES AND MARK ROWLAND
Control of Chagas Disease 129
YOICHI YAMAGATA AND JUN NAKAGAWA
Human African Trypanosomiasis: Epidemiology and Control 167
E.M. FÈVRE, K. PICOZZI, J. JANNIN, S.C. WELBURN AND I. MAUDLIN
Chemotherapy in the Treatment and Control of Leishmaniasis. 223
JORGE ALVAR, SIMON CROFT AND PIERO OLLIARO
Dracunculiasis (Guinea Worm Disease) Eradication 275
ERNESTO RUIZ-TIBEN AND DONALD R. HOPKINS

Intervention for the Control of Soil-Transmitted Helminthiasis in the Community 311
MARCO ALBONICO, ANTONIO MONTRESOR, D.W.T. CROMPTON AND LORENZO SAVIOLI
Control of Onchocerciasis 349
BOAKYE A. BOATIN AND FRANK O. RICHARDS, JR.
Lymphatic Filariasis: Treatment, Control and Elimination 395
ERIC A. OTTESEN
Control of Cystic Echinococcosis/Hydatidosis: 1863–2002 443
P.S. CRAIG AND E. LARRIEU
Control of *Taenia solium* Cysticercosis/Taeniosis 509
ARVE LEE WILLINGHAM III AND DIRK ENGELS
Implementation of Human Schistosomiasis Control: Challenges and Prospects 567
ALAN FENWICK, DAVID ROLLINSON AND VAUGHAN SOUTHGATE

Colour plate section

Plate 1.2 An example of treatment of a sparse data set. Records of the presence of Rift Valley Fever (RVF) in Africa and the near Middle East are modeled here in three different ways. (A) Maximum likelihood discriminant analysis applied to records for RVF presence ($n = 62$) and 2000 randomly selected points for RVF absence within 0.5 to 10.00 degrees of any presence point; (B) mean predictions of 100 bootstrap samples from these data, of 200 presence and 200 absence points sampled at random from the training set, with replacement; (C) after expansion of the environmental envelope using the predictions of the bootstrap samples of Figure 2B and re-assignment of 'absence' to 'presence' status of training set absence points that were consistently classified as presence points. A single model was then produced, as for Figure 2A. The records of RVF presence used in all models are shown as blue points. These records were obtained from querying two sources: (a) the PubMed reference database (National Center for Biotechnology Information (NCBI) of the National Library of Medicine (NLM), MD, USA; url: http://www.ncbi.nim.nih.gov/entrez/query.fcgi) for the period 1920–2005 and (b) the ProMED database (url: http://www.promedmail.org/) for the period January 1994–December 2004. In each case, geographical point locations were extracted wherever possible from the original literature or from geo-locatable place names referenced in the GEONet Names Server of the National Geospatial-Intelligence Agency's and the US Board of Geographic Names' databases (http://earth-info.nga.mil/gns/html/), the Encarta 2005 Premium Suite (Microsoft Corporation, WA, USA), the Alexandria Digital Library Gazetteer (http://middleware.alexandria.ucsb.edu/client/gaz/adl/index.jsp) or the Getty Thesaurus of Geographic Names Online (http://www.getty.edu/research/conducting_research/vocabularies/tgn/index.html). I thank Dr Simon Hay for providing these data for modelling.

The environmental data were derived from the Advanced Very High Resolution Radiometer (AVHRR) on board the National Oceanographic and Atmospheric Administration (NOAA) satellites and cover the period from 1982 to 1999. These data are described in detail in Hay *et al.* (this volume, pp. 37–77). Monthly maximum-value composited data from the AVHRR middle infra red, the derived land surface temperature (LST) and the normalized difference vegetation index (NDVI) data were temporal Fourier processed to extract annual, bi-annual and tri-annual seasonal signals, which were captured as separate images showing the amplitudes and phases or timing of the first peak of each of the three signals (Rogers, 2000). In addition the signal means, maxima, minima and variances were also available, as was a single digital elevation surface (DEM) derived from the GTOPO30 coverage (http://edcdaac.usgs.gov/gtopo30/gtopo30.asp). All AVHRR data were originally produced and made available at a spatial resolution of 8×8 km in the Goode's Interrupted Homolosine projection and after Fourier processing were projected back to latitude/longitude format by bi-linear interpolation to 0.10 degree spatial resolution. The DEM data at an original 30 arc second resolution (1/120th of a degree) were similarly resampled to 0.1 degree resolution by averaging. Satellite and DEM data were later extracted for each of the disease 'presence' and 'absence' points and these data formed the training sets for model construction.

All models used a non-linear discriminant analysis approach (Rogers, 2000). Training set data were first clustered into 2 presence and 3 absence clusters and, for each model, 10 variables were selected by step-wise inclusion using the minimum AIC_c values as the selection criterion (Burnham and Anderson, 2002). Models assumed equal prior probabilities and the output risk maps shown here are of posterior probabilities (averaged values in the case of the bootstrapped models).

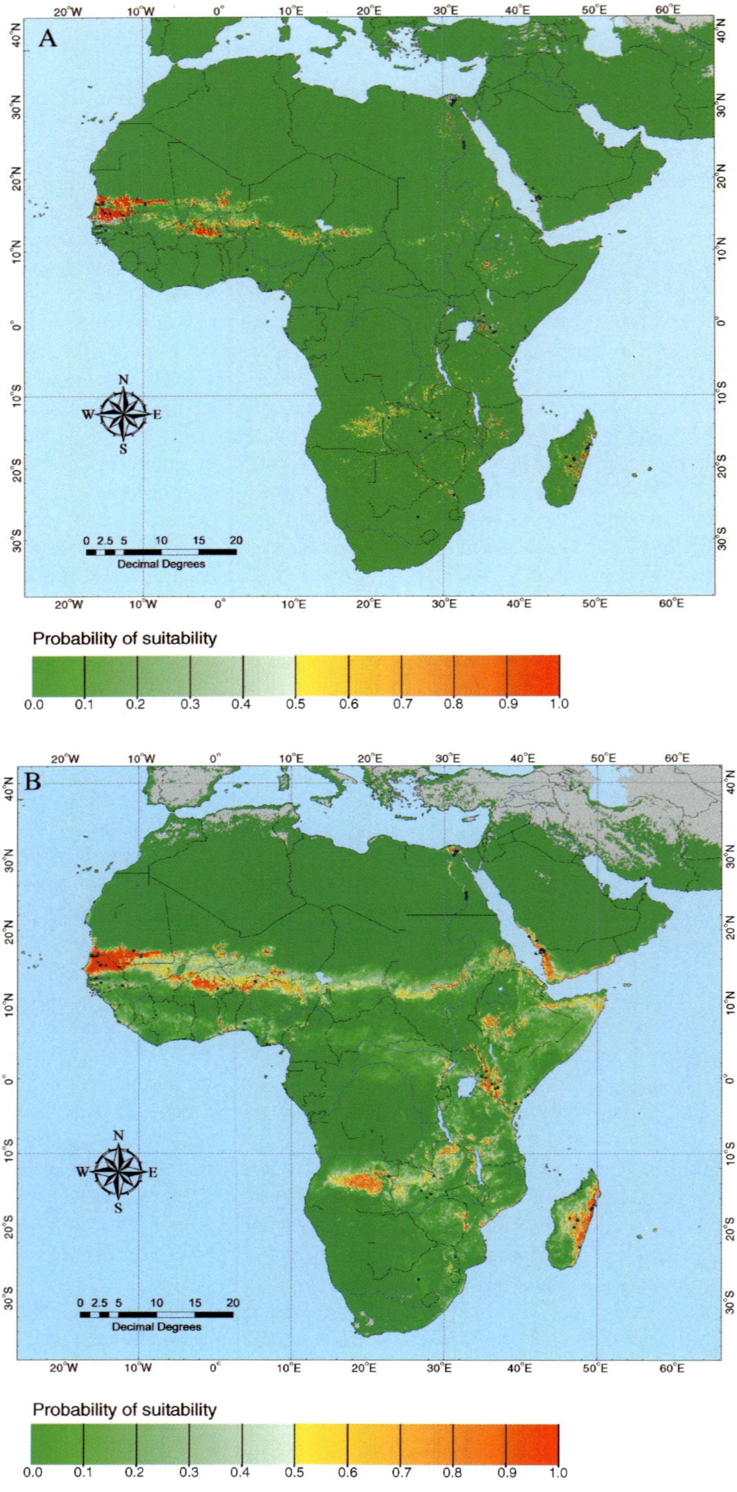
A
B
20°W
10°W
0°
10°E
20°E
30°E
40°E
50°E
60°E
40°N
30°N
20°N
10°N
0°
10°S
20°S
30°S
N
S
E
W
0 2.5 5 10 15 20
Decimal Degrees
Probability of suitability
0.0 0.1 0.2 0.3 0.4 0.5 0.6 0.7 0.8 0.9 1.0

Plate 1.2

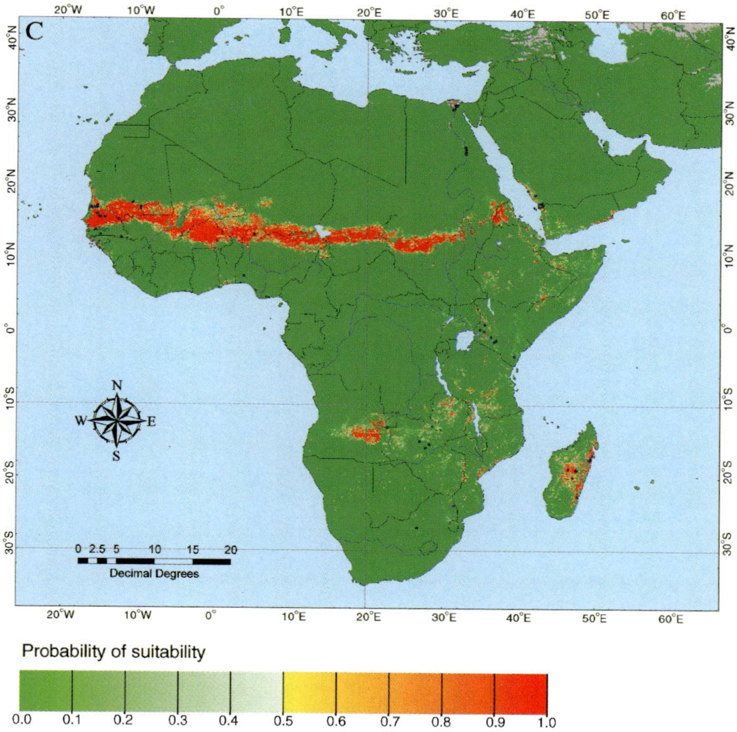

Plate 1.2 (continued)

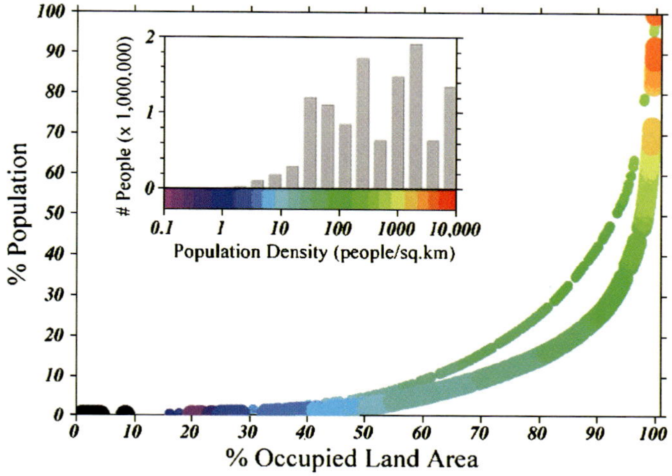

Plate 4.7 Spatial Lorenz curve for the population distribution compared to the land area of Ecuador, 2000 (with insert indicating the non-cumulative distribution of population density).

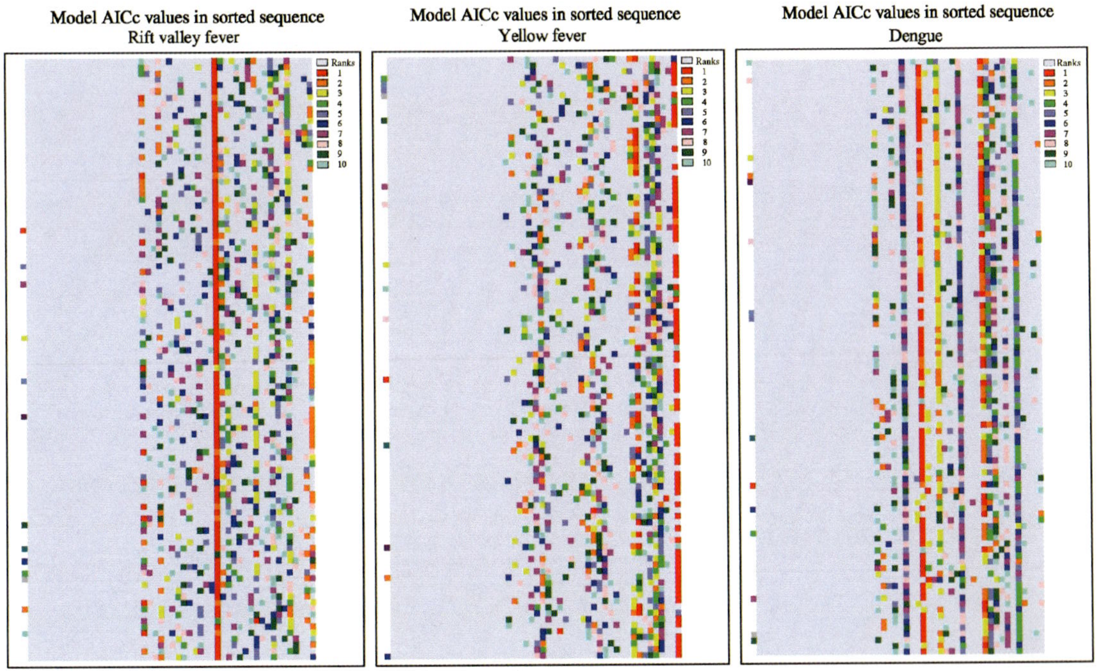

Plate 1.3 Results from the 100 bootstrap models for (left) Rift Valley Fever, (middle) Yellow Fever and (right) Dengue. Each row in the image refers to one of the models, which are arranged in rank order, with 1 (lowest AIC_c value) at the top and 100 (highest AIC_c value) at the bottom. Each of the 31 columns on the right of the image indicates one of the satellite predictor variables available to describe the disease. The first column of these 31 columns is for the digital elevation layer or DEM. There then follow three sets of 10 columns referring to the Fourier-processed AVHRR MIR, LST and NDVI imagery. These layers are in the following order: mean, phase of annual cycle, amplitude of annual cycle; phase of bi-annual cycle, amplitude of bi-annual cycle; phase of tri-annual cycle, amplitude of tri-annual cycle; maximum of fitted Fourier cycles (summed annual to tri-annual), minimum of fitted Fourier cycles and variance of the original signal. In any single model (row) the top (i.e. first selected) predictor variable is coloured red, the second most important variable is coloured orange and so on according to the rainbow colour scale to the right of the image. Variables not chosen in any model are not coloured at all in that row. The red line down the first image indicates variable 14 in the variable list, which is the annual amplitude of LST. This variable is consistently chosen in all RVF models, and is usually the most important variable, but there is no other single variable which is consistently chosen second. (The left-most column refers to the model number in the sequence; this, and the grey area to the left of the variable columns should be ignored.) The other two images may be similarly interpreted (see Rogers *et al.*, this volume, pp. 181–220, for more details).

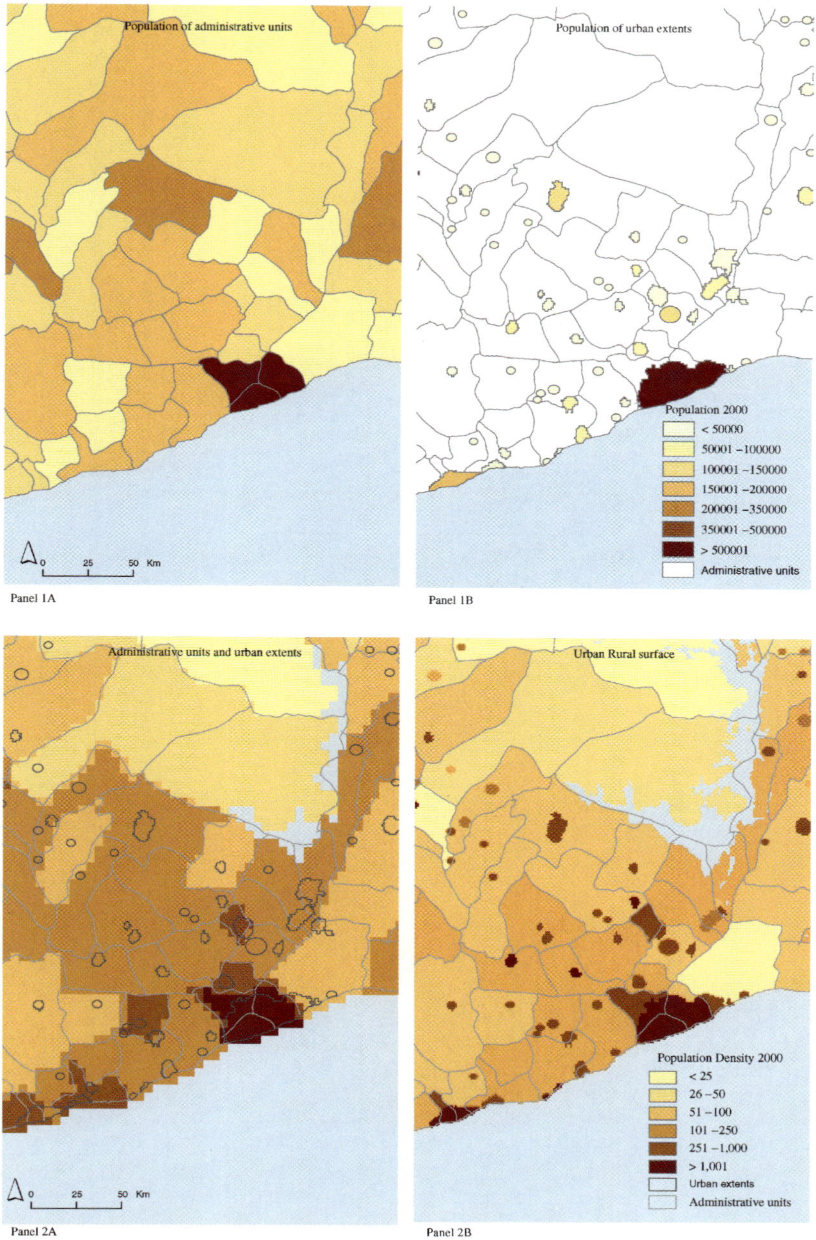

Plate 4.3 Process by which GRUMP population surface is constructed, illustrated for southern Ghana. Panel 1 shows inputs side by side with their population counts. Panel 1A is identical to the inputs to GPW, panel 1B shows the additional urban areas used in GRUMP. In panel 2, the inputs are merged, first illustrated as an overlay of the urban footprints over the administrative polygons in panel 2A, and the final grid, in panel 2B (with administrative and urban) boundaries overlaid (density/square km).

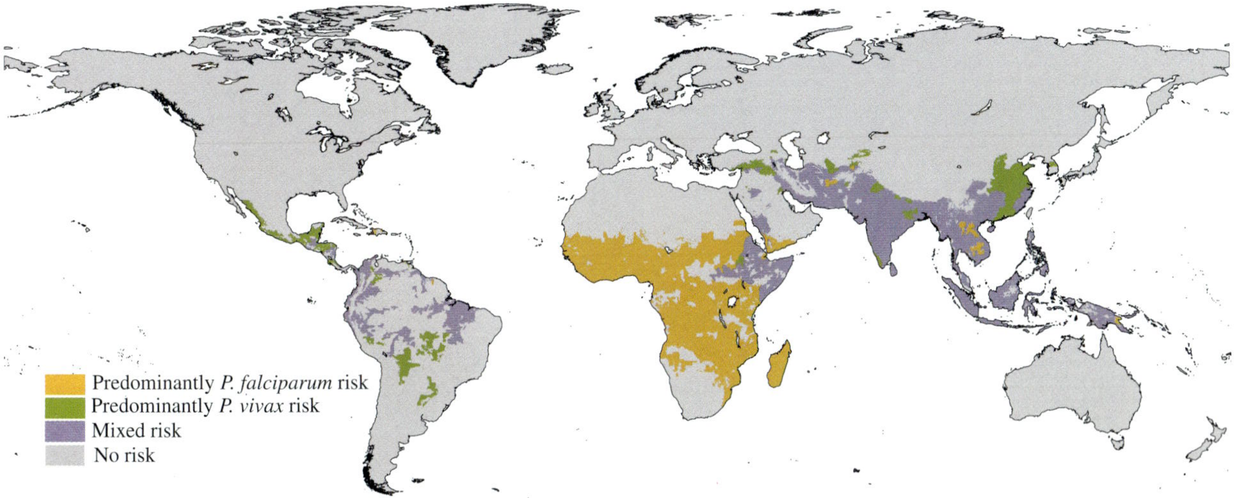

Plate 5.1 Malaria distribution in 2005 after altitudinal and population exclusions indicating areas at risk according to species of *Plasmodium*.

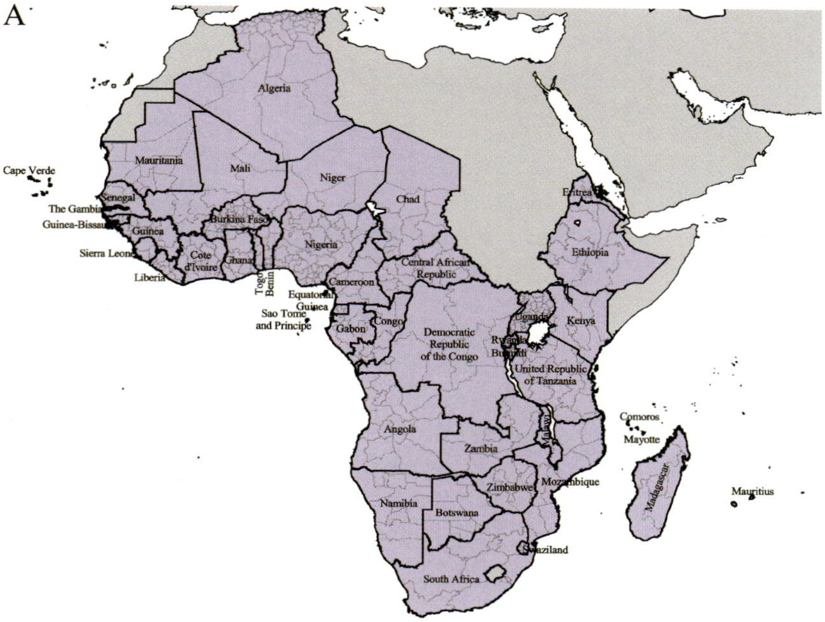

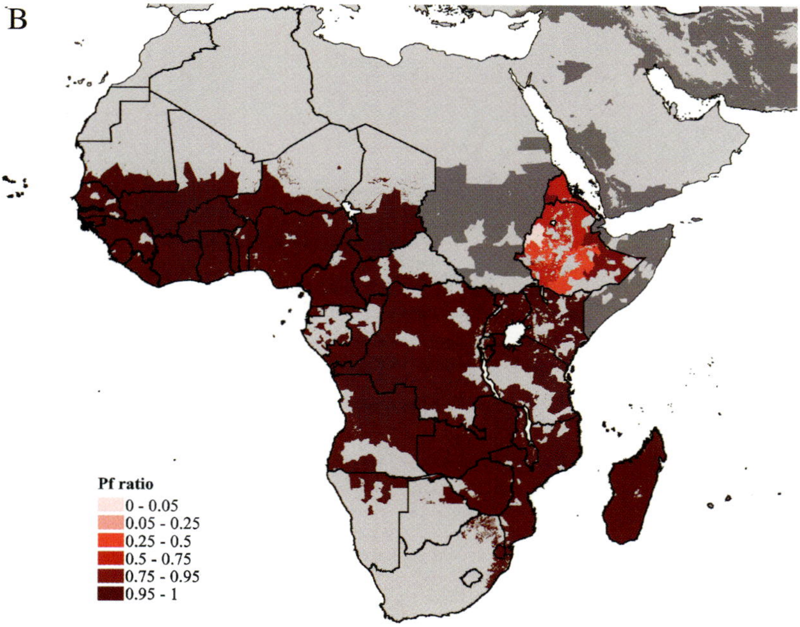

Plate 5.2 A–R First level administrative division boundaries and *P. falciparum* (Pf) and *P. vivax* (Pv) ratios by WHO region (A–C: AFRO; D–F: AMRO; G–I: EMRO; J–L: EURO; M–O: SEARO; P–R: WPRO). Outlined in thick black line are countries belonging to each region. Malarious countries are filled in light blue with a light grey thin line representing sub-national boundaries. Dark grey areas represent malaria distribution outside the given region and light grey ones are malaria-free areas.

C

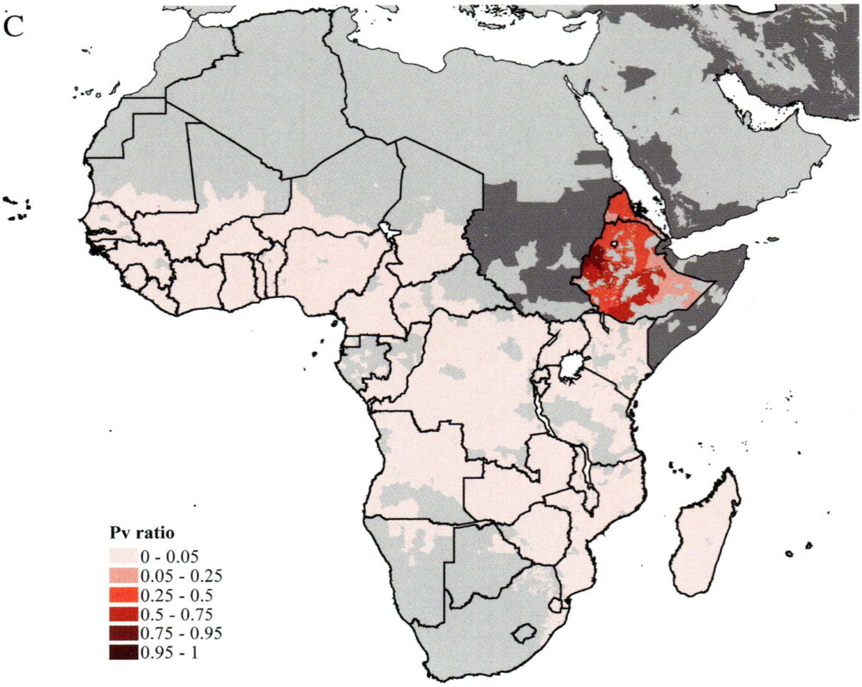

D

Plate 5.2 (continued)

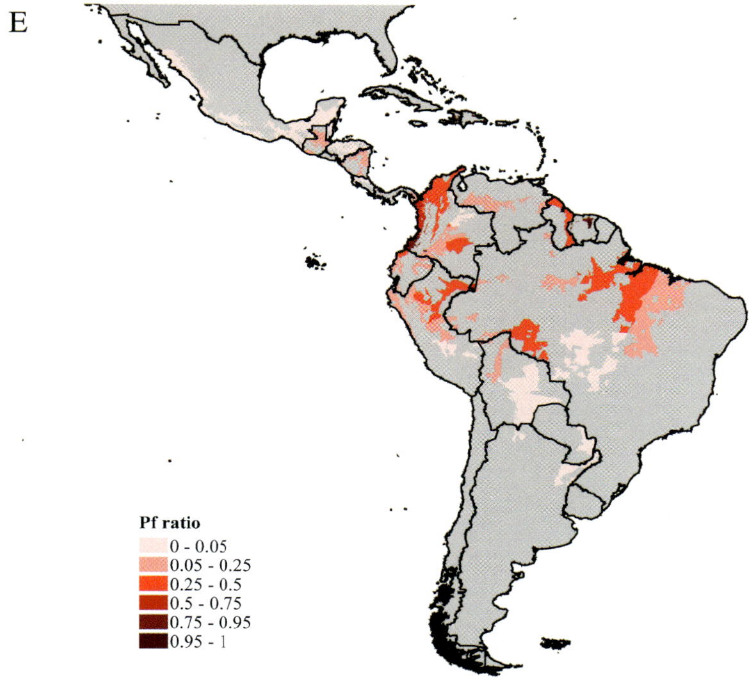

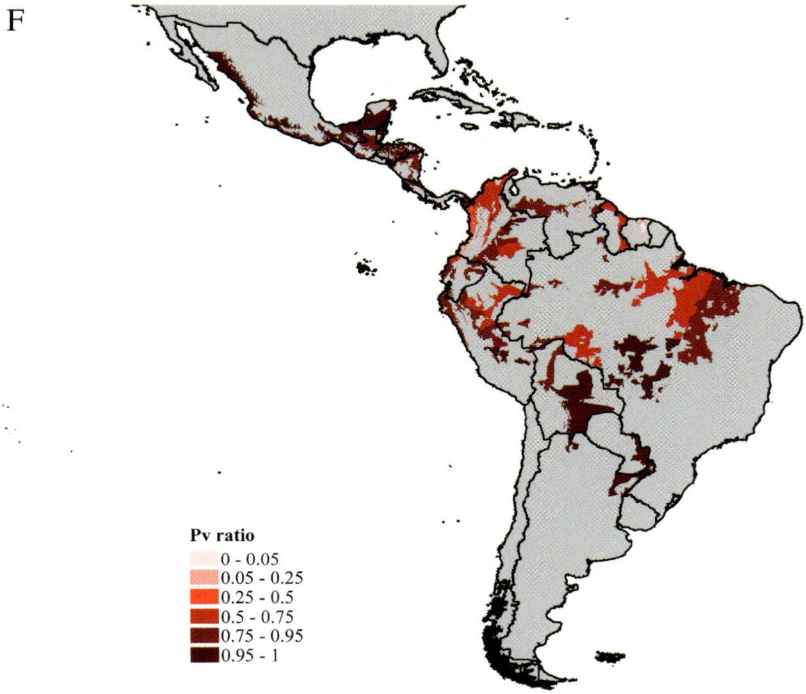

Plate 5.2 (continued)

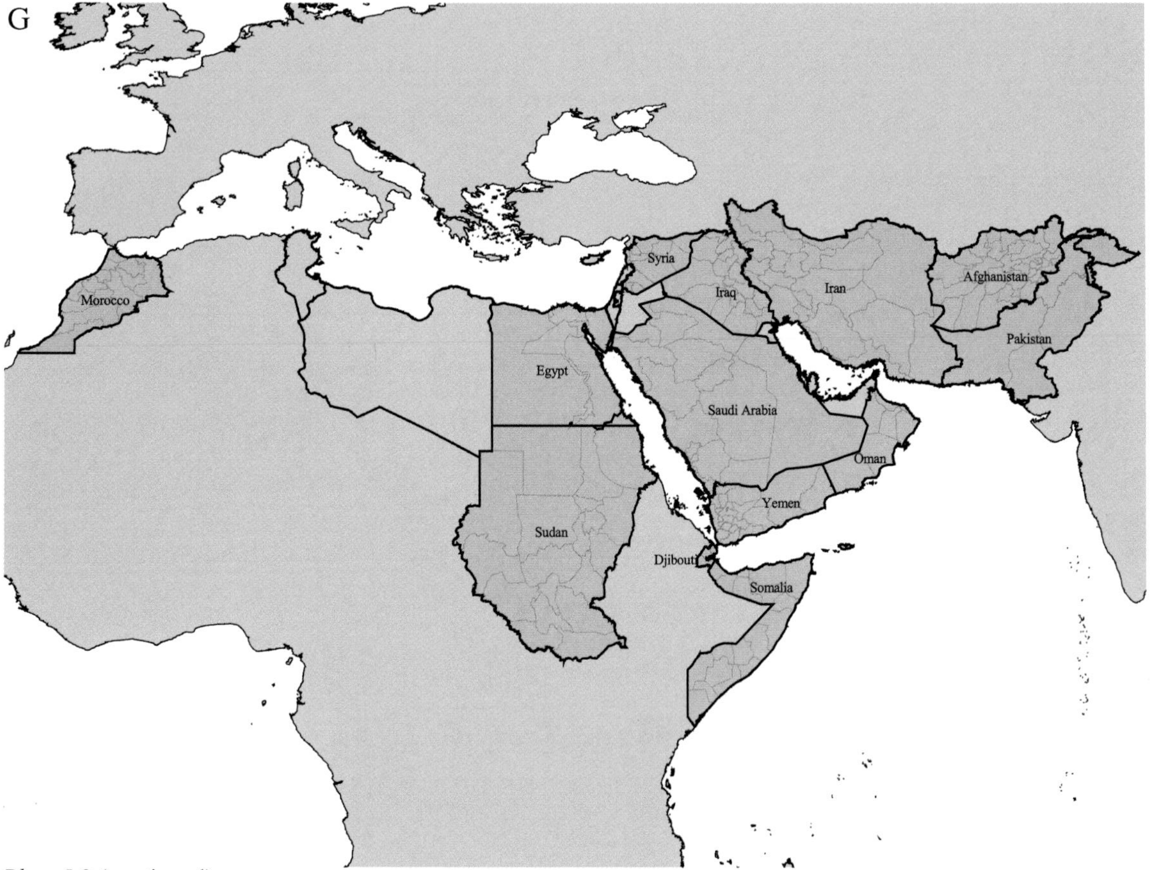

Plate 5.2 (continued)

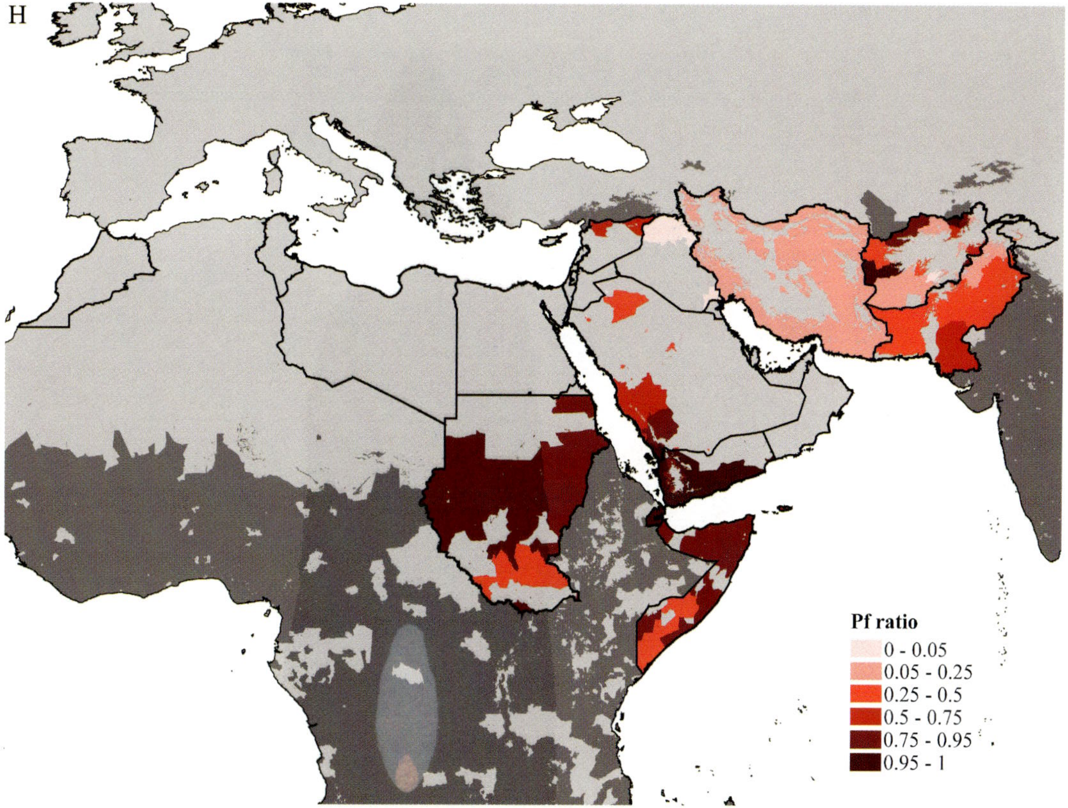

Plate 5.2 (continued)

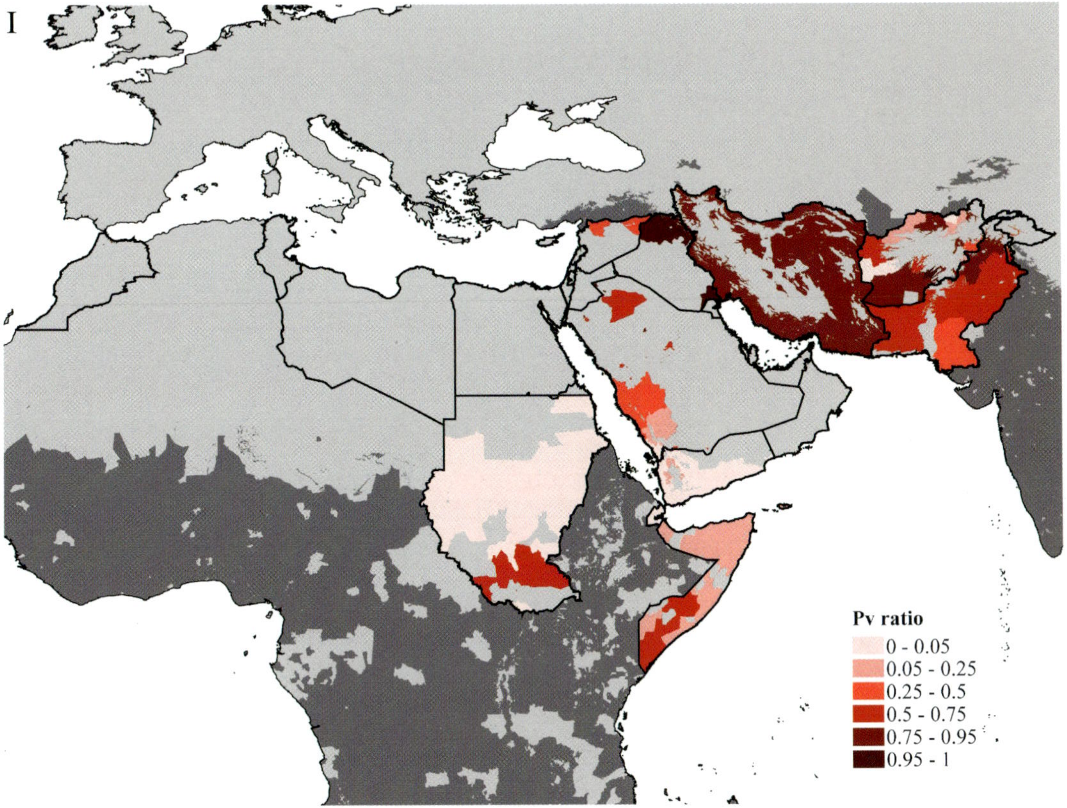

Plate 5.2 (continued)

J

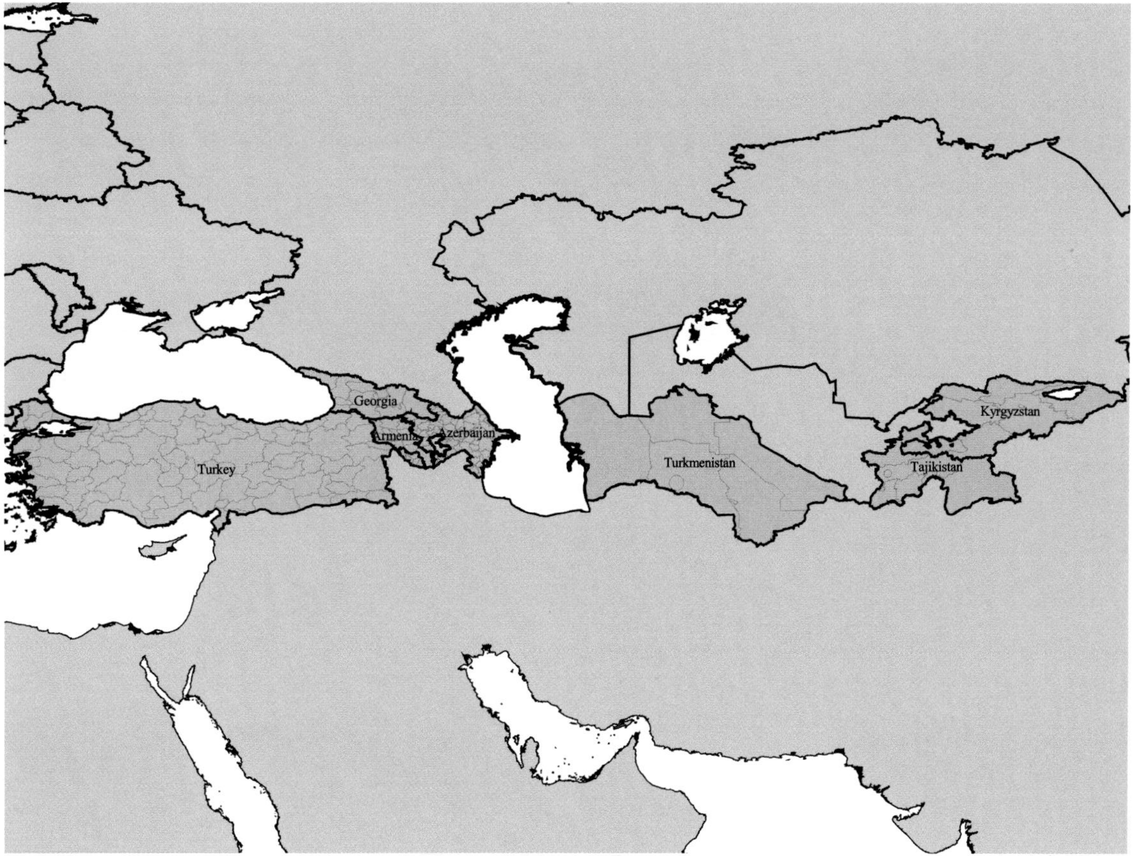

Plate 5.2 (continued)

K

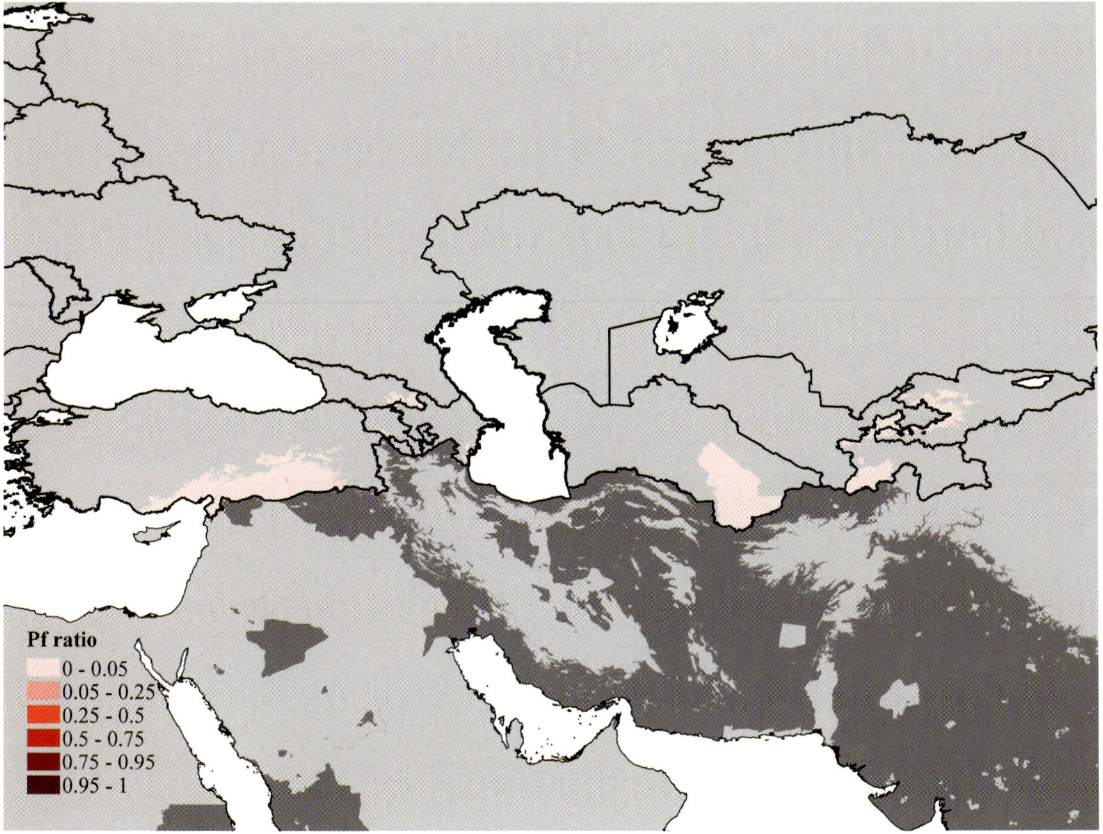

Plate 5.2 (continued)

L

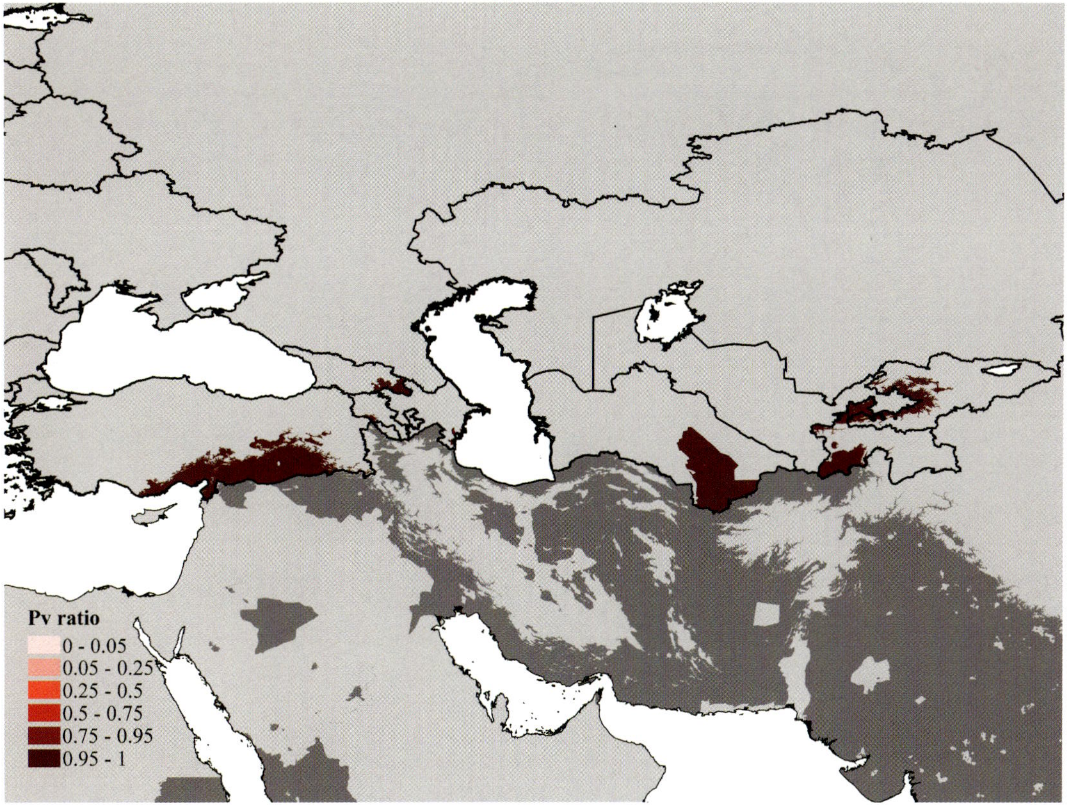

Plate 5.2 (continued)

M

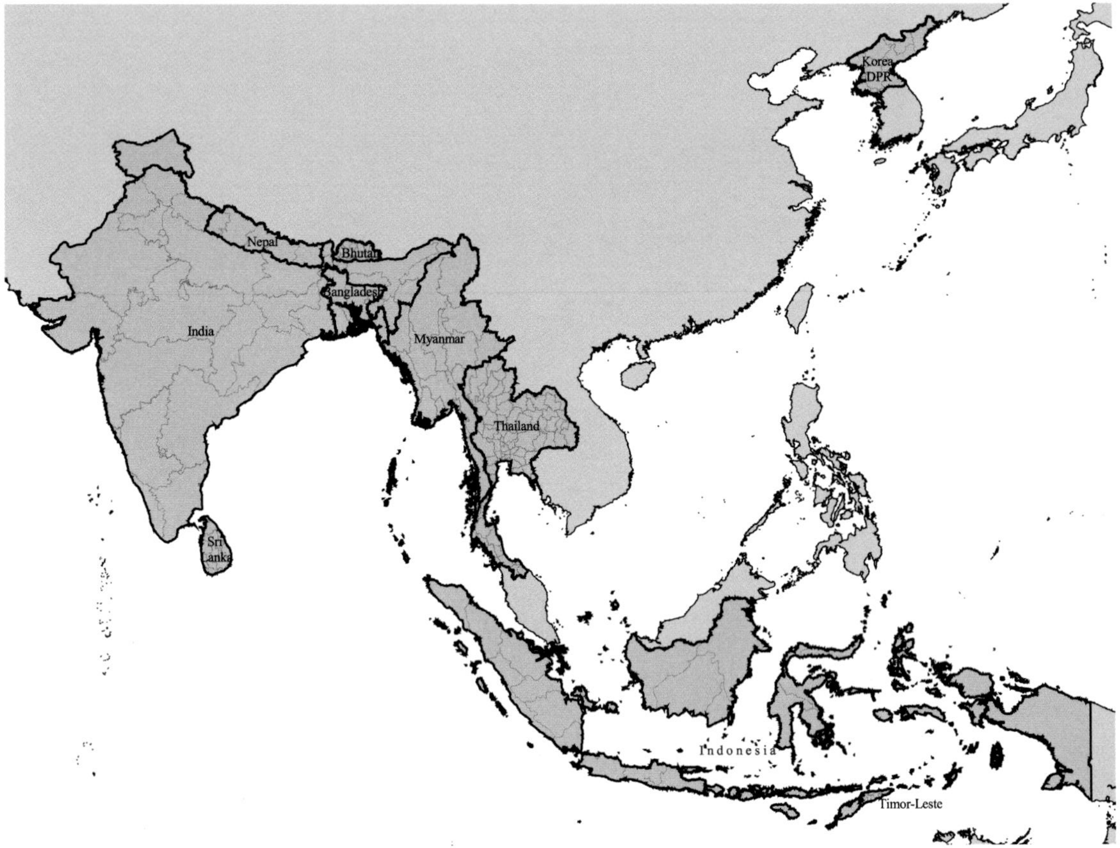

Plate 5.2 (continued)

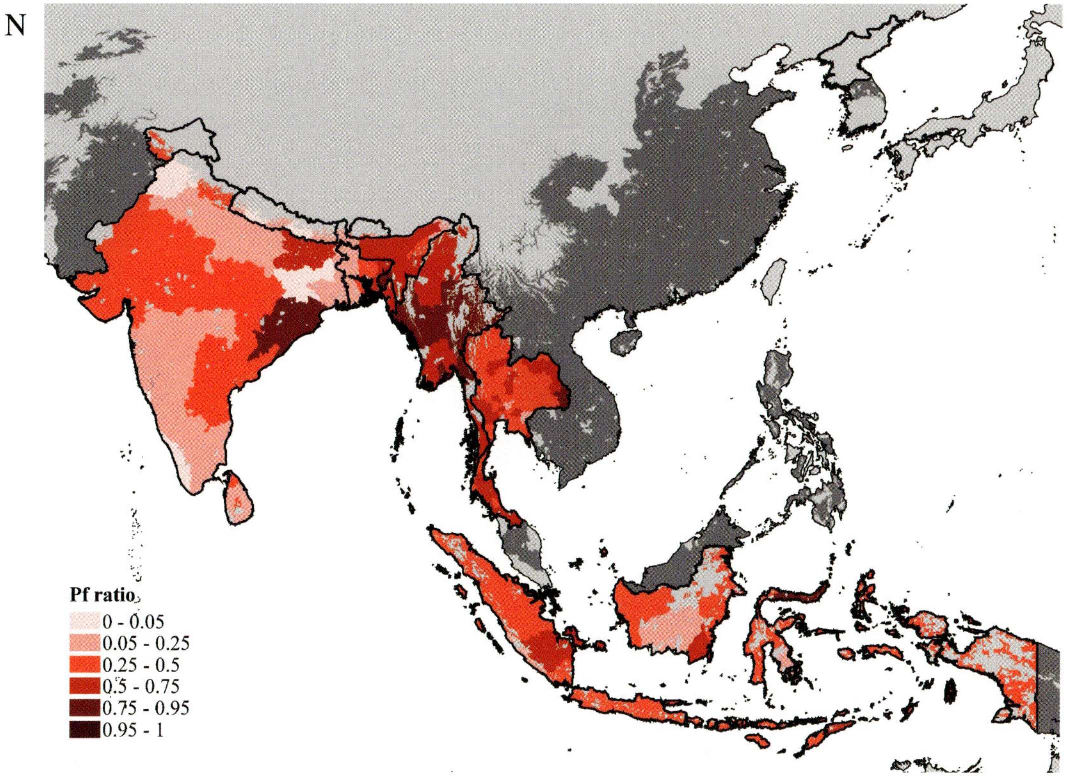

Plate 5.2 (continued)

O

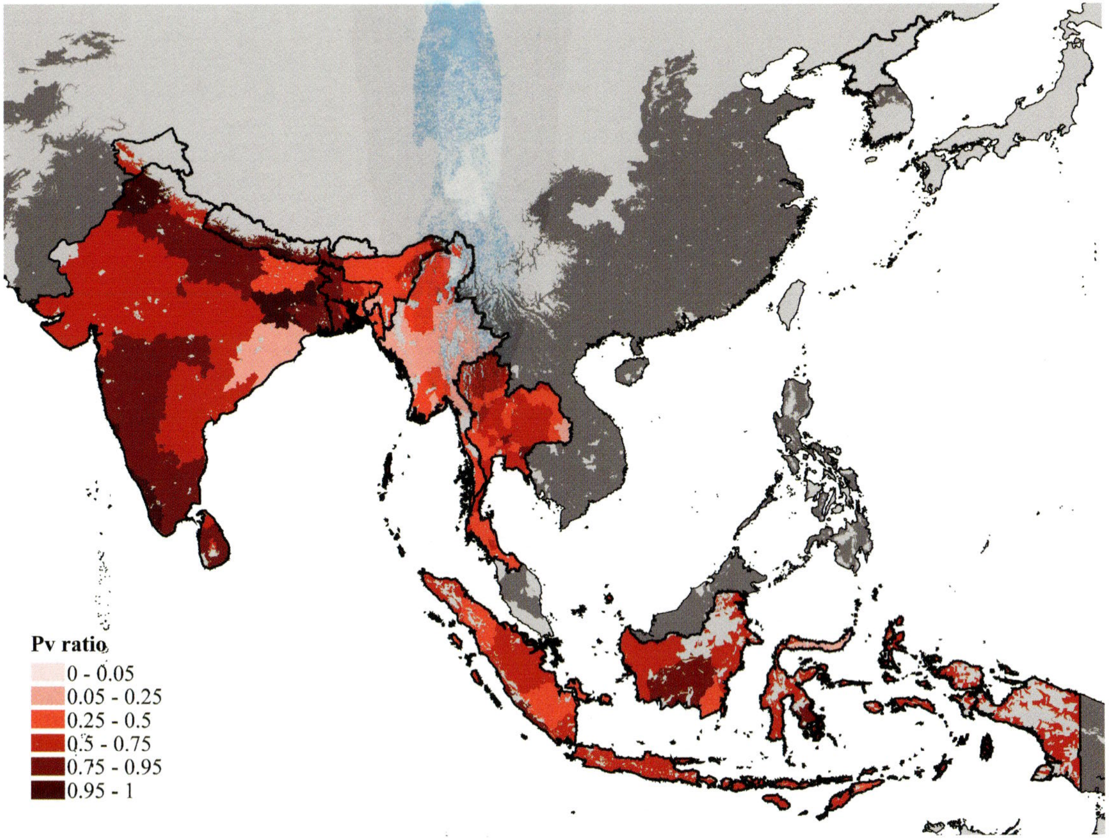

Plate 5.2 (continued)

P

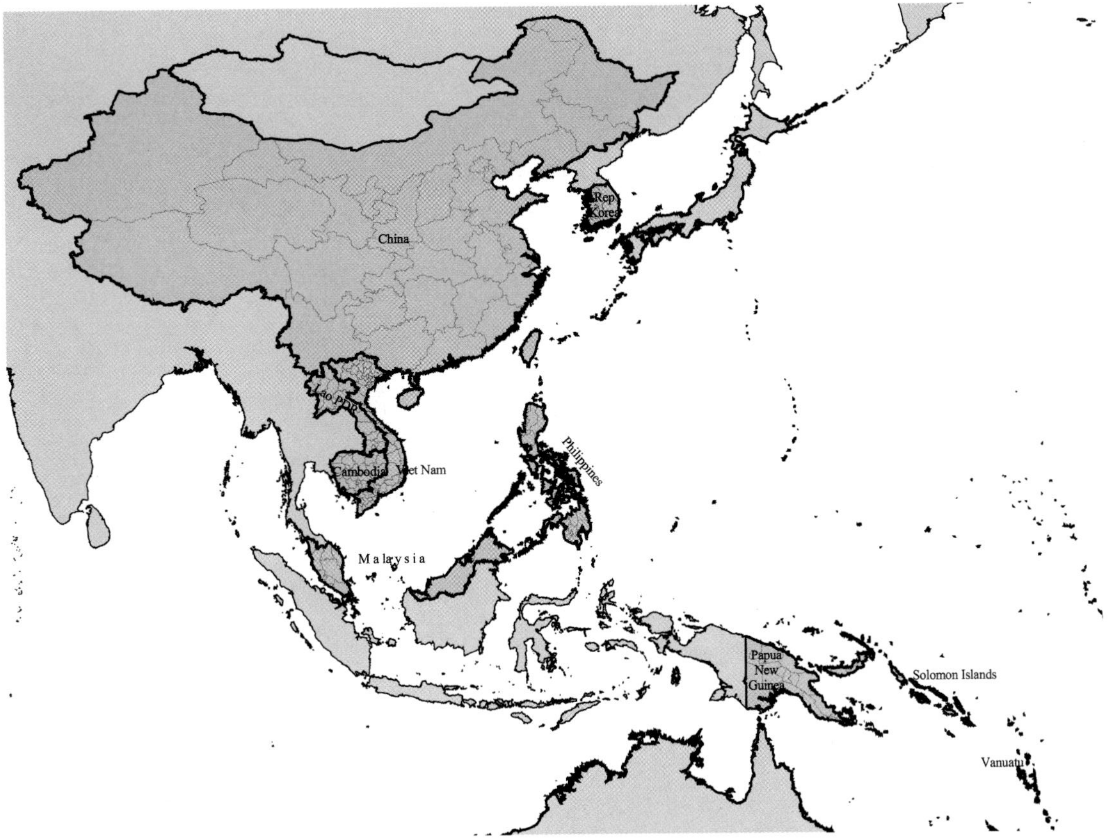

Plate 5.2 (continued)

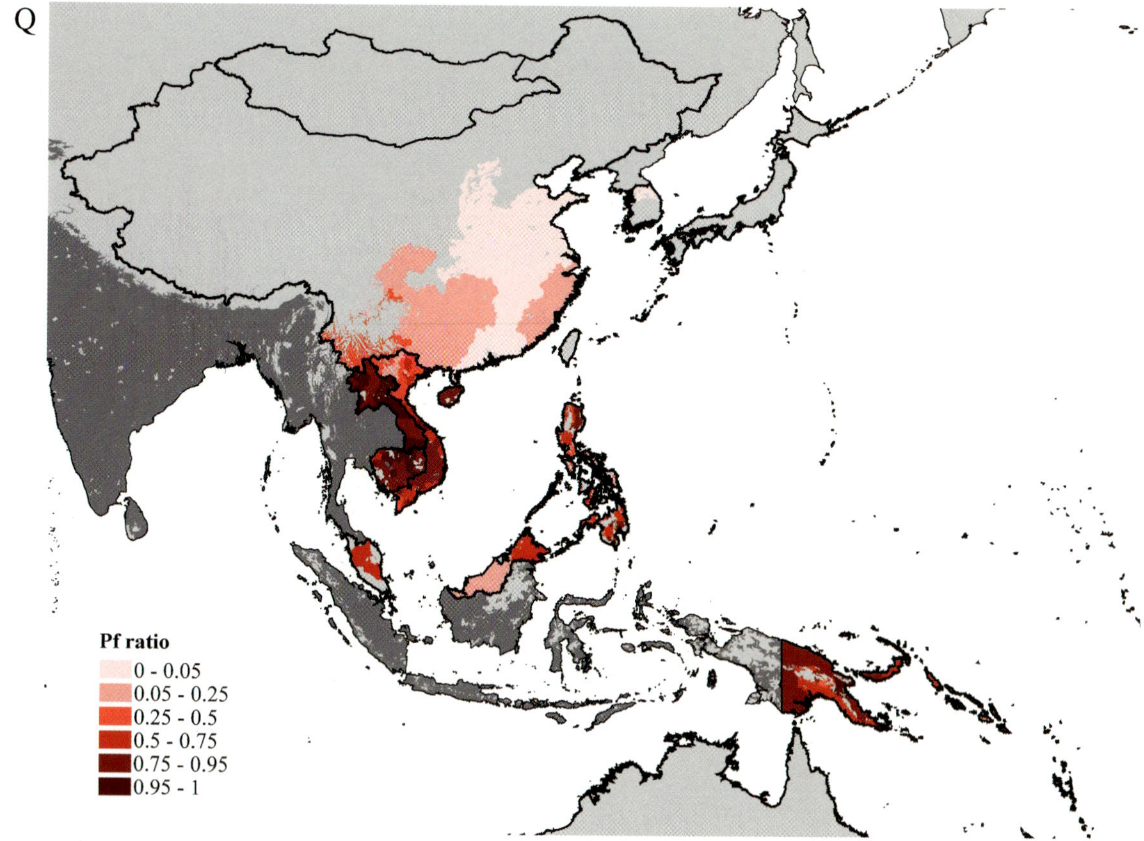

Plate 5.2 (continued)

R

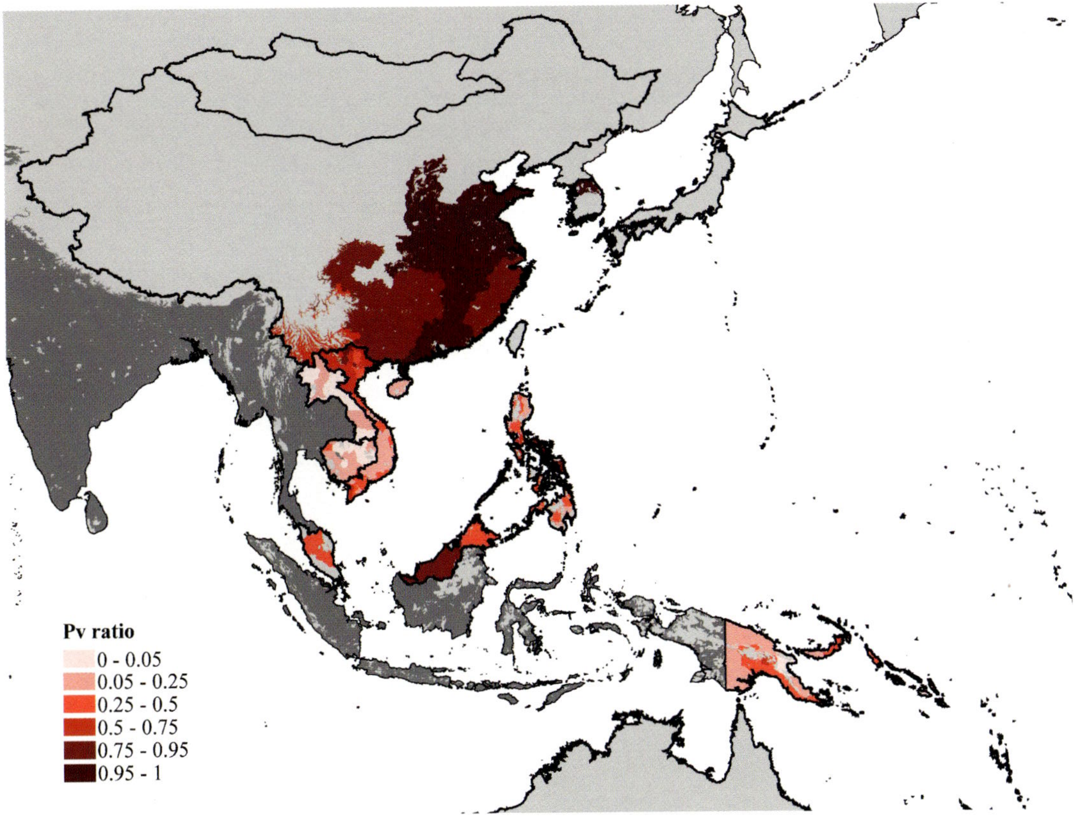

Plate 5.2 (continued)

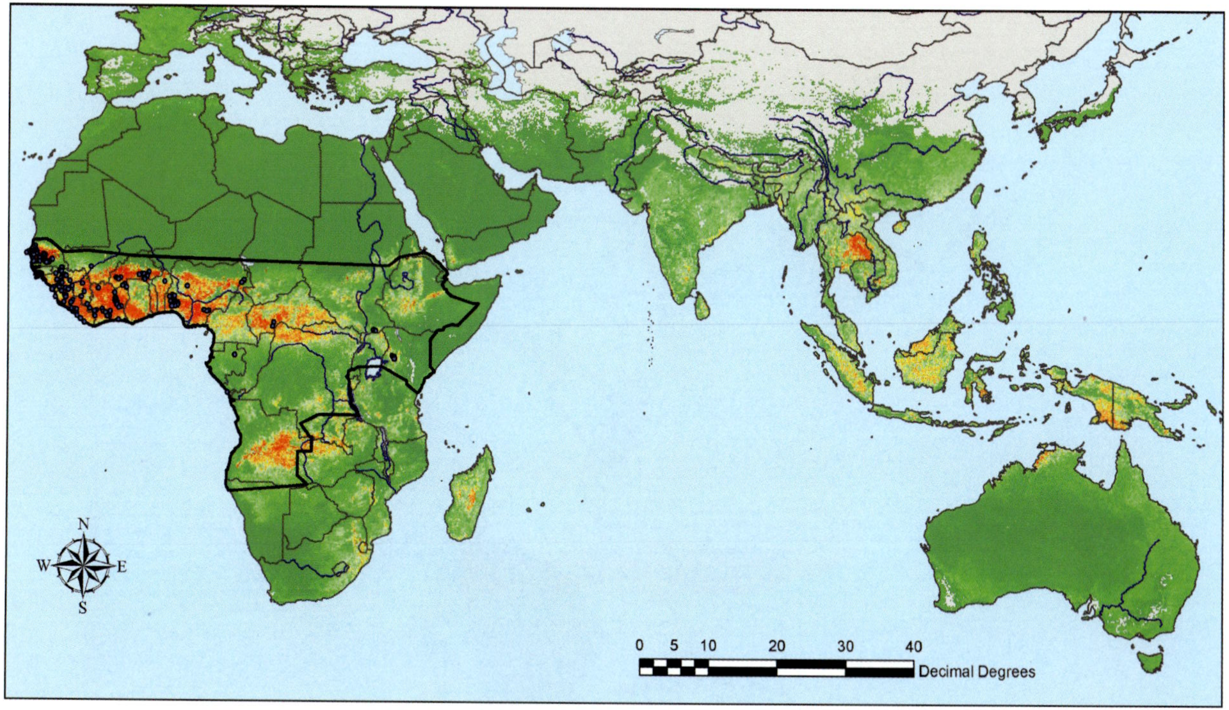

Plate 6.7 Risk map for yellow fever. This risk map is the average of 100 bootstrap models each based on a sample of 300 presence and 300 absence pixels selected at random with replacement from the training set for this disease. Risk is on a probability scale from zero to 1.0. Probabilities from 0.0 to 0.49 are coloured green (darker to lighter) and indicate conditions not suitable for the disease (*i.e.* predicted absence of disease). Probabilities from 0.50 to 1.0 are coloured yellow through to dark red, indicating conditions increasingly suitable for the disease. The database observations of presence are indicated by the blue dots and the WHO 2003 map for yellow fever by the thick black outline.

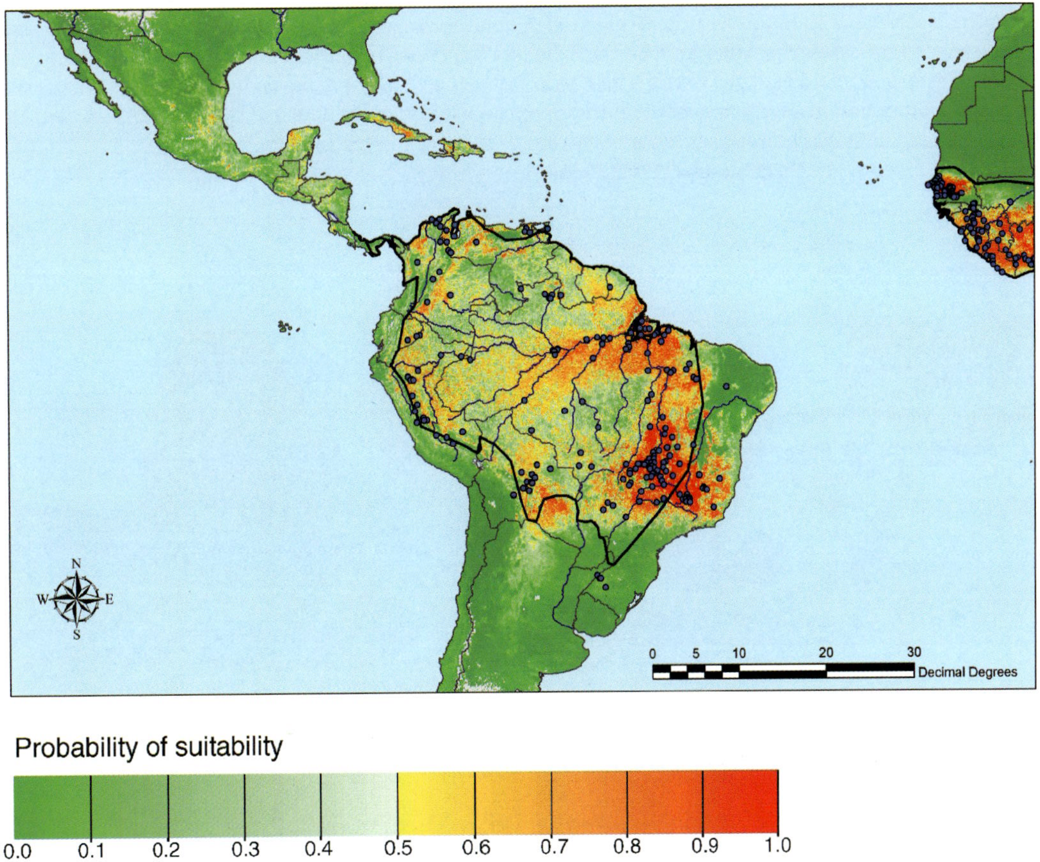

Plate 6.7 (continued)

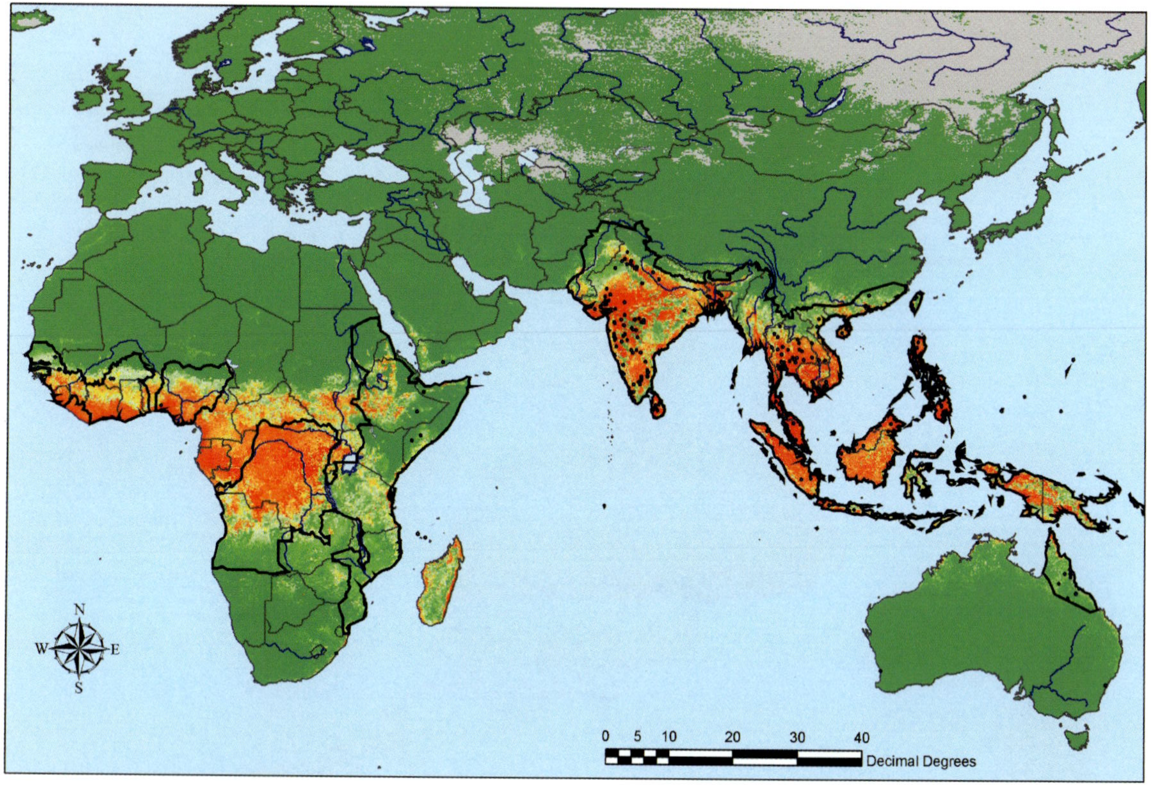

Plate 6.8 Risk map for dengue. This risk map is the average of 100 bootstrap models each based on a sample of 900 presence and 900 absence pixels selected at random with replacement from the training set for this disease. Risk is on a probability scale from zero to 1.0. Probabilities from 0.0 to 0.49 are coloured green (darker to lighter) and indicate conditions not suitable for the disease (*i.e.* predicted absence of disease). Probabilities from 0.50 to 1.0 are coloured yellow through to dark red, indicating conditions increasingly suitable for the disease. The database observations of presence are indicated by the blue dots and the WHO 2003 map for dengue by the thick black outline.

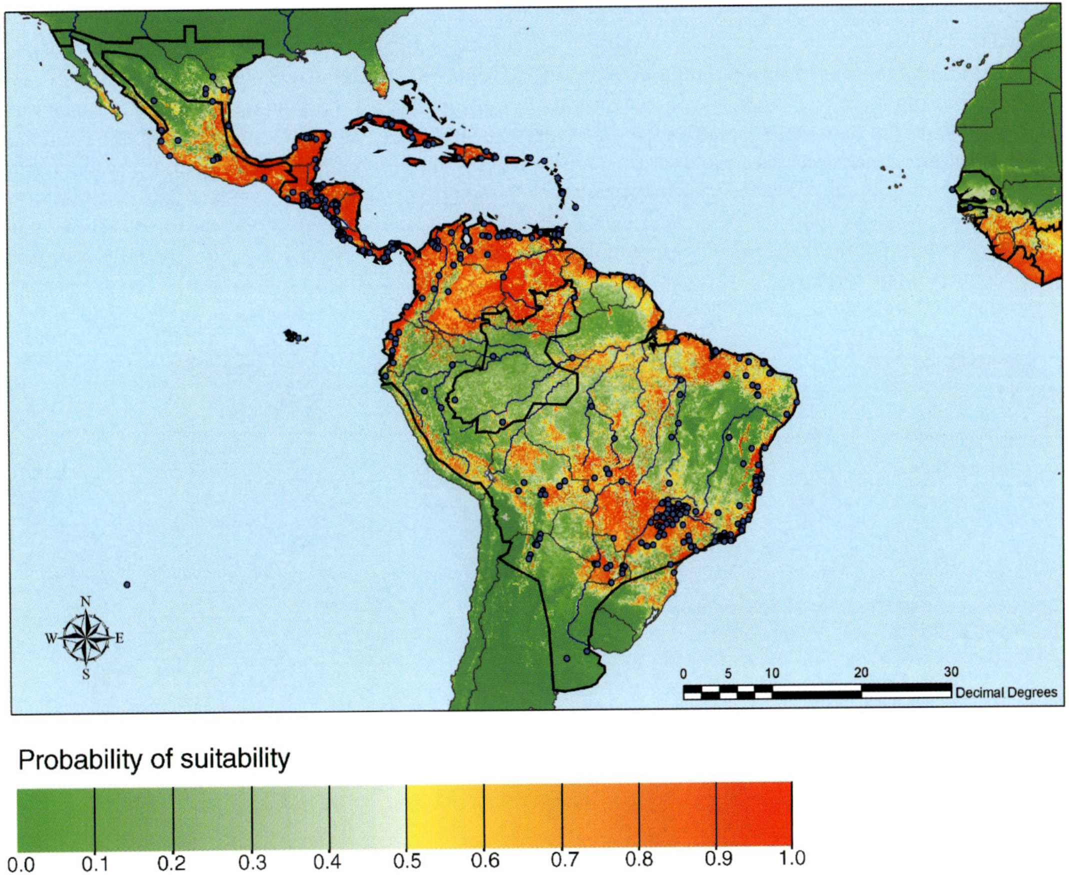

Plate 6.8 (continued)

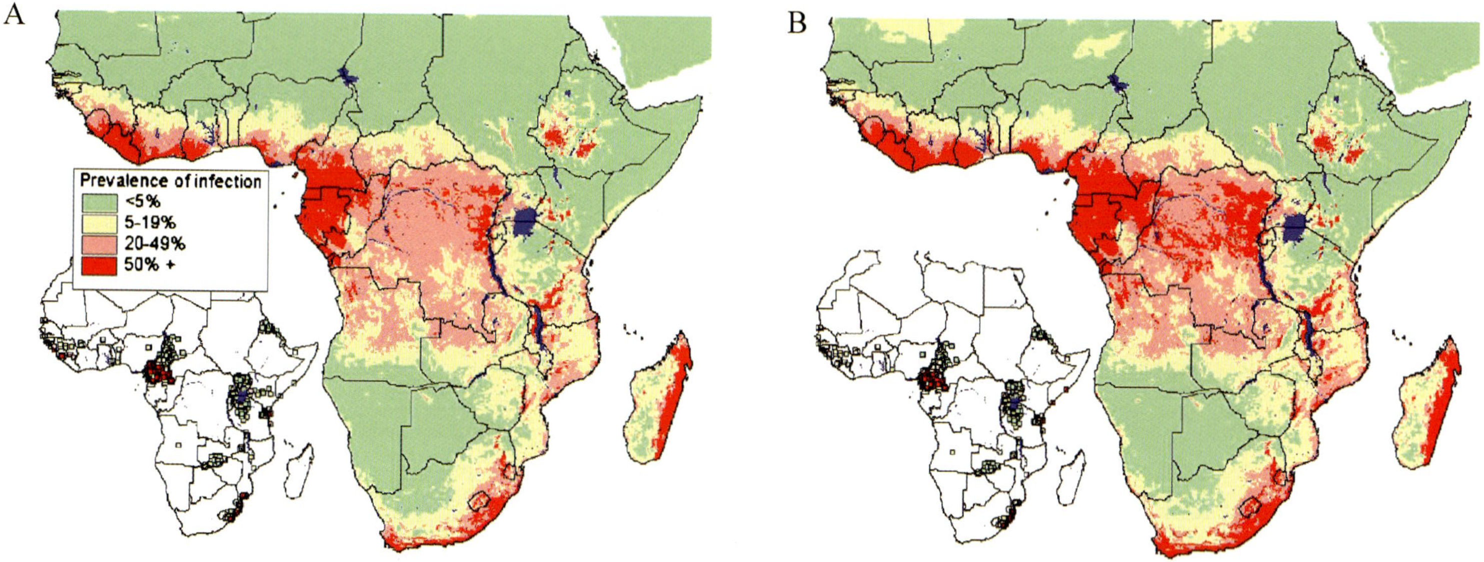

Plate 7.2 Predicted prevalence of (A) *A. lumbricoides*, (B) *T. trichiura* and (C) hookworm, based on relationships between observed prevalence of infection among school-aged children (insert) and AVHRR satellite data (see Hay *et al*., this volume for details) and elevation obtained from an interpolated digital elevation model from the Global Land Information System (GLIS) of the United States Geological Survey (http://edcwww.cr.usgs.gov/landdaac/gtopo30/). Prevalence data are available for 1172 sites across SSA including 84 412 children. All surveys were conducted using similar diagnostic techniques (direct smear, typically using the Kato–Katz method) and based on random samples of children in areas where no control measures have previously been undertaken. Due to non-linear relationships between observed prevalence and predictor variables, the predictors were categorized before being entered into the models. The coefficients from these models were then applied to the categories of the predictor variables to generate a predicted prevalence of infection. Validation statistics including area under the curve (AUC), optimal prediction threshold and sensitivity, specificity, positive predictive value (PPV) and negative predictive value (NPV) at the optimal prediction threshold are presented for observed prevalence thresholds of 5% and 50%.

C

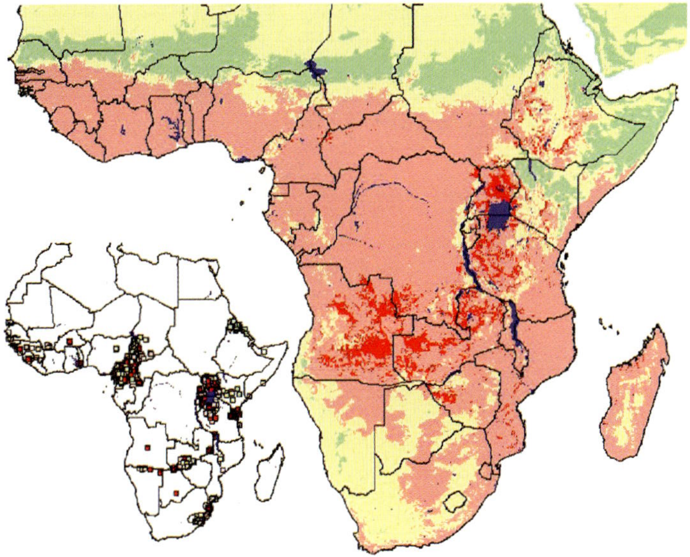

	Prevalence threshold	
(A) Validation statistic	5%	50%
AUC	0.90	0.91
Optimal prediction threshold	0.13	0.28
Sensitivity (%)	86.2	89.5
Specificity (%)	80.9	81.9
PPV (%)	79.9	60.0
NPV (%)	86.9	96.3

(B) Validation statistic	5%	50%
AUC	0.88	0.92
Optimal prediction threshold	0.12	0.32
Sensitivity (%)	84.8	87.7
Specificity (%)	77.5	85.0
PPV (%)	76.0	68.0
NPV (%)	85.9	95.0

(C) Validation statistic	5%	50%
AUC	0.76	0.70
Optimal prediction threshold	0.31	0.36
Sensitivity (%)	78.6	67.7
Specificity (%)	79.7	68.5
PPV (%)	88.7	50.8
NPV (%)	68.9	81.5

Plate 7.2 (continued)

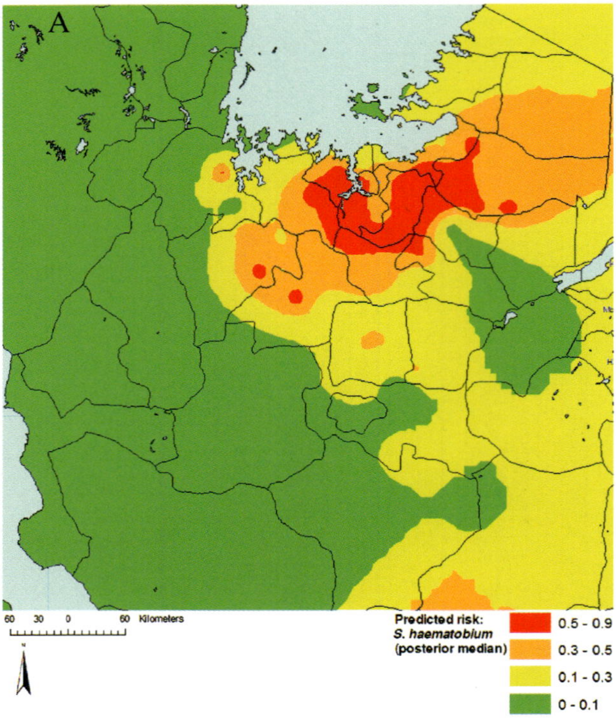

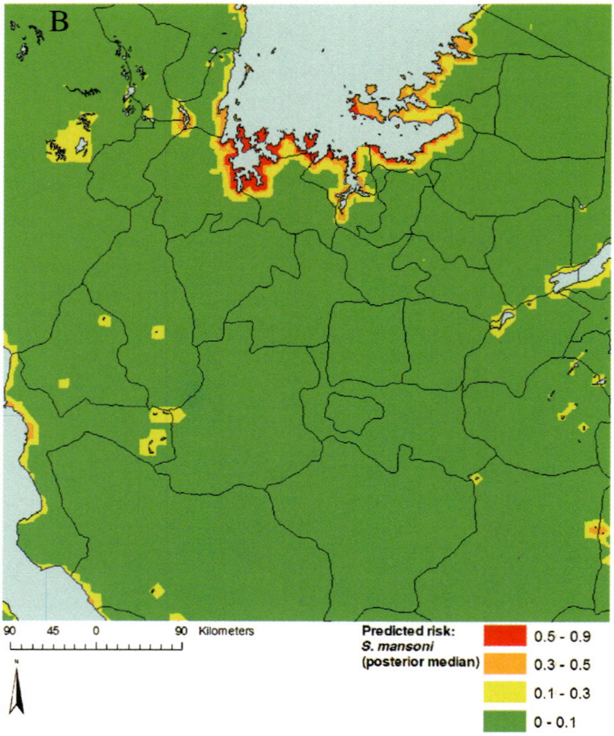

Plate 7.6 (continued)

C

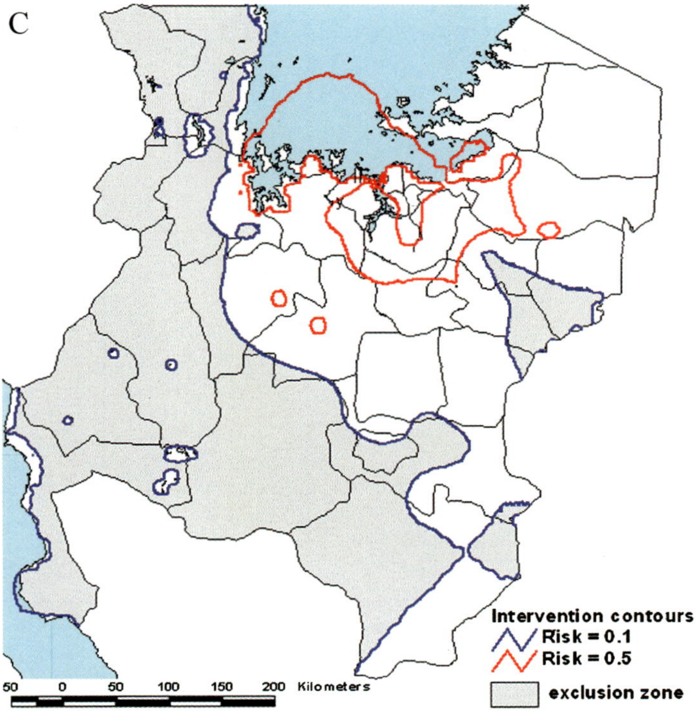

Plate 7.6 (A) Risk prediction surface for prevalence of *S. haematobium* infection in northwest Tanzania. Values presented are interpolated median posterior risk estimates from a Bayesian geostatistical binomial logistic regression model. Model parameters were: α (intercept) $= 2.3$ (95% Bayesian CI -0.7 to 5.9), κ (smoothing parameter) $= 0.9$ (95% Bayesian CI 0.6 – 1.2), φ (decay of spatial correlation) $= 0.2$ (95%Bayesian CI 0.1 – 0.5) and σ (overall variance) $= 4.8$ (95%Bayesian CI 2.7 – 7.6). (B) Risk prediction surface for prevalence of *S. mansoni* infection in northwest Tanzania. Values presented are interpolated median posterior risk estimates from the Bayesian geostatistical binomial logistic regression model. Model parameters were: α (intercept) $= -12.3$ (95%Bayesian CI -18.8 to -4.5), coefficient for distance to perennial water body, <0.04 decimal degrees $= 4.1$ (95%Bayesian CI 2.8 – 5.4), coefficient for distance to perennial water body, 0.04 – 0.1 decimal degrees $= 2.3$ (95%Bayesian CI 1.2 – 3.4), coefficient for distance to perennial water body, 0.1 – 0.4 decimal degrees $= 1.1$ (95%Bayesian CI 0.1 – 2.0), coefficient for annual minimum temperature $= 0.4$ (95%Bayesian CI 0.0 – 0.8), κ (smoothing parameter) $= 0.8$ (95%Bayesian CI 0.5 – 1.3) $= \varphi$ (decay of spatial correlation) $= 2.8$ (95%Bayesian CI 1.0 – 5.7) and σ (overall variance) $= 1.2$ (95%Bayesian CI 0.8 – 1.9). (C) Intervention contour map overlying districts of northwest Tanzania. Areas outside the 0.1 risk contour will be excluded from the mass treatment programme and PQZ will be made available in health centres. Areas between the 0.1 and 0.5 risk contour will receive mass treatment, targeted at school-age children. Areas within the 0.5 risk contour are priority areas where mass treatment will be targeted at school-age children and other high-risk groups.

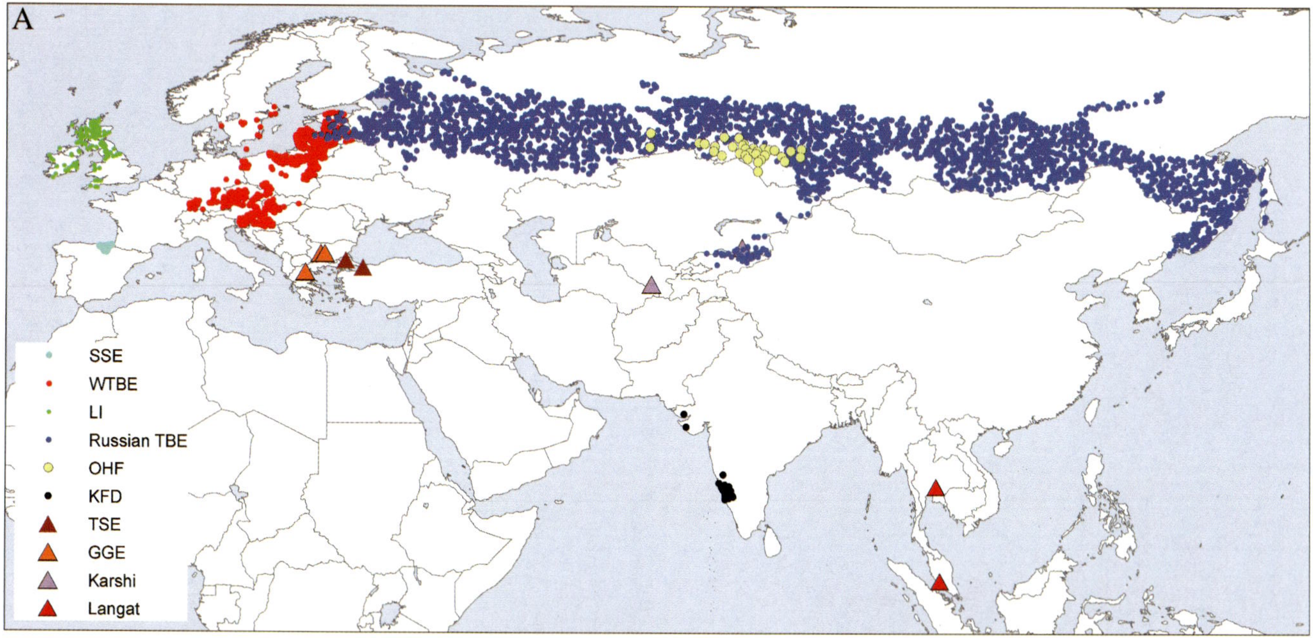

Plate 8.3 (A) The distribution of geo-referenced sites of known presence of TBE complex flaviviruses used in this analysis. In addition to those shown associated with the phylogenetic matrix (circles), see 1–3 sites for each of TSE (brown triangle) in Turkey, GGE (orange triangles) in Greece and Bulgaria, Karshi (mauve triangles) in Uzbekistan and Langat (red triangles) in SE Asia. For sources, see text. (B) The distribution of six tick-borne flaviviruses predicted in a single exercise of discriminant analysis, based on satellite-derived climatic variables. Each virus is represented by a colour that matches those in Figure 3A: LI, green; SSE, turquoise; WTBE, red; all Russian TBE, blue; OHF, yellow; KFD, black

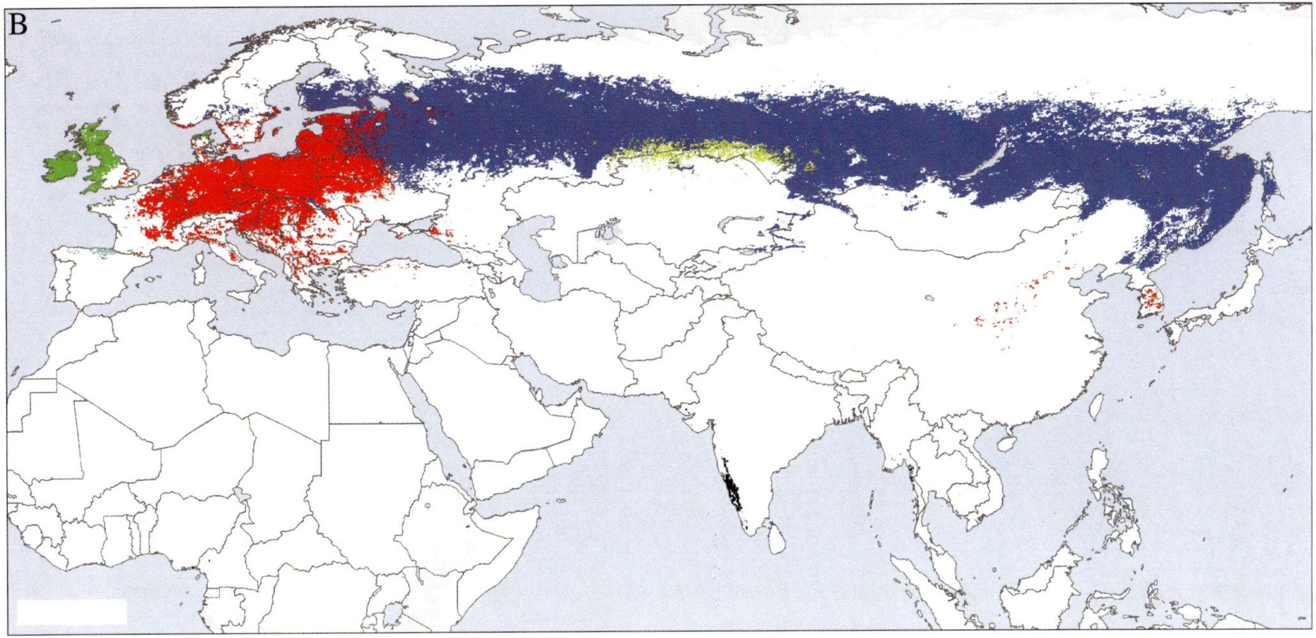

Plate 8.3 (continued)

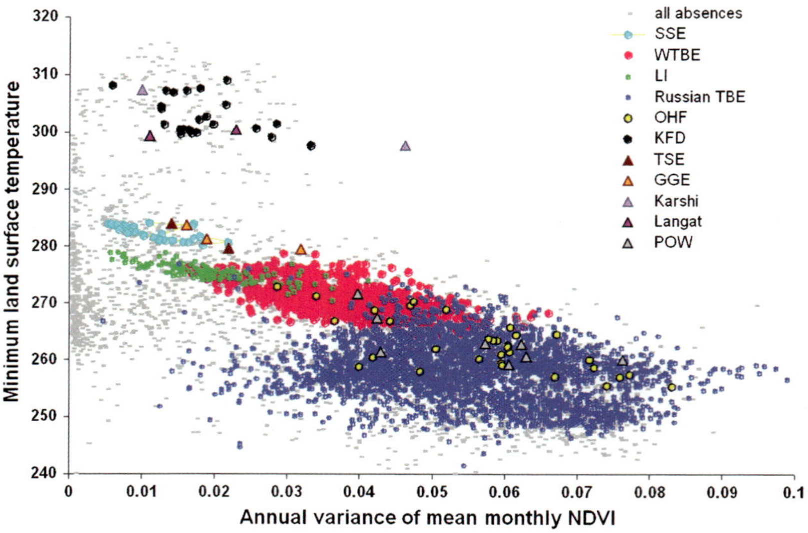

Plate 8.4 Each virus of the tick-borne encephalitis complex occupies a distinct 'eco-climatic' space, illustrated here in bi-variate space defined by two of the most significant climatic variables that predict the distribution of each virus. NDVI (normalized difference vegetation index) is an indirect measure of moisture conditions.

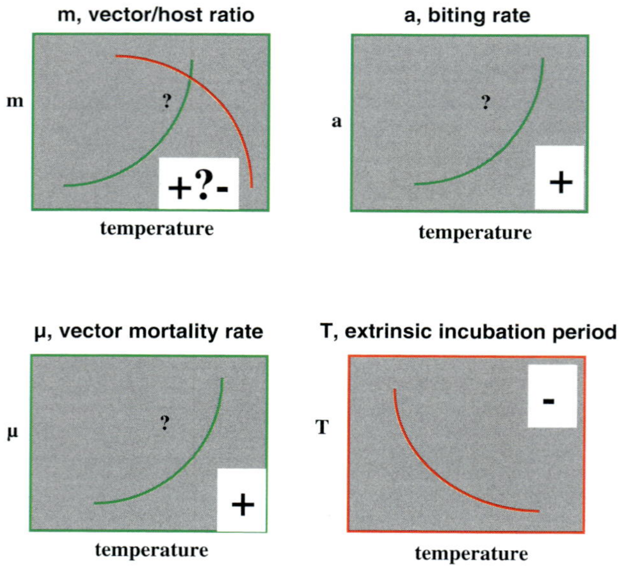

Plate 10.1 Likely effects of increasing temperature on the variables and parameters of the R_0 equation. The net effect is indicated by the positive or negative symbol within each panel. Notice that a positive effect here might decrease transmission (e.g. the effect on μ) or increase it (e.g. the effect on a).

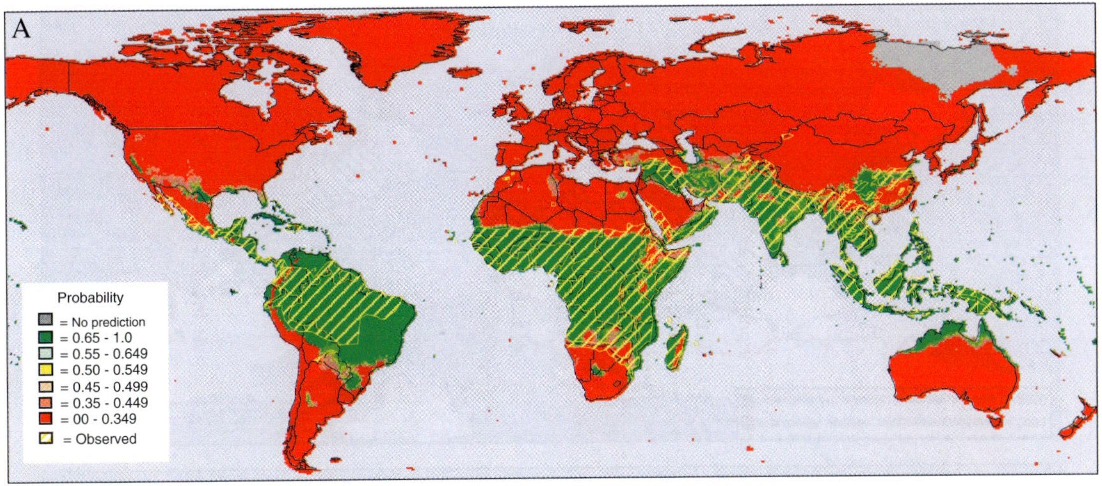

Plate 10.2 (a) Global map of malaria distribution according to the WHO (1997) (yellow cross-hatching) and predicted distribution made from 1961–1990 global climate norms. Predictions were made using a discriminant analysis approach (Chapter 1 and (Rogers, 2000)) and are on a probability scale from zero (coloured red) to 1.0 (coloured green) (see inset legend) (model results: 78% correct with 14% false positives and 8% false negatives). (b) Predicted global distribution of malaria in 2050 under the HadCM2 High scenario of global warming. The model from (A) was run using these climate predictions to produce an estimate of malaria distribution in 2050 (colour scale as in A). The WHO map of malaria is shown for reference (yellow cross-hatching). (c) The difference between Plate 10.2A and B reveal the predicted changes in global malaria distribution in 2050. Areas coloured red are presently suitable for malaria but will become unsuitable (generally because of higher temperature or lower rainfall). Areas coloured green are presently unsuitable but are predicted to become suitable. All areas of no change (i.e. suitable or unsuitable, now and in 2050) are coloured white. Plates 10.2A–C from Rogers and Randolph (2000), with permission.